Resolving Human–Wildlife Conflicts

The Science of Wildlife Damage Management

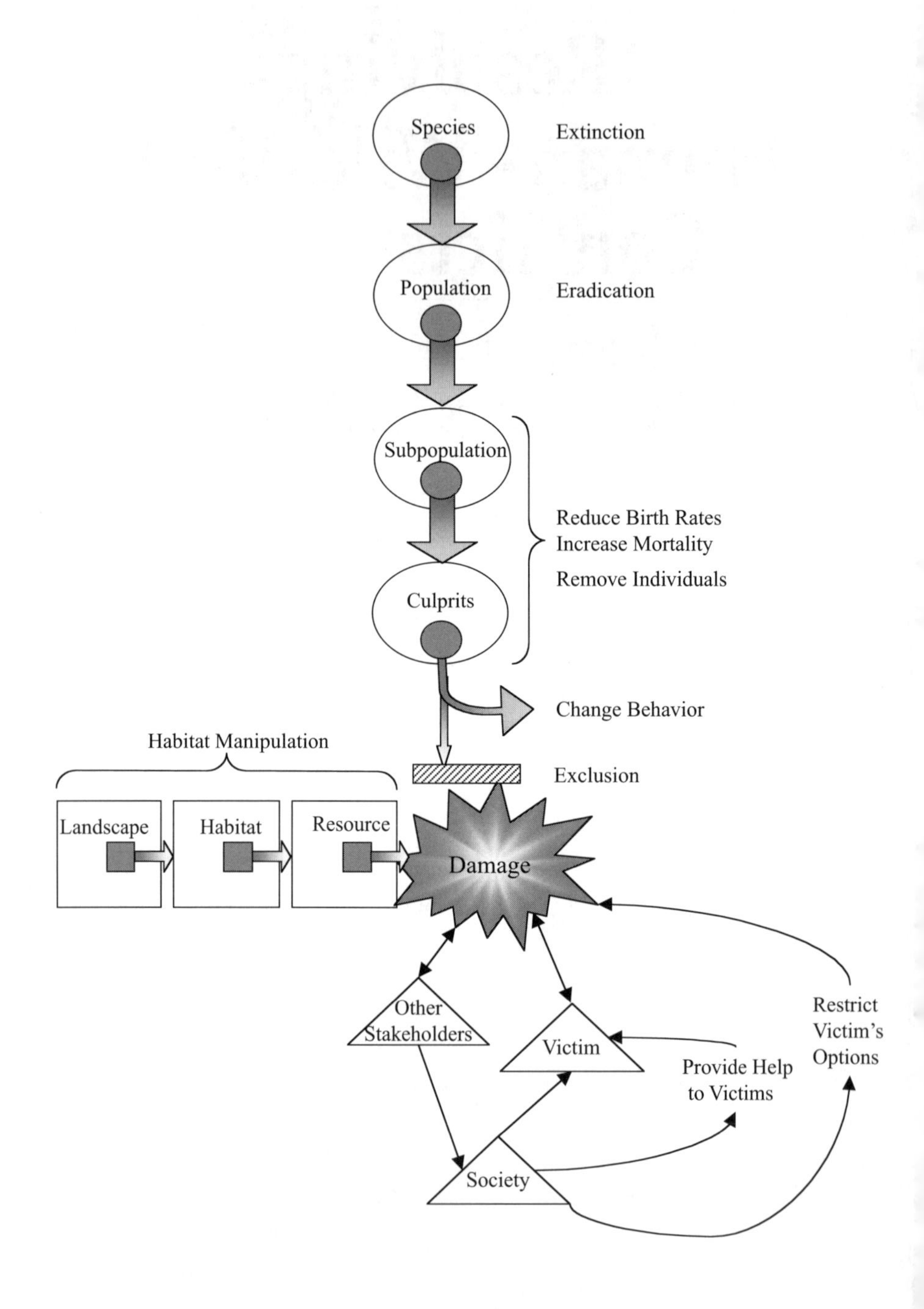
Species
Extinction
Population
Eradication
Subpopulation
Reduce Birth Rates
Increase Mortality
Remove Individuals
Culprits
Change Behavior
Habitat Manipulation
Exclusion
Landscape
Habitat
Resource
Damage
Other Stakeholders
Victim
Restrict Victim's Options
Provide Help to Victims
Society

Resolving Human–Wildlife Conflicts

The Science of Wildlife Damage Management

Michael Conover

LEWIS PUBLISHERS

A CRC Press Company

Boca Raton London New York Washington, D.C.

Library of Congress Cataloging-in-Publication Data

Conover, Michael R.
Resolving human–wildlife conflicts : the science of wildlife damage management/ Michael Conover.
p. cm.
Includes bibliographical references and index.
ISBN 1-56670-538-X
1. Wildlife management. 2. Wildlife depredation. I. Title.

SK355 .C66 2001
639.9—dc21 2001032672

Direct all inquiries to CRC Press LLC, 2000 N.W. Corporate Blvd., Boca Raton, Florida 33431.

Visit the CRC Press Web site at www.crcpress.com

Lewis Publishers is an imprint of CRC Press LLC

International Standard Book Number 1-56670-538-X
Library of Congress Card Number 2001032672
Printed in the United States of America 6 7 8 9 0
Printed on acid-free paper

Dedication

I dedicate this book to Jack H. Berryman
— the man and his institute.

Jack H. Berryman — The Man

Figure 1 Jack H. Berryman.

Jack H. Berryman was born on July 28, 1921, in Salt Lake City, Utah (Figure 1). He earned a B.S. degree in science and an M.S. degree in ecology from the University of Utah, then started working for the Utah Fish and Game Department in 1947. In 1950, he moved to New Mexico, where he worked for the U.S. Fish and Wildlife Service. He then moved to Minnesota and became the assistant regional supervisor in the Branch of Federal Aid of the U.S. Fish and Wildlife Service. From 1959 to 1965, he served as an associate professor and extension wildlife specialist at Utah State University. After that, he worked for the U.S. Fish and Wildlife again, as chief of the Division of Wildlife Service in Washington, D.C., where he strove to develop new strategies for reducing wildlife damage. From 1974 to 1977, he was chief of the Division of Technical Services and from 1977 to 1979, chief of the Office of Extension Education. He then served as the executive vice president of the International Association of Fish and Wildlife Agencies, from 1979 to 1988, and as counselor emeritus for the International Association of Fish and Wildlife Agencies after that.

Jack was president of The Wildlife Society in 1965 and received the Aldo Leopold Award from The Wildlife Society in 1993. He authored over 100 publications on natural resource policy and management, extension education, outdoor recreation, wildlife damage management, and related topics.

Jack was a visionary leader and a man of great integrity, respected nationally and internationally. His contributions to the fields of wildlife management, natural resource conservation, and extension education helped shape the course of the wildlife profession. He cared deeply for wildlife and people, had strong ethics and values, and inspired those around him to strive for excellence in their profession. The Jack H. Berryman Institute, created at Utah State University in 1993, was named after Jack to honor his lifelong accomplishments.

Jack H. Berryman — The Institute

Berryman Institute

Figure 2 Logo of the Jack H. Berryman Institute.

The Jack H. Berryman Institute is a national organization based in the Department of Fisheries and Wildlife at Utah State University (Figure 2). It is named after Jack H. Berryman to honor his distinguished career in wildlife management. The Berryman Institute is dedicated to improving human–wildlife relationships and resolving human–wildlife conflicts through teaching, research, and extension.

As human and wildlife populations increase, conflicts between humans and wildlife are increasing in frequency and severity. These conflicts also have become more numerous due to humans moving into remote areas to live and recreate and to wildlife moving into urban–suburban areas to survive. A human–wildlife conflict occurs whenever the interests of humans and wildlife collide. When this happens, both wildlife and humans are losers.

The mission of the Berryman Institute is to help create a world where neither humans nor wildlife have an adverse impact upon the other. Specific objectives of the Berryman Institute are to

- increase our understanding of wildlife and of human–wildlife conflicts
- develop new techniques to reduce wildlife damage that are effective, cost-efficient, humane, socially acceptable, and environmentally benign
- enhance the positive values of wildlife and thereby increase human tolerance of wildlife
- provide information about wildlife damage management issues and techniques to wildlife professionals and the general public
- assist state and federal agencies in their important mission of managing our majestic natural resources for the benefit and enjoyment of everyone
- promote communication among wildlife damage management professionals, natural resource managers, wildlife extension specialists, academicians, and stakeholders affected by wildlife management issues
- create educational opportunities for future and current professionals in the field of wildlife damage management

To learn more about the Berryman Institute, please visit its website at www.BerrymanInstitute.org

Preface

"If we can see further than others, it is because we are standing on the shoulders of giants."

— Isaac Newton

Our knowledge on how to resolve human–wildlife conflicts, which is the science of wildlife damage management, has advanced tremendously since the first Vertebrate Pest Conference was held in 1962. This book is, in essence, a celebration of this remarkable advance. I invite readers to thumb through the references at the end of each chapter to gain some appreciation for the hundreds of people who have devoted their lives to increasing our knowledge base. Two groups of scientists stand out for the magnitude of their contributions to the advancement of this field. The first is the employees of the USDA/APHIS/Wildlife Services' National Wildlife Research Center. The second is wildlife extension specialists employed by the many land-grant universities throughout the United States. It is my hope that this book, the first compendium of our collective wisdom about how to resolve human–wildlife conflicts, will provide a sense of accomplishment to the many people who have labored in this field and will stimulate future research. It is the goal of wildlife damage management to create an ideal world where humans and wildlife can coexist without either having an adverse impact upon the other.

Organizing a book on the resolution of human–wildlife conflicts can take many directions. I could have included a chapter on each wildlife species or one on each type of wildlife problem. Instead, this book is organized on the basis of the fundamental concepts and principles upon which the field of wildlife damage management rests. I believe that by understanding these basic concepts, readers will be able to apply the information to any specific wildlife problem. For instance, the use of supplemental feeding is based on the optimal-foraging theory. The opportunities and challenges involved with using supplemental feeding are the same whether we are trying to divert deer from feeding along roads, bears from girdling trees, cranes from eating newly planted corn seeds, or foxes from depredating waterfowl nests.

In the first six chapters, I define the field of wildlife damage management, discuss its philosophy and history, and examine how wildlife threatens human health and safety, our economy, and the environment. I describe in chapters 7 to 13 how human–wildlife conflicts can be managed by reducing wildlife populations (lethal control and fertility control) or removing individual animals (lethal control and translocation) or by changing animal behavior (fear-provoking stimuli, chemical repellents, diversion, and exclusion). I then point out that we also can resolve human–wildlife conflicts by changing the resource so that it is less vulnerable to wildlife damage (chapter 14) or by changing people's perceptions about wildlife (chapter 15). In the final chapter, I provide three examples of how human–wildlife conflicts can be alleviated by using an integrated approach.

I think readers will discover that there are no panaceas for the resolution of human–wildlife conflicts, but rather there are many methods that can be used to address these conflicts. Although no single technique can help resolve all of them, hopefully, there will always be at least one technique which can be used to alleviate any specific wildlife problem.

A book of this size is usually a team effort, and I have many people to thank for contributing to this work. Anne Brown wrote the rough draft of the habitat manipulation, translocation, and part of the zoonoses chapters. Jaimi Butler-Curl wrote the rough draft of the fertility control chapter. Anne and Jaimi also edited all of the other chapters. Anne Brown and Ben West drew all the uncredited figures. Robert McLean and Kathleen Fagerstone provided up-to-date information for the zoonoses and chemical repellent chapters, respectively. Comments by Fred Knowlton and Robert Schmidt helped improve the lethal control and human dimensions chapters. Terry Messmer contributed to many of the ideas expressed in this book. Many students in my Principles of Wildlife Damage Management class helped edit the book, including Olin Albertson, Kimberly Anderson, Jarom Bangerter, Sophia Bates, Adam Bronson, Clint Brunson, Michael Burrell, Breta Campbell, Joe Caudell, Ruth Cecil, Trisha Cracroft, Justin Crump, John Curl, David Dahlgren, Joel Dunlap, Nicki Frey, Eriek Hansen, Ryan Hillyard, Jeremy Johnson, Tamara Johnson, Adrienne Marler, Damon McRae, Kevin Nettleton, Grizz Oleen, Landon Olson, Robert Peterson, Tana Pickett, John Reichert, Shane Ross, Dixie Sadlier, Ryan Shaw, Todd Sullivan, Maria Torres, John Treanor, Sharon Ward, and Edward Zakrajsek. To all of these people, my deepest thanks.

Michael Conover

Logan, Utah

June 2001

Contents

CHAPTER 1

Philosophy

"Wonder is the feeling of a philosopher, and philosophy begins in wonder."

Plato

"The point of philosophy is to start with something so simple as not to seem worth stating, and to end with something so paradoxical that no one will believe it."

Bertrand Russell

"Conscience implies a sense of duty, and this infers a moral obligation and a human responsibility towards animals; not only to those species which have been bred and reared to be of service to man, but also to those which have been affected by man's environmental modification."

Charles Hume

Wildlife damage management is a field within the discipline of wildlife management. To understand the former, we need to know how it is related to the broader goals of wildlife management, why we manage wildlife, and how wildlife damage management helps achieve societal goals. Wildlife damage management is built upon the fundamental knowledge gained in the fields of ecology, behavior, population dynamics, and wildlife management. The same principles apply whether we are trying to increase or decrease a wildlife population, regardless of our motives.

DEFINITIONS

Wildlife is one of those terms we use every day without worrying about its precise definition. It is a misnomer to say that wildlife means wild animals because the term "wild" has several meanings. Wild sometimes describes an individual or animal that exhibits erratic, unpredictable, or uncontrollable behavior. One synonym for it is "crazy." Certainly when we use the term wildlife, we do not mean crazy animals. By wild, we really mean animals whose behavior and movements are not controlled by humans, i.e., undomesticated animals. To avoid confusion over the different interpretations of "wild," many scientific journals prefer the term "free-ranging animals" rather than wild animals. It is unlikely that the term "wildlife" will ever be replaced by "free-ranging life," but the latter is a more correct term. Moreover, not all free-ranging animals are considered wildlife. Excluded are insects and other invertebrates. Historically, the term meant free-ranging vertebrates, but it was unclear whether fish were included in this term. To reduce ambiguity, people started using the term "fish and wildlife" when referring to both. By doing so, the term "wildlife" became restricted to vertebrates excluding fish. Hence, the commonly accepted definition of wildlife is free-ranging vertebrates other than fish.

Wildlife is often referred to as a resource. By definition, a resource provides a benefit to people. Resource is a human construct; without people, wildlife exists, but it is not a resource. Resources are the physical, observable, and conceivable goods and services in the environment that have use and value for humans (Brown 1984, Giles 1990). A value is the ability to cause an effect. An effect might benefit people, i.e., have a positive value, or harm people, i.e., have a negative value. If there were no people, wildlife could not affect them and therefore could not be assigned a value. The lives of wildlife and humans are so entwined that wildlife affects people in many different ways. That is, free-ranging animals have many values, some of which are positive and some negative.

Wildlife management can be defined as the act of influencing or modifying the wildlife resource to meet human needs, desires, or goals. Wildlife management is a science. Wildlife biologists use the scientific method of asking a question, devising hypotheses or potential answers to that question, and then conducting experiments to try to disprove the hypotheses. Because wildlife management is a science, we can try some form of management, observe the effects of our action, and learn from our mistakes. Through this process, we can determine how best to achieve our goals for managing wildlife.

PHILOSOPHIES OF WILDLIFE MANAGEMENT

Philosophy can be defined as a study of the general principles or beliefs which underlie a particular branch of knowledge or subject. These beliefs cannot be scientifically tested but instead are believed to be true. The major philosophical questions in wildlife management include: Why do we try to manage wildlife? What are our goals for managing wildlife? We must be trying to accomplish something very important because we spend vast amounts of time and money to achieve our objectives. In the 1990s, U.S. agricultural producers collectively spent $2.5 billion and 120 million hours annually to manage wildlife on their farms and ranches (Conover 1998). During the same period, U.S. metropolitan households spent $5.5 billion and 1.6 billion hours yearly to manage wildlife around their homes (Conover 1997).

We can draw several conclusions from these large expenditures to manage wildlife. First, we as individuals and as a society are trying to achieve some goal regarding wildlife that we believe is important. Second, we believe that our goal will not be achieved without our active intervention. Third, we believe that our goal will be achieved if we do intervene. The question remains, what are we, as a society, trying to accomplish when we manage wildlife?

That question has been debated for centuries, and the answer has changed over time, as we discuss in the next chapter. One philosophical debate centers on whether we should be managing wildlife for the benefit of the wildlife resource itself (Sidebar 1.1) or for the benefit of humanity (Sidebar 1.2).

Sidebar 1.1 Developing a Land Ethic — the Views of Aldo Leopold

"All ethics so far evolved rest upon a single premise: that the individual is a member of a community of interdependent parts. His instincts prompt him to compete for his place in that community, but his ethics prompt him also to co-operate (perhaps in order that there may be a place to compete for). The land ethic simply enlarges the boundaries of this community to include soils, waters, plants, and animals, or collectively: the land.

"This sounds simple: do we not already sing our love for and obligation to the land of the free and the home of the brave? Yes, but just what and whom do we love? Certainly not the soil, which we are sending helter-skelter downriver. Certainly not the waters, which we assume have no function except to turn turbines, float barges, and carry off sewage. Certainly not the plants, of which we exterminate whole communities without batting an eye. Certainly not the animals, of which we have already extirpated many of the largest and most beautiful species. A land ethic of course cannot prevent the alteration, management, and use of these 'resources,' but it does affirm their right to continued existence, and at least in spots, their continued existence in a natural state.

"In short, a land ethic changes the role of *Homo sapiens* from conqueror of the land-community to plain member and citizen of it. It implies respect for his fellow-members, and also respect for the community as such" (Leopold 1949).

Sidebar 1.2 Saving Nature, but Only for Man — the Views of Charles Krauthammer

"Environmental sensitivity is now as required an attitude in polite society as is, say, belief in democracy or aversion to polyester.... But how are we to choose among the dozens of conflicting proposals, restrictions, projects, regulations, and laws advanced in the name of the environment? Clearly not everything with an environmental claim is worth doing. How to choose?

"There is a simple way. First, distinguish between environmental luxuries and environmental necessities. Luxuries are those things it would be nice to have if costless. Necessities are those things we must have regardless. Then apply a rule. Call it the fundamental axiom of sane environmentalism: combating ecological change that directly threatens the health and safety of people is an environmental necessity. All else is luxury....

"A sane environmentalism is entirely anthropocentric: it enjoins man to preserve nature, but on the grounds of self-preservation. It does not ask people to sacrifice in the name of other creatures. After all, it is hard enough to ask people to sacrifice in the name of other humans…." (Krauthammer 1991).

My philosophy of wildlife management, which I believe is shared by most people, has two main tenets. First, I adhere to the concept that we do not inherit natural resources from our parents but borrow them from our children. Hence, we have a right to "live off the interest" of the wildlife resource but not to "spend the capital." This means that no generation, including ours, has a right to take an action that diminishes the value of the wildlife resource for future generations. The problem each generation faces is not knowing what component of the natural resources might become important to a future generation. As we see in the next chapter, former generations viewed wildlife very differently than we do today. Their interest was focused almost exclusively on wildlife species of economic value, e.g., beaver, deer, elk, and bison. Most could not have dreamed that our generation would care if a frog or warbler became extinct. Our wildlife goals have changed in a way that was unforeseen by former generations. How then can we foresee the interests of future generations to ensure that our actions do not prevent them from achieving their goals for the wildlife resource? The answer is that we cannot. Hence, we need to manage the wildlife resource cautiously and should not take any action that is irrevocable, nor should we fail to act and allow an irrevocable loss to occur.

An example of an irrevocable loss is the erosion of soil from our agricultural fields. This loss causes a permanent reduction in the land's productivity. Another is the loss of a species through extinction. During the 1700s, one of the most abundant birds in North America was the passenger pigeon. It also was one of the most popular game birds. However, former generations allowed this species to become extinct, and hence, we and future generations will not have the pleasure of seeing or hunting this magnificent bird.

My philosophy combines elements of both Leopold's and Krauthammer's philosophies (Sidebars 1.1 and 1.2). Like Leopold, I believe that society needs to develop an environmental ethic which constrains us from causing environmental degradation. Like Krauthammer, my philosophy is based on Protagoras' maxim that "man is the measure of all things" and that the focus of environmental protection should be to protect ourselves and our children. However, our goals of wildlife management should extend beyond preserving wildlife for future generations.

The second major tenet in my philosophy of wildlife management is that we manage wildlife to increase its value. As individuals, we manage wildlife to benefit ourselves and our families. Some landowners manage their property to reduce wildlife damage to their crops or to grow trophy deer or elk so that they can lease the hunting rights to others. As a society, we manage wildlife to increase the value of the wildlife resource for all members of society. In summary, my philosophy is that the goal of wildlife management should be to increase the net value of the wildlife resource for society while avoiding those actions which might cause the irrevocable loss of any part of this resource. In the next section, we discuss the positive, negative, and net values of wildlife.

WHAT POSITIVE VALUES ARE PROVIDED BY WILDLIFE?

The benefits or positive values provided to humans by the wildlife resource can be grouped into the following categories:

1. physical utility,
2. monetary,
3. recreational,
4. scientific,
5. ecological,
6. existence,
7. historic values (Giles 1978).

Physical utility is the use of wildlife for food and clothing or to meet other necessities for human survival. Before the advent of agriculture, this was the main use of wildlife. Even today, many people in the world meet at least part of their subsistence needs by harvesting wildlife.

Economic or monetary benefits of the wildlife resource are those which can be exchanged for money. The economic value of wildlife is captured by individuals, governments, and communities. By harvesting animals, individuals capture the value of furs, hides, or meat, which can be sold. Landowners receive a monetary benefit from wildlife when they charge hunters a fee for access to their property. Hunting guides earn money by providing services to hunters. Government agencies capture some of the monetary value of wildlife by taxing wildlife commodities (such as furs) or by requiring people to buy licenses for the right to hunt or trap. Communities gain an economic benefit from wildlife when hunters or tourists spend money for meals, lodging, or supplies while recreating in the area.

Recreational value can be viewed as the enjoyment people receive from recreational activities involving wildlife. Many hunters derive great pleasure from the sport and often consider it to be one of the most fulfilling activities in their lives. Other people enjoy watching birds and other wildlife and would consider trips to parks or other outdoor activities less worthwhile if they could not see them.

The scientific value of wildlife results from the role it serves in the advancement of science. Much of what we know about ecology and behavior came from studying wildlife. Some types of wildlife serve as sentinel species and are used to monitor environmental health. For instance, spotted owls are used by environmentalists to monitor whether we have preserved enough old-growth forests in the Pacific Northwest of the U.S. Because spotted owls have large home ranges, they are one of the first species to be affected when old-growth forests become scarce. Hence, we reason that if there are sufficient old-growth forests to support a healthy population of spotted owls, then there should be a sufficient amount of forests to meet the needs of other species.

The ecological value of wildlife stems from its importance in maintaining the ecosystem. Wildlife is a vital component of a functioning natural ecosystem. For instance, if we want a healthy cougar population, we need a healthy deer population to sustain it.

Therefore, one benefit of deer is that they sustain cougar populations. Likewise, without predation from cougars, deer may become too abundant and may overbrowse some plants and cause those plant species to disappear. Consequently, an ecological value of cougars is their ability to keep the deer population from becoming too large.

Wildlife has an existence value, which is its potential to become valuable in the future. Even the most useless land in the world can be sold for money because of its future potential to become valuable. Likewise, wildlife species may have future values which we fail to recognize today. For instance, armadillos and humans are the only two species known to be susceptible to leprosy. If we could not use armadillos for medical research, we would know much less about how to treat leprosy in humans. Each wildlife species has a unique combination of genes and proteins that can never be reproduced and may someday prove useful to humans.

Wildlife has a historic value that stems from mankind's fear of change. We are uneasy about how the actions of man might be changing the environment, and we long for the way things used to be. Thus, we gain a greater sense of well-being from knowing that there are still wilderness areas and that healthy wildlife populations exist as they have for thousands of years. One example of this value is the satisfaction many people feel from the reintroduction of wolves into Yellowstone National Park. People have a sense that the world has been made right and that things have been returned to the way they were.

Some authors have considered historic value synonymous with existence value. It is, however, different. One way to illustrate this is to ponder why we reintroduced wolves into the Yellowstone ecosystem. It was not to preserve wolves, because large populations exist elsewhere. If we wanted to maximize the existence value of wildlife, we should have released a predator that is threatened with extinction, perhaps the Siberian tiger. This Asiatic species is much more endangered than wolves and may thrive in the Yellowstone ecosystem, but the idea of having Siberian tigers in Yellowstone is disquieting. Why? The answer is that historically they were not present. If we placed tigers there, the wildlife resource in Yellowstone would increase in existence value, but the loss in historic value would far outweigh the gain.

Some of these values are tangible, meaning that we can assign a dollar value to them. Others are intangible, meaning that we cannot assign a dollar value to them. Physical utility, monetary, and recreational values are tangible. Scientific, ecological, existence, and historical values are usually intangible, but intangible values are just as real and important as tangible ones.

WHAT IS WILDLIFE DAMAGE MANAGEMENT?

The values listed above are positive values of wildlife, meaning that they provide a benefit to members of society. Wildlife also cause many problems for people, and these can be considered negative values of wildlife. Negative values would include loss of agricultural productivity or destruction of property due to wildlife damage and human injuries or fatalities caused by wildlife-related diseases or wildlife–automobile collisions. To determine the true value of wildlife for society, we need to add up all of the positive values of wildlife and then subtract from them all of the negative values. This determines the net value of wildlife. The goal of wildlife management is, in essence, to increase this net value of wildlife for society.

Most efforts in wildlife management attempt to do this by increasing the positive values of wildlife. We try, for instance, to increase the value of game species by managing them to produce the maximum sustainable yield so that hunters can obtain the maximum recreational value. On the other hand, the field of wildlife damage management tries to increase the net value of wildlife for society by reducing the negative values of wildlife. Thus, wildlife damage management can be defined as the science and practice of increasing the value of the wildlife resource by reducing the negative values of wildlife (Sidebar 1.3).

Sidebar 1.3 Mission and Vision Statement of the U.S. Department of Agriculture's Wildlife Services

"The mission of Wildlife Services (WS) is to provide federal leadership in managing problems caused by wildlife. WS recognizes that wildlife is a significant public resource, greatly valued by the American people. By its very nature, however, wildlife is a highly dynamic and mobile resource which can cause damage to agricultural and industrial resources, pose risks to human health and safety, and impact other natural resources, including endangered species. WS fulfills a federal responsibility for helping to solve problems which occur when human activity and wildlife are in conflict with one another.

"WS's vision recognizes that the entire field of wildlife management is in a period of great change. Wildlife damage management must increasingly take into account a wide range of legitimate public interests which seriously contend with each other. These interests include wildlife conservation, biological diversity, and the welfare of animals as well as the use of wildlife for purposes of enjoyment, recreation, and livelihood.

"WS strives to develop and utilize wildlife damage management strategies which are environmentally, socially, and biologically sound. In its vision, WS' strategies will be designed to prevent any loss of human health, safety, or the resource base while minimizing any loss of wildlife. This vision represents the future towards which WS is moving. In charting this course, WS must continuously improve upon and modify damage management strategies which are constrained by current technologies, knowledge or resources and do not reach this high standard" (U.S. Dept. of Agriculture 1994).

By this definition wildlife damage management is a very large field. Anything that a wildlife species does that causes human injuries or illnesses, economic-productivity loss, physical damage, or a reduction in a person's quality of life or well-being would be considered wildlife damage. Wildlife damage management also includes the negative impacts of one wildlife species upon another, like the impact of mammalian predation on nesting ducks. Because people like ducks and want large duck populations, wildlife damage occurs when duck populations are reduced by mammalian predation on nesting ducks or their eggs.

WHY WORRY ABOUT HUMAN–WILDLIFE CONFLICTS?

A human–wildlife conflict occurs whenever an action by humans or wildlife has an adverse impact upon the other. Human–wildlife conflicts occur when coyotes kill sheep, raccoons destroy someone's garden, a beach is closed because it is littered with goose feces, or mice chew a hole in a cereal box. Human–wildlife conflicts also occur when humans do something that has an adverse impact on wildlife. For instance, a human–wildlife conflict occurs when a wildlife habitat is converted into an asphalt parking lot. Humans should be concerned about the impact of their actions on wildlife even from a purely anthropocentric standpoint (see Sidebar 1.2) because wildlife provide many positive benefits to society. In actuality, whenever a human–wildlife conflict occurs, both parties (humans and wildlife) lose. Consider a collision between an automobile and a deer. Both the driver and society lose because of the economic damage to the car and also because of the risk of human injury or death. But the deer also loses because most deer struck by vehicles are killed. Society also suffers from the loss of the deer because the future pleasure and enjoyment that the deer might have brought someone are lost. Likewise, when coyotes kill sheep, the rancher suffers losses, as does society through higher food costs. But the coyote will also lose because the rancher will respond by trying to kill it.

One approach to human–wildlife conflicts is to create preserves, wildlife refuges, or parks where human impact on wildlife is minimized. Although this approach is well intended, it does little to resolve human–wildlife conflicts because societal demands for natural resources are so great that only a small fraction of the environment can ever be set aside in parks. There is also the problem that wildlife do not respect our boundary lines and will not stay inside parks. In fact, the vast majority of wildlife live outside parks — the same place people live. Wildlife populations thrive in our most densely settled cities. Clearly, if human–wildlife conflicts are going to be resolved, ways must be found for humans and wildlife to coexist harmoniously without either having an adverse impact on the other. This is the goal of the field of wildlife damage management.

CONTRIBUTIONS OF WILDLIFE DAMAGE MANAGEMENT TO THE LARGER FIELD OF WILDLIFE MANAGEMENT

In recent decades, the field of wildlife damage management has become an integral part of the wildlife profession. As such, it has had a profound impact on the entire field of wildlife management. Historically, wildlife management was concerned with increasing game populations for hunters by protecting the animals from overharvesting and by protecting and managing wildlife habitats. This approach has been widely successful, and many wildlife populations have soared above historic levels. Some people would say that wildlife managers have been too successful and that the current problem is not too few animals but too many. Several authors have suggested that if wildlife managers are going to maintain their relevance to society and to survive as a profession, they must change their

Figure 1.1 How people perceive a human–wildlife conflict depends on their perspective. A wood duck hunter sees this beaver pond as a great place to hunt, but a timber producer sees a disaster because of the trees killed by the impoundment. (Photo courtesy of Gary San Julian.)

traditional emphasis of managing populations to enhancing the value of the wildlife resource for society (Minnis and Peyton 1995; Decker et al. 1996; Messmer 2000). A major difficulty in achieving this optimization of values is that the benefits and liabilities of wildlife do not fall evenly upon everyone in society (Figure 1.1). Because of this, some people want more wildlife and others want less. Trying to reach management decisions when stakeholders cannot agree on what is in the best interest of society is difficult. However, wildlife damage managers have been doing this for years, and the procedures they have developed can serve as a guide for the rest of the wildlife profession.

With changes come opportunities. Wildlife managers have long been accused of managing the wildlife resource for hunters, who compose only a small and decreasing segment of society. While most people do not hunt, many are confronted with wildlife damage and are unable to solve their wildlife problems by themselves (Conover 1997). Clearly, this is a call for help that governmental wildlife agencies can answer. In doing so, wildlife agencies will become more relevant to all segments of society.

ALTERNATIVE DEFINITIONS FOR WILDLIFE DAMAGE MANAGEMENT

Some people have defined wildlife damage management as the science and management of overabundant species (Wagner and Seal 1991), but this definition is too narrow. All wildlife species act in ways that harm human interests. Thus, all species cause wildlife damage, not just overabundant ones. One interesting example of this involves endangered peregrine falcons in California, which prey on another endangered species, the California least tern. Certainly, we would not consider peregrine falcons as being overabundant, but we wish that they would not feed on an endangered species. In this case, one of the negative values associated with a peregrine falcon population is that its predation reduces the population of another endangered species. The goal of wildlife damage management in this case would be to stop the falcons from eating the terns without harming the falcons.

Another problem with defining wildlife damage management as the management of overabundant species is that it implies that wildlife damage is occurring because there are too many individuals of a particular species. It also implies that the way to stop the problem is to reduce the wildlife population to an acceptable level. As we will see in later chapters, wildlife damage is rarely resolved by reducing wildlife populations.

Some people also have defined wildlife damage management as the science and management of vertebrate pests. A pest population can be defined as one whose negative values outweigh its positive values. I believe that few wildlife species are pests, because the positive values of a wildlife population almost always outweigh its negative values. I can only think of two groups of animals that might be considered pests, and both groups are exotic species. The first group is commensal rodents, such as mice and rats, especially when they inhabit our homes and buildings (Figure 1.2). The second group consists of exotic species that have escaped into a new area and are causing great environmental damage. An example would be the brown tree snake in Guam, which has already caused the extinction of most of the island's avifauna. However, even for these species, only certain populations may be considered pests, rather than the entire species. For instance, brown tree snakes are protected in Australia, where they are native. For this reason, I doubt that any wildlife species can accurately be labeled as a pest species. Is there a wildlife species for which humans would rejoice if it were driven to extinction? I do not think so. Pest species do occur, most frequently as organisms which cause human diseases. Few people would regret it if the viruses which cause AIDS, rabies, or the common cold disappeared from the face of the earth. As one clear example of this, the World Health Organization spent millions of dollars to drive the species that caused smallpox in humans to extinction.

WHAT IS IN A NAME?

There is a constant debate about whether the field of wildlife damage management should have a different name. One example of the evolution of names is seen in the changing names for the federal agency responsible for wildlife damage (U.S. Department of Agriculture 1994). When created, the federal agency was called Economic Ornithology, but was later renamed Economic Ornithology and Mammalogy as the group started addressing problems caused by mammals. In fact, as work on mammal problems intensified, the next name change was to Predator and Rodent

Figure 1.2 Commensal mice and rats are among the few species which might be called "pest." (From Thompson, H. V., 1990. Proceedings of the Vertebrate Pest Conference. With permission.)

Control. This name was then changed to Animal Damage Control, a significant change that signaled a recognition that the field's responsibility was to control damage, not necessarily to control populations. Most recently the name was changed to Wildlife Services to reflect the agency's mission to provide services to people. The change from "Animal" to "Wildlife" was designed to indicate that the agency's jurisdiction is wildlife, rather than all animals.

I like the name "resolution of human–wildlife conflicts," but it is too long. The most widely accepted name for the field is wildlife damage management. For instance, that is the name the Wildlife Society has given to its working group dealing with these issues. I favor that name because it reflects our concern about wildlife, as opposed to other animals, and that we are trying to manage damage, not necessarily the animals themselves. Lastly, I prefer wildlife damage management to wildlife damage control because "control" implies a level of success that we rarely achieve in wildlife management. That is, while we manage damage, and hopefully reduce it to an acceptable level, we cannot control it. Another problem is that the word "control" sometimes connotes "kill." I do not think we should use the name "animal control" because lethal control should be the last resort in trying to resolve a problem. However, the debate over the best name for this field continues. Again, names used for different conferences on wildlife damage management indicate the debate, such as Vertebrate Pest Conference, Eastern Wildlife

Damage Control Conference, Great Plains Wildlife Damage Control Workshop, and British Pest Conference. Problem wildlife management is another common name for the field (Dorrance 1983). Perhaps our only agreement that there is no ideal name.

WHAT ARE THE NECESSARY INGREDIENTS FOR DAMAGE BY WILDLIFE?

For wildlife damage to occur, three ingredients are required: a wildlife species doing the damage, an object being damaged, and a person being adversely affected. The objects being damaged may be agricultural crops, livestock, buildings, humans, timber, wildlife, or other natural resources. The affected person often is the owner of the object being damaged, but it may be anyone in society. Examples of people adversely affected by wildlife include a farmer suffering crop damage, a homeowner with a raccoon living in his chimney, a hunter unable to shoot a pronghorn because most of the fawns have already been killed by coyotes, and a person saddened because a bird population is declining as a result of vegetation loss from deer over-browsing. It is important to note that if no person is being adversely affected by wildlife, then wildlife damage is not occurring (Figure 1.3).

However, wildlife damage can occur even if no physical object appears to be damaged. For instance, if people are afraid to walk in their yards because they fear poisonous snakes (even if that fear is irrational), then wildlife damage has occurred because those people have suffered a loss of security and a reduction in the quality of their lives (i.e., their fear keeps them from enjoying life to its fullest). In this

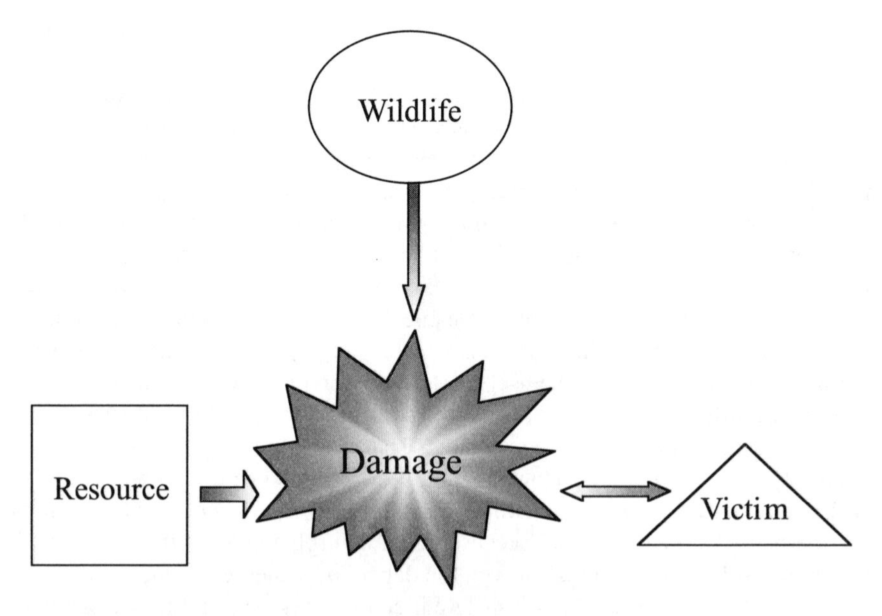

Figure 1.3 For wildlife damage to occur, there must be a combination of wildlife, a resource being damaged, and a person suffering a loss.

case, the "object" that has been damaged by their fear of snakes is their perception or mental image of their yard. That is, when they picture their backyard, they do not see a pleasant and reassuring place; rather, they see a terrifying site where a deadly poisonous snake may be lurking.

This book explores several options for resolving human–wildlife conflicts. We can try to remove the animal causing the problem or change its behavior, we can try to modify the object being damaged to reduce its vulnerability, or we can change human behavior or perceptions so that people are willing to tolerate the damage. Any of these approaches can be used to resolve wildlife damage problems.

THE ROLE OF GOVERNMENT IN WILDLIFE MANAGEMENT

Two premises of a free society that are applicable to wildlife damage management are, first, that individuals should be allowed to do whatever they want so long as their actions do not infringe on another person's right, and second, that an individual's behavior can be restricted when that behavior is detrimental to others. Governments are created by society to administer its will, to take action on its behalf, to accomplish tasks too large to be tackled by individuals, and to ensure that people coexist harmoniously. Governments decide what is in the best interest of society, when the rights of neighbors outweigh the rights of an individual, and what is not acceptable human behavior.

It is difficult for one person to manage wildlife because mammals and birds are so mobile that their home ranges often encompass lands owned by many people. Because wildlife management actions by one person affect others, the management of wildlife cannot be left entirely in individual hands. The role of government in wildlife management is to regulate the harvest of wildlife by people, to restrict human behavior that would be detrimental to the wildlife resource, to conduct large-scale management activities, and to manage wildlife for the benefit of society.

Wildlife management in North America is complicated because most land is owned by individuals who can restrict public access or manage it as they see fit, so long as their actions do not adversely impact other people. Wildlife, however, is not owned by individuals or landowners but by society. This is the result of English common law, which stated that wildlife was owned by the crown. When English colonies were established in America, colonial governments administered on the crown's behalf, and they owned the wildlife (Matthews 1986). When the colonies gained their independence following the American Revolution and became states, they continued to own wildlife but now on behalf of "the people" who had replaced the crown as the ultimate authority on which government was based. Consequently, in North America, we have this interesting phenomenon in which society owns and manages the wildlife, but private individuals own and manage the land and habitat on which wildlife depend (Figure 1.4). Clearly a partnership is required between landowners and government if the wildlife resource is going to be managed wisely.

Given that society is made up of individuals, who vary in their perceptions of wildlife and how they want the wildlife resource managed, society must decide what goals should be adopted in wildlife management. In the U.S., this is

Figure 1.4 Landowners do not own the deer and other wildlife that may live on their property. (Photo by Mark McClure.)

accomplished through our elected officials who assemble yearly and decide what is in the best interest of society. The process of determining what is the greatest good for the greatest number of people is more complicated than just taking a vote. The problem occurs when an issue is of varying importance to different members of society. For instance, consider the question of whether a deer population inside a hypothetical national forest should be reduced by encouraging the harvest of does. This question would be extremely important to the 117 farmers surrounding the forest who are facing financial ruin because the deer are eating their crops; it also would be important to the 1500 people who hunt deer in the forest. At the same time, there may be 250 million people in the U.S. who have never heard of the place and will never visit it, but if you ask them, most will say they do not want deer to be hunted there.

The problem is how to balance the rights of a small group of people who are deeply affected by an issue or care deeply about it, with the rights of the majority who do not really care very much about the issue. In our hypothetical example, deer hunters and farmers might lobby their legislator, who has to decide what he or she thinks is in the best interests of his or her constituents. If this is a crucial issue for the constituents, it may affect the legislator's chances of being reelected. Therefore, the legislator may use the power of persuasion to gain the support of other legislators. The legislator also may engage in vote trading in which he or she will support a bill important to other legislators if they will support his or her bill. If voters elsewhere do not really care, other legislators might go along and the view of the minority will prevail. If, however, the issue is important to voters elsewhere, the view of the majority will carry. Through this process of give and take, our elected officials reach decisions on what is in the best interest of society. If, however, they make the wrong decision, voters will select someone else to represent them in the next election and the law will be changed. This process of lawmaking may seem inefficient, chaotic, and full of abuses. But it results in a system that allows government to decide what is the greatest good for the greatest number of people. It is also worth noting that no other system of government has been devised which does a better job of ensuring that the will of society prevails.

THE ROLE OF GOVERNMENT WILDLIFE BIOLOGISTS

We also need to consider the respective roles played by elected government officials and government wildlife biologists. It is the duty of legislative bodies to translate values into public policies that drive wildlife management programs and to answer the major questions concerning the management of the wildlife (Wagner and Seal 1991). It is their responsibility to decide what the goals of wildlife management are and what is in the best interest of society. For instance, the government should decide whether the best use of deer is to allow them to be hunted or whether the interests of hunters should be sacrificed to please those people who oppose hunting.

The job of government wildlife biologists is twofold: 1. to provide the scientific and management information which the legislative bodies and our citizens need to make informed decisions, and 2. to use their expertise about wildlife to decide how best to achieve the goals set by elected officials. For instance, if elected officials decide that the deer population should be managed for the maximum sustained yield and that deer should be hunted, wildlife managers have the responsibility of determining how best to accomplish this. This includes making decisions on how long the hunting season should be, how many deer each hunter should be allowed to kill, what hunting methods should be legal, and how public land should be managed to enhance deer hunting opportunities. They also have the responsibility to ensure that the laws are obeyed. In sum, elected officials make strategic decisions, while wildlife biologists employed by the government make tactical decisions. At the same time, wildlife biologists are members of society like everybody else and have a right as citizens to make their views known and to try to influence policy. They should do so, however, in the role of citizens and not as government employees.

SUMMARY

The wildlife resource provides many benefits for society and has many types of positive values — physical utility, monetary, recreational, scientific, ecological, existence, and historic. Wildlife also causes many problems for people, and these can be considered negative values of wildlife. Anything that wildlife does that causes human injuries or illnesses, loss of economic productivity, physical damage, or a reduction in a person's quality of life or well-being is wildlife damage. The goal of wildlife management is to increase the value of the wildlife resource for society while protecting it for the benefit of future generations. Wildlife damage management, which is an integral part of the field of wildlife management, is devoted to solving problems caused by wildlife and thereby increasing the net value of wildlife by reducing its negative values. Its goal is resolving human–wildlife conflicts so that humans and wildlife can coexist without either having an adverse impact on the other.

Wildlife management in the U.S. is complicated because government owns and controls wildlife but wildlife habitats are owned and controlled by landowners. Legislative bodies have the responsibility to make strategic decisions about how wildlife should be managed to achieve the greatest good for the greatest number of people. Wildlife biologists who work for the government use their scientific knowledge about wildlife to make tactical decisions about how best to achieve the broad wildlife goals set by the government.

LITERATURE CITED

Brown, T. C., The concept of value in resource allocation. *Land Econ.,* 60, 648–656, 1984.

Conover, M. R., Wildlife management practices of metropolitan residents in the United States: practices, perceptions, costs, and values. *Wildl. Soc. Bull.,* 25, 306–311, 1997.

Conover, M. R., Perceptions of American agricultural producers about wildlife on their farms and ranches. *Wildl. Soc. Bull.,* 26, 597–604, 1998.

Decker, D. J., C. C. Krueger, R. A. Baer, Jr., B. A. Knuth, and M. E. Richmond, From clients to stakeholders: a philosophical shift for fish and wildlife management. *Hum. Dimen. Wildl.,* 1, 70–82, 1996.

Dorrance, M. J., A philosophy of problem wildlife management. *Wildl. Soc. Bull.,* 11: 319–324, 1983.

Giles, R. H., Jr., *Wildlife Management*. W. H. Freeman, San Francisco, 1978.

Giles, R. H., Jr., A new focus for wildlife resource managers. *J. For.,* 88, 21–26, 1990.

Krauthammer, C., Saving nature, but only for man. *Time*, June 17, 1991, 82, 1991.

Leopold, A., *A Sand County Almanac and Sketches Here and There*. Oxford University Press, New York, 1949.

Matthews, O. P., Who owns wildlife? *Wildl. Soc. Bull.,* 14, 459–465, 1986.

Messmer, T. A., The emergence of human–wildlife conflict management: turning challenges into opportunities. *Int. Biodeterioration Biodegradation,* 45: 97–102, 2000.

Minnis, D. L. and R. B. Peyton, Cultural carrying capacity, modeling a notion, in *Urban Deer: A Management Resource?*, J.B. McAninch, Ed., North Central Section, The Wildlife Society, St. Louis, Missouri, 19–34, 1995.

Thompson, H. V., Animal welfare and the control of vertebrates. In *Vertebrate Pest Conf.,* 14, 5–7, 1990.

U.S. Department of Agriculture, *Animal Damage Control Program — Final Environmental Impact Statement*, Volume Two. U.S. Dept. of Agriculture, Washington, D.C., 1994.

Wagner, F. H. and U. S. Seal, Values, problems, and methodologies in managing overabundant wildlife populations: an overview. p. 279–293 in *Wildlife 2001: Populations*, D. R. McCullough and R. H. Barrett, Eds., Elsevier Applied Science, New York, 279–293, 1991.

CHAPTER 2

History

"History may not repeat itself, but it does rhyme."

Mark Twain

"The Knights of the Round Table may have been the world's first professional NWCOs (nuisance wildlife control operators). After all, they were the ones who were always called upon when there was trouble with fire-breathing dragons."

Anonymous

"Americans will always do the right thing.... After they have exhausted all the alternatives."

Winston Churchill

Our perceptions of wildlife and our wildlife philosophy are products of history. Current wildlife management practices and beliefs make sense only when viewed in the context of historical development. Knowledge of the wildlife attitudes of past generations and how wildlife was historically managed may provide insight into how wildlife management might change in the future. In this chapter, I focus on the history of wildlife damage management and the broader field of wildlife management in the United States and throughout the world.

PREHISTORIC WILDLIFE MANAGEMENT

We describe early humans as hunter–gatherers, but they did much more gathering than hunting and probably did not venture very far from safety. Foraging for seeds, scavenging carcasses, or killing small prey would have provided protein. However, as humans learned how to use tools (e.g., clubs, stones, spears, and fire), to communicate with each other, and to coordinate the behavior of large numbers of people

Figure 2.1 A pictograph from Newspaper Rock in Utah showing a Native American on a horse hunting two desert bighorn sheep.

in communal hunts, they became efficient predators. For the first time, they were able to protect themselves from other predators and kill large mammals for food (McCade and McCade 1984).

So efficient were these prehistoric hunters that they often suppressed the wildlife populations they hunted. For instance, Native Americans hunted ungulates with such skill (Figure 2.1) that large populations of ungulates occurred only in "no-man zones," which existed along the territorial boundaries of warring tribes (Kay 1998). In such areas, hunting was limited because of the danger of being ambushed by enemies. West (1995) believed that bison would not have survived on the central Great Plains without the presence of these no-man zones.

Given the importance of wildlife for food, it is not surprising that early man burned the vegetation at regular intervals to produce more game and to concentrate animals for easier hunting. Burning by Native Americans was so extensive that fire-tolerant vegetation and early successional plant communities existed in large parts of North America when the European colonists first arrived.

We know that early humans managed prey populations, but did they also take steps to reduce competition from other predators? If so, they were engaging in wildlife damage management. Native Alaskans apparently kept bear populations at low levels (Birkedal 1993 as cited by Kay 1998), but they might have been hunting these bears for food, furs, sport, or prestige, making it unclear whether they were engaging in wildlife damage management. However, a clear example of wildlife damage management by Native Americans can be found at the American Museum of Natural History in Washington, D.C., where there is a display of Inuit (Eskimo) artifacts that were certainly used for wildlife damage management. These devices were made by taking baleen from whale carcasses and cutting it into thin pieces, 10 to 15 cm in length, and sharpening each end to a point. These dagger-like tools were folded back upon themselves repeatedly to create "S" shapes and placed in pieces of blubber which were allowed to freeze solid. The blubber was then distributed in areas where wolves or polar bears would find it. After the blubber was swallowed, the baleen would spring back to its original shape, piercing the predator's digestive system. A predator so injured would take several days to die, and Inuits would not have an opportunity to find the carcass. Hence, the motivation for using these items could not have been a quest for meat or fur. Instead, the sole function of these items was to kill predators, which the Inuits feared either as competitors for food or as a potential predators of humans.

Elsewhere, Native Americans have used poisons and traps since prehistoric times to kill troublesome animals. For instance, the first English colonists in New England reported that the Native Americans were using pitfall traps to catch wolves (Conover and Conover 1987, 1989).

WILDLIFE DAMAGE MANAGEMENT IN THE ANCIENT WORLD

When people started growing crops and raising animals, probably 10,000 to 15,000 years ago in the Middle East, they gained reliable sources of food throughout the year but faced new threats of wildlife damage. For farmers, wildlife posed a threat both during the growing season and after the crops were stored. In response, humans developed several methods to reduce wildlife damage. Oral tradition indicates that scarecrows and poisons were used prior to recorded time to reduce wildlife damage. One of Aesop's fables, written about 570 B.C., describes a farmer using a net to capture and kill cranes depredating his grain field.

Rodents posed a serious threat to stored grain, and some of the first known attempts at wildlife damage management were directed at killing mice and rats. During the third millennium B.C. (4000 to 5000 years ago), Egyptians used cats (Keeler 1931, as cited by Hygnstrom 1990), poison (Fitzwater 1990), and traps to reduce the numbers of mice and rodents (Drummond 1992). During the same period, Chinese were making ceramic mouse traps (Anonymous 1967), and inhabitants of the Indus River Valley were using choker traps designed to strangle animals (Drummond 1992).

When humans first started farming, fields were small because crops had to be planted and cared for by hand, aided only by primitive tools. Consequently, even a small number of animals could quickly destroy a year's worth of work. Information on how Native Americans stopped birds from eating their corn is available from paintings by two early European explorers. These paintings show tall platforms, often with small huts, erected in the mature corn fields. These platforms reportedly were occupied from dawn to dusk by children, whose job it was to throw stones at birds and scare them out of the corn fields.

In the Old World, children were also the main deterrent against bird damage to crops. This practice lasted in Europe until the nineteenth century. For instance, William Lawrence, who later won fame during the 1815 battle of Waterloo, came from a poor English family and had to start working full time when he was still a young boy. His first job was to scare birds at a local farm (Howarth 1968).

Wildlife damage also was a difficult problem for early pastoralists. Captive animals would have been tempting targets to large predators, especially during periods of food scarcity. Learning how to keep predation on captive herds within tolerable levels was undoubtedly one of the major breakthroughs that allowed humans to switch their dependency from hunting to herding.

The Bible describes the life of the tribes of Israel, who were pastoralists and raised sheep and goats in the Middle East. Not surprisingly, the Israelis' herds were often attacked by large predators, especially lions and bears. Lions also attacked humans; in fact, one of the prophets of God was killed by a lion (1 Kings 13: 24–25). Shepherds were the main defense used to prevent predation on livestock; these individuals constantly stayed with the flocks and attempted to scare off or kill predators by hurling stones at them with slings. In his youth, David was a shepherd and killed both lions and bears with his sling—the same weapon he used later to kill Goliath (1 Samuel 17: 41–51). Samson also is credited with killing a young lion using only his bare hands (Judges 14: 5–6).

WILDLIFE DAMAGE MANAGEMENT IN MEDIEVAL EUROPE

During the Middle Ages, there was considerable use of toxicants to control rodents. One formula called for "…a paste made of honey, copperas, and ground glass…" Another called for adding powdered mercuric chloride to a paste made from sweet butter, oatmeal, apple pulp, flour, and sugar. Other popular poisons included finely ground hemlock and oleander. The common names of some plants also speak of their uses as toxicants. Wolfsbane literally means "the scourge of wolves" and was used to kill wolves as well as foxes and rodents. Ratsbane ("scourge of rats") is a plant that contains arsenic and was used to kill rats (Fitzwater 1990).

Lethal means were also used to reduce damage by birds. Laws were established as early as 1424 in Scotland calling for the killing of rooks. In England, an act was passed in 1533 ordering local parishes to set nets to capture jackdaws, crows, and rooks. Noncompliance was punished by a heavy fine. This act apparently did not solve the problem because in 1566 another act was passed which ordered church wardens to pay bounties for a list of birds and mammals described as vermin (Wright 1980).

The Middle Ages also saw the development of several ingenious traps to catch mice and rats. Some of these were illustrated and described in Leonard Mascall's book published in 1590 (Drummond 1992). One trap called a "mill to take mice" was a pinwheel that was placed at the edge of a table over a bucket of water. The vanes on the wheel were baited and the idea was that when a mouse reached up to grab the bait, its weight would cause the pinwheel to move and the mouse would fall into the bucket and drown (Figure 2.2A). Another trap, the "fall for rats," was a deadfall trap where a heavy wood block was guided downward by two posts which fit through it (Figure 2.2B). The block was designed to fall and

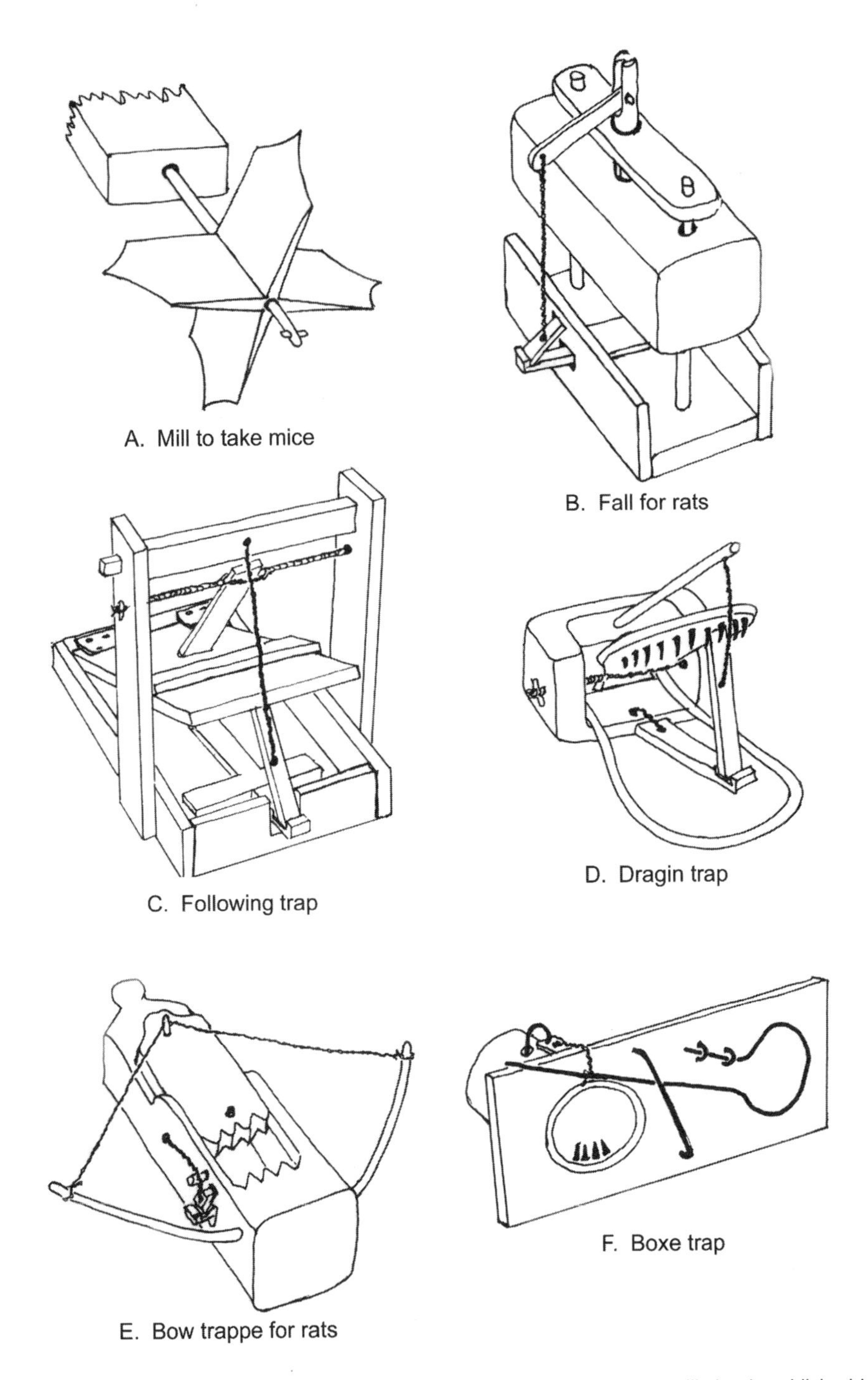

Figure 2.2 Some of the many rat and mouse traps pictured in Mascall's book published in 1590. (From Drummond, D.C., 1992. Proceedings of the Vertebrate Pest Conference. With permission.)

crush any rat which stepped on the treadle positioned beneath the block. It had been in use at least since the 1400s (Drummond 1992). The "following trap" was a more sophisticated deadfall trap because the falling block was powered by both gravity and a twisted-cord spring (Figure 2.2C). This trap had a long history; it appeared in an early-1400s painting by the Master of Flemalle and also was illustrated during the 1780s by Roubo (Drummond 1992). The "dragin trap" was an early example of the snap trap (Figure 2.2D). It consisted of a paddle-shaped piece of wood armed with metal teeth. This paddle was released downward when a rodent stepped on a treadle sited beneath the paddle. Extra force was delivered to the paddle by running the back end of it through a twisted rope that functioned as a spring. When set, the trap looked like a dragon's open mouth; this was the origin of the trap's name. This type of snap trap was very successful and had a long history of use extending from the ancient Egyptians through modern times (Drummond 1992). The "bow trappe for rats" was similar except that it was powered by a bow (Figure 2.2E). The "boxe trap" was a guillotine-type trap that was activated when a mouse put its head in a hole to reach the bait (Figure 2.2F). This released a spring-powered rod that struck the mouse on the head and killed it (Drummond 1992).

During the Middle Ages, people often tried to reason with rodents. One practice was to write a "rat letter" asking the rodents to vacate the premises. These were then left where rodents would find them and were always left with the printed side up so that it would be easier for rats to read. If this approach failed to work, another option was the use of scare tactics. Sometimes when a rodent was captured, it was tortured rather than killed and then released to warn the others to leave (Fitzwater 1990).

Many medieval Europeans were superstitious. When all else failed, they relied on magical potions and spells to solve intractable problems, such as those caused by wildlife. The Irish were supposed to rhyme rats to death (Fitzwater 1990). Music was also used to resolve rat problems in the Middle Ages. The story of the Pied Piper of Hamelin is based on a true incident that occurred in Hamelin (Hameln), Germany, during the fourteenth century. According to the legend, the town agreed to pay a pied piper if he would rid the town of rats. The piper used his flute to lure the rats to their deaths. Unfortunately, the story does not have a happy ending because once the rats were gone, town officials refused to pay the piper for his work. In revenge, the pied piper returned a few days later, but this time he used music to lure the children of Hamelin into following him. Neither the piper nor the children were ever seen again.

Sidebar 2.1 Wildlife Damage Management's Greatest Accomplishment in Medieval Europe

According to medieval literature, the most serious wildlife damage problems during the Middle Ages were caused by fire-breathing dragons and lamias. The latter were four-legged, scaly monsters which had the face of a woman and "large and comely-shaped breasts" (Figure 2.3). These were used by the monster to lure men into approaching close enough to be captured and devoured (Topsell 1607 as reported

Figure 2.3 The "lamia" is described by Topsell as a monster that uses its great beauty to attract men (obviously this must not be a good drawing of it), who are then captured and devoured. (From Fitzwater, F.D., 1990. Proceedings of the Vertebrate Pest Conference. With permission.)

by Fitzwater 1990). If the reader will excuse me for having a little fun, I would like to point out that medieval wildlife damage managers were very successful in their efforts to control these wildlife species. After all, there have been no complaints about wildlife damage caused by dragons or lamias in several centuries.

I now shift from a focus on just wildlife damage management to the broader questions of 1. what the European colonists who settled America thought about the New World's wildlife, and 2. how this wildlife should be managed. I discuss how their views changed over the centuries and, more importantly, how our own beliefs about wildlife and human–wildlife conflicts are a product of our past.

WILDLIFE MANAGEMENT IN COLONIAL AMERICA FROM 1620 TO 1776

English immigrants who first colonized North America clearly viewed the New World and its wildlife as hostile. The colonists were accustomed to the orderly world of Europe with its patchwork of fields, pastures, and woods (Cronin 1983). The vast wilderness of the New World filled them with awe and dread. Their perceptions were described by William Bradford, leader of the Plymouth Bay colony, who called the area "a hideous and desolate wilderness, full of wild beasts and wild men" (Geller 1974).

The New England colonists believed that the wilderness was a place of evil and hardship that had to be subdued, conquered, or vanquished before they could create their utopian world (which was their reason for coming to North America). In diaries, addresses, and memorials of the period, the colonists articulated this need to transform and eradicate the wilderness, in other words, to "tame" it. Two such targets of eradication were the Native Americans and "bad" wildlife, which were those animals that threatened the colonists' health, food supplies, or property (Nash 1979; Reed and Drabelle 1984; Conover and Conover 1987, 1989).

For both moral and practical reasons, the colonists "made war" on wildlife. In these early years, starvation was a basic concern of colonists. Any threat to their subsistence, particularly predation of livestock, was very serious. By destroying predators that threatened their livestock, the Puritans were protecting a food source upon which their lives depended. Livestock's importance to the early English settlers was indicated in the journals of John Winthrop, who was the leader of the Massachusetts Bay colony, and William Bradford; they recorded the arrival of every shipment of sheep, goats, swine, and cattle (Walcott 1936; Conover and Conover 1987, 1989).

These attitudes toward predators were translated into action by means of bounties that colonies paid for dead wolves, cougars, and other predators. For example, one of the first laws passed by the Puritans in the New Haven colony established a bounty on wolves and foxes. The intention of the colonists was not merely to manage predator populations but to eradicate them. As wolf populations declined, bounties increased dramatically to encourage the killing of the last few wolves (Conover and Conover 1987, 1989).

Hunting with dogs and trapping were the primary means of predator control in the 1600s. This was demonstrated in the 1648 order to all towns making up the Massachusetts Bay colony to use "so many hounds as they thinke meete [sic]...that so all meanes may be improved for the destruction of wolves" (Trumbull 1850). Other methods of predation control included habitat destruction. In particular, swamps were drained and cleared as a means of eliminating predators (Trumbull 1850; Hoadly 1857; Conover and Conover 1989). The English colonists succeeded in the quest to reduce the threat that large predators posed to their livestock. Wolves, the main predation threat, were practically eliminated from southern New England and many of the mid-Atlantic colonies by the end of the colonial period, although they remained on the frontier (Sidebar 2.2).

Sidebar 2.2 How Wolf Depredation Helped Americans Win the Revolutionary War

A turning point of the American Revolution was General Burgoyne's ill-fated attempt to divide the American colonies by marching his army from Canada down the Hudson River to New York City. To reach the latter, Burgoyne's army had to cross the wilderness which lay between Canada and Albany, New York. Burgoyne's soldiers were from Europe and were unprepared for a march through the frontier. As the march continued, primitive conditions and a fear of the "hostile and alien" wilderness led to an increasing sense of foreboding among his soldiers and a steady erosion of their morale. Still, Burgoyne's army made slow but steady progress and had almost succeeded in making its way to Albany when the path was blocked by

American forces at Saratoga. A series of battles ensued as Burgoyne's army tried to dislodge the Americans. Several of Burgoyne's Hessian soldiers remarked in letters and diaries that the horror of this battle was compounded by wolves coming out onto these battlefields to feed on the dead bodies. These scenes unsettled Burgoyne's men and helped convince them that they were trapped in a hideous wilderness where their only hope for survival was to surrender. Lacking the will to continue the fight or to retreat back through the wilderness to Canada, Burgoyne surrendered his army to the Americans on October 17, 1777. This defeat not only deprived the English government of many of its best troops in North America, but also convinced France to enter the war on the American side (Doughty 1996). Hence, one can argue that the gruesome behavior of New York's wolves terrorized Burgoyne's men and increased their desire to surrender.

Birds, particularly "sterlings" or red-winged blackbirds, also threatened the colonists' food supplies. As John Winthrop reported:

> Crowes, Sterlings, and other Birds ... come in greate flights into the fields, when the Eare beginneth to be full, before it hardneth, and being allured by the Sweetness of the Corne, will sitt upon the stalke, or the Eare it selfe, and so pick at the Corne through the huske at the top of the Eare (for there it is tenderest) and not cease that worke toill they have pulled away some of the huske that they may come at the Corne which will be plucked out so far as they can come at it (Mood 1937).

Colonial governments responded to this threat by offering bounties as an incentive for killing the birds. For instance, New Haven in 1648 offered 10 shillings for every thousand blackbirds killed. Colonial farmers also targeted passenger pigeons because they would come by the thousands to forage in the grain crops (Hoadly 1857; Walcott 1936).

From the perspective of the English colonists, however, not all wildlife was bad. Those species which had commercial value or provided food were "good." Consequently, wildlife management by the English colonies consisted of trying to eradicate "bad species" and to harvest "good species." Beaver was especially important to early New England settlers, owing to the monetary value of its pelts when shipped back to Europe. Therefore, the beaver was considered a "good species." As William Bradford noted in 1623, his settlers had "...no other means to procure them foode [sic] which they so much wanted, and cloaths allso [sic]" than by acquiring beaver pelts for commercial exchange. Beaver pelts in New England, like tobacco in the Chesapeake colonies, were such an important commodity that they were used as legal tender for a time (Conover and Conover 1989). However, the beaver supply was soon exhausted in New England, and the fur trade declined. In Connecticut, the beaver population dwindled within a few decades after the arrival of the first English colonists (Conover and Conover 1989). Like beaver, deer were overhunted because their hides were coveted colonial exports and venison was an important food source (Dexter 1917; Nettles 1927). Despite various belated management efforts by the colonial leadership to maintain deer populations, deer were practically eliminated from southern New England before the American Revolutionary War.

English colonists, in summary, had a utilitarian view of wildlife. Animals were judged by their economic impact. Some were good because their pelts could be sold; others were bad because they caused damage. Given the constant risk of starvation, colonists were not tolerant of wildlife damage and viewed wilderness and wildlife as obstacles that had to be overcome before the colonists could create their utopian civilization.

WILDLIFE MANAGEMENT IN THE U.S. FROM 1776 TO 1880

From the beginning of the United States as an independent nation through the post–Civil War years, American philosophy and attitudes toward wildlife scarcely changed from colonial days. The dominant philosophy continued to be an optimistic one of man creating a better world by taming the wilderness and creating a civilized world. Wildlife retained its dual function for Americans: a source of food or revenue or an obstacle or hindrance to be eliminated. Westward expansion, the predominant theme in American history from the 1770s to the 1880s, guaranteed that this colonial pattern of ambivalence and overexploitation of natural resources would continue as settlers moved across the North American continent.

An important driving force in westward expansion was the desire for beaver pelts, which led to repeated overexploitation of beaver populations (Figure 2.4). The constant need to locate unexploited populations led trappers further and further west. Their descriptions of the trans-Mississippi West, in turn, fueled interest among settlers to move there (Trefethen 1975; Anderson 1991).

The westward-bound American farmers, who followed trappers to the frontier, continued to detest "bad" wildlife. They held the dominant Anglo-American view that the "wilderness" must be conquered. In this dominant mind-set, wolves, cougars, coyotes, and other predators served as "symbols of the savage wilderness" that early Americans sought to tame (Kellert and Berry 1980; Kellert and Westervelt 1982; Feldman 1996).

Consider the American experience during the early nineteenth century when Ohio was the frontier. Insight into the views held by this new wave of American settlers has been provided by historian Stephen Ambrose, who wrote:

> Getting rid of it —with "it" meaning anything or anyone who stood in the way of progress — was a universal American passion and a commonplace experience for all those living in the Old Northwest.
>
> This assault on nature ... owed much to sheer need, but something also to a compelling desire to destroy conspicuous specimens of the fauna and flora of the wilderness....

What was the result of this Anglo-American move into Ohio? Writes Ambrose (1975), "The Ohio Valley today has neither trees nor animals to recall adequately the splendor of the garden of the Indian which the white man found and used so profligately."

One example of this prevailing mind-set that Ambrose described is provided by General Philip Sheridan, Civil War hero and, in the post–Civil War era, commander of the military department of the Southwest. His aim was to eliminate Native Americans by eliminating bison upon which they depended (Sidebar 2.3). Sheridan's efforts to eliminate both were largely successful.

Figure 2.4 Fur trappers were often the vanguard of western expansion in North America. (Courtesy of U.S. Bear River Migratory Bird Refuge.)

Sidebar 2.3 Why Bison Should Be Exterminated — The Views of General Philip Sheridan

In late 1870, General Philip Sheridan, commander of military forces in the Southwest and Civil War hero, traveled to Austin to address the Texas Legislature, which was debating a bill to protect bison herds. According to one source, Sheridan warned the Texas legislature:

> "…that they were making a sentimental mistake by legislating in the interest of the buffalo. He told them that instead of stopping the hunters they ought to give them a hearty, unanimous vote of thanks, and appropriate a sufficient sum of money to strike and present to each one a medal of bronze, with a dead buffalo on one side and a discouraged Indian on the other."

Specifically, Sheridan said:

> "These men [the buffalo hunters] have done more in the last two years and will do more in the next year, to settle the vexed Indian question, than the entire regular army has done in the last thirty years. They are destroying the Indians' commissary…. Send them [bison hunters] powder and lead, if you will; but, for the sake of a lasting peace, let them kill, skin and sell until the buffaloes are exterminated. Then your prairies can be covered with speckled cattle and the festive cowboy, who follows the hunter as a second forerunner of an advanced civilization" (Marcus and Burner 1991).

WHY THE CLOSING OF THE FRONTIER AND THE INDUSTRIAL REVOLUTION SPARKED A NEW PHILOSOPHY OF WILDLIFE MANAGEMENT

Even as Sheridan and others continued to espouse the traditional rhetoric about wildlife, other people's views of wildlife were beginning to change. Consider the words of the editors of the newly created magazine *Forest and Stream*, who stated in 1873 that their objective was to promote a "healthful interest in outdoor recreation and … a refined taste for natural objects." It was hoped that the readers of *Forest and Stream* would become "familiar with the living intelligences that people the woods and the fountains" (*Forest and Stream* 1873). Clearly such had not been the typical attitude of Americans toward wildlife in past decades. While the utilitarian view remained, a new view of wildlife was emerging.

Several factors accounted for the emergence of this new attitude toward wildlife, including the closing of the frontier and the excesses of the Industrial Revolution. By 1890, America had ceased to be a "frontier" country. As the national census announced, the frontier had been closed — wilderness had finally been conquered. The goal of Americans for 250 years had been obtained. But rather than celebrating or having a sense of accomplishment, Americans began to consider what had been lost. During this period, the Industrial Revolution was in full swing and America surpassed Britain as the world's leading industrial power. The Industrial Revolution and concomitant urbanization created many problems: long working hours, dangerous working conditions, low wages, child labor, squalor, and poverty. In retrospect, Americans were beginning to question whether "progress" was such a good idea and to view the lost wilderness as an object of beauty and reverence.

Along with these new forces of modernization came the clear realization that wildlife populations were not inexhaustible. The visible overexploitation of natural resources would help transform attitudes and result in new policies for the management of America's resources. Underlying the earlier mentioned Sheridan-Texas legislature debate on the protection of bison (Sidebar 2.3) was the realization that in just a few years, from 1872 to 1874, nearly 4 million bison were slaughtered. Even earlier, in the late 1850s, the Ohio legislature had debated a bill to protect the passenger pigeon, a bird whose numbers had once seemed unlimited but would become extinct by the twentieth century (Trefethen 1975; Marcus and Burner 1991).

During the latter half of the 1800s, sport hunting became popular and helped create a more positive attitude toward wildlife. The transformation of hunting from a commercial or life-sustaining activity to a sport was, according to Dunlap (1988), one of the first steps toward wildlife preservation. The greatest advocate of this new view of hunting was Henry William Herbert, who wrote under the pseudonym of Frank Forester. He was an English writer who had moved to the U.S. in the mid-1800s. In his writings, he urged fellow Americans to hunt only game animals and only by "sporting methods" (e.g., not shooting sitting ducks). He also urged hunters to treat their dogs and horses humanely; cruelty to animals, in Herbert's view, indicated that a man was not "a true sportsman and gentleman."

Forester's advocacy of hunting and sportsmanlike conduct began to spread among the upper class, who began to appreciate wildlife and to adopt a more positive attitude towards it. Sportsmen's clubs began to appear, most prominent of which was the Boone and Crockett Club, founded in 1887 by Grinnell, editor of *Forest and Stream*, and Theodore Roosevelt, future U.S. president. Roosevelt and others like him felt that hunting, like warfare, provided an arena for forming and testing the character of Americans that would substitute for the now vanishing frontier (Reiger 1975; Belanger 1988; Dunlap 1988).

Meanwhile, to save their sport as the supply of game declined rapidly, hunters began to take action. They organized and called upon local, state, and federal governments to save wildlife by outlawing such unfair or "unsporting" activities as jacklighting, hunting deer with dogs, and using baits. Other helpful regulations included lowering bag limits, shortening the length of hunting seasons, and restricting the kind of firearms that hunters could use. Finally, these hunting organizations wanted these new laws enforced, preferably by a professional set of wardens under the direction of a state game commission (Dunlap 1988). As a result of these actions, a conservation effort slowly began to emerge at the state, and then national, level.

In response to changes in American attitudes toward wilderness and wildlife, the federal government initiated some important policy changes for managing the nation's natural resources. The most famous change was the establishment of Yellowstone National Park in 1872. Meanwhile, numerous forest reserves were established to manage and protect America's timber resources. Yet another indication of policy change was the federal government's creation in 1885 of a wildlife agency, the Division of Economic Ornithology and Mammalogy, in response to pressure from the American Ornithologists Union (Anderson 1991). This agency would ultimately evolve into the U.S. Fish and Wildlife Service and the U.S. Department of Agriculture Wildlife Services (Sidebar 2.4).

Sidebar 2.4 History of the U.S Department of Agriculture Wildlife Services Program

The first involvement of the U.S. government in wildlife damage management occurred in 1885 when the newly created U.S. Department of Agriculture Branch of Economic Ornithology sent questionnaires to farmers about bird damage (Di Silvestro 1985). The next year, the Branch was renamed the Division of Economic Ornithology and Mammalogy. The agricultural commissioner stated that the new division's responsibility would be to educate farmers about birds and mammals affecting their interests to minimize the destruction of useful species. Efforts to educate farmers included conducting research and demonstrating wildlife damage control techniques. In 1905, the division's name was changed to the Bureau of Biological Survey, which ultimately evolved into the U.S. Fish and Wildlife Service. The first congressional appropriation for predator control occurred in 1915 when $125,000 was given to the Bureau of Biological Survey for that purpose. In 1920, an eradication methods laboratory was established in Albuquerque, NM, to improve methods and techniques for the control of predators and rodents. A year later, this lab was moved to Denver and much later renamed the Denver Wildlife Research Center (U.S. Department of Agriculture 1994). During the 1990s, the center was moved to Fort Collins, CO, and given the new name: National Wildlife Research Center.

In 1930, the American Society of Mammalogists opposed the $1 million federal predator and rodent control program, and this nearly ended it. Supporters of the program rallied, however, and the Animal Damage Control Act of March 2, 1931 was passed, giving the federal government authority to conduct wildlife damage control activities and to enter into cooperative agreements with state and local governments (U.S. Department of Agriculture 1994).

President Franklin D. Roosevelt transferred the Bureau of Biological Survey from the U.S. Department of Agriculture to the Department of the Interior in 1939 as part of efforts to consolidate all wildlife activities in one department. The federal Animal Damage Control (ADC) program, especially its use of poisons, came under close public scrutiny during the environmental awareness movement of the 1960s. In response, Secretary of the Interior Steward Udall appointed a group to investigate the federal program. Their findings, called the Leopold Report, contained six recommendations:

1. create an advisory board, 2. reassess the goals of the ADC program, 3. revise ADC's guidelines, 4. increase research efforts, 5. obtain legal control for the use of certain toxicants, and 6. change the organization's name (Hawthorne et al. 1999). Udall accepted the report as a "guidepost," Jack H. Berryman was selected as the new chief, and the ADC program was modified (Hawthorne et al. 1999).

Despite the changes, however, criticism continued. In 1971, another advisory committee was created and published the Cain Report which also was critical of the use of poisons as inhumane and nonselective and favored the use of leg-hold traps (Wade 1980). President Richard M. Nixon signed an executive order in 1972 banning the use of toxicants for predator control by federal employees or on federal lands. This order was amended by President Gerald R. Ford to allow the use of sodium cyanide in a predator control device called the M-44. President Ronald W. Reagan revoked Nixon's executive order and Ford's amendment.

During the 1980s, agricultural interests were concerned that the ADC program was not being adequately funded by the U.S. Fish and Wildlife Service and made their opinions known to their congressional delegations. Congress responded by directing the secretaries of agriculture and interior to assess the pros and cons of moving the ADC program back to the U.S. Department of Agriculture. In 1985, Congress made its decision and moved all ADC programs back to the U.S. Department of Agriculture (U.S. Department of Agriculture 1994). In 1997, the name of the ADC program was officially changed to "Wildlife Services."

Despite America's expanded consciousness about wildlife, the division of animals into "good" and "bad" groups continued. The difference was that "good" animals were now those species that could be hunted or provided sport. "Bad" animals were those that preyed upon or competed with the "good" animals. Hence, government policy was still dualistic; actions were taken to protect some species and to eradicate others. In particular, wolves and cougars were still targeted as "threats" to be removed through the same methods used since colonial times: trapping, hunting, and poisoning.

However, another philosophy was beginning to emerge, which held that nature and wilderness possessed special values of beauty and spiritualism and should be preserved for their own sake, independent of any value which they might provide man. This idea can trace its origins to the antebellum period, when European romanticism inspired writers, such as Henry David Thoreau and Ralph Waldo Emerson, to view nature (and wildlife) in spiritual terms. This aesthetic appreciation of nature grew among writers and artists during the post-Civil War period (Trefethen 1975; Anderson 1991). But America's growing appreciation of nature and increased concern for the environment did not constitute a smooth process. The world wars would bring different changes.

CONSEQUENCES OF THE WORLD WARS AND THE GREAT DEPRESSION ON WILDLIFE MANAGEMENT

The world wars had important repercussions for America's wildlife policy because European economies were forced to emphasize war production over agriculture and to send much of their agricultural labor force to the military. The result was worldwide shortages of food and soaring food prices. As America tried to feed both itself and its allies, concern for livestock waxed and concern for wildlife waned. Throughout history, wildlife damage becomes more important during poor economic times or when human food supplies are threatened; the world wars were no exception. During World War I, Congress allocated $125,000 to deal with predatory animals, and for the first time, the federal government hired professional hunters to kill predators (Anderson 1991). The government, justifying these actions on economic grounds, met with little opposition (Dunlap 1988; Feldman 1996). Likewise, during the Great Depression, economic needs and the desire to create new jobs took precedence over concern for the environment.

This pattern of sacrificing wildlife interests for economic ones is repeated across time and throughout the world. People, either as individuals or as a society, almost always place their own survival over environmental protection. The bottom line is that many people view wildlife as a luxury that becomes important when they can afford it but not when the necessities of human life are not being met. Therefore, concerns about the environment are more prevalent in wealthy nations than in poor ones.

Sidebar 2.5 How Wildlife Damage during World War II Changed the Course of History

During World War II, armed forces were sent around the globe by the warring nations and often faced wildlife damage problems for which they were unprepared. This problem was particularly acute for Nazi Germany because, unlike the United States, Germany did not have the equivalent of the federal U.S. Department of Agriculture Wildlife Services program which they could call upon to help resolve these problems. This oversight helped produce the turning point of the war — the surrender of the German Sixth Army during the battle of Stalingrad.

During this battle, Russian forces surrounded the Sixth Army and placed them in a serious predicament. Rather than allow the Sixth Army to break through the encircled Russian forces to safety, Hitler ordered them to remain in Stalingrad and ordered the 48th Panzer Corp to break through to relieve them. However, the relief effort failed because many of its tanks had been immobilized prior to the battle by mice eating through the tanks' electrical insulation (Clark 1985). With the surrender of the Sixth Army, Germany lost 13 infantry divisions, 3 panzer (tank) divisions, 3 motorized divisions, and 1 anti-aircraft division — a loss of 250,000 men and all of their equipment (Clark 1985).

WILDLIFE MANAGEMENT IN MODERN AMERICA

Owing to an era of unprecedented prosperity following World War II, Americans predictably became more interested in their nation's wildlife. Americans had more money and leisure time, which they increasingly spent on outdoor recreation. By 1960, there were 30 million hunters and fishermen who spent a total of nearly $4 billion in pursuit of wildlife. Better highways and affordable cars gave more Americans the opportunity to travel to the nation's national parks. By 1970, 128 million people participated in outdoor recreation, not only hunting and fishing but also nature walking, bird watching, and wildlife photography. Clearly, the wildlife conservation movement was drawing an increasingly diverse clientele (Belanger 1988).

A new invention, television, also elevated interest in wildlife as people across the country could watch and marvel at the beauty of the nation's wildlife from their living rooms. Television produced a national constituency for wildlife, making wildlife problems no longer just local issues. Now, people in New York City could follow and care about the fate of a wildlife population a thousand miles away. But as local ideas about how wildlife should be managed clashed with the concerns of distant citizens, tensions began to mount concerning wildlife management. Opinions often differed between the expanding urban population and the declining rural one. Especially publicized were the constant struggles between local commodity interests in the West and national environmental interests. Those who espoused the "commodity" point of view included representatives of the livestock, mining, oil and gas, and timber industries. Supporting the opposing viewpoint, or environmental interests, were the Friends of the Earth, the National Wildlife Federation, the Natural Resources Defense Council, the Sierra Club, and the Wilderness Society (Satchell 1990; Reiger 1992; Cawley 1993).

Battle lines also were drawn between hunters and antihunters. Although the major conservation organizations — the National Audubon Society, Wilderness Society, Wildlife Society, American Forestry Association, Sierra Club, and National Wildlife Federation — still supported sport hunting and considered it to be an important tool of wildlife management, American public opinion was shifting against it. The media helped to fuel the flames (Belanger 1988; Dunlap 1988). An early example of this occurred in November 1969, when NBC television aired a program "The Wolf Man" that showed the slaughter of wolves by bounty hunters in Alaska. Thousands of television watchers were motivated by the grisly scenes and sent letters of protest to the U.S. Department of the Interior (Feldman 1996).

Beginning in the 1960s, polarization also increased when some Americans experienced a paradigm shift in how they perceived the environment and their role in it. The new view was that the environment was fragile with many interconnected features and that changes brought about by man could have serious and unexpected consequences. Helping to lead the change was Rachel Carson's book *Silent Spring*, which promoted the adoption of an "ecologist" mindset. The spread throughout the country of this mind-set led to the establishment of events such as "Earth Day" in 1970 (Feldman 1996; Norton et al. 1996).

Still, this new environmental consciousness was not universally accepted. Rural residents continued to hold more utilitarian perspectives than urban ones. Rural dwellers relied more directly on the land than urban residents, and they traditionally worked in more extractive occupations (farming, logging, trapping, etc.) than did urbanites. Given their economic dependence on natural resources, many rural Americans maintained the traditional perspective of their pioneer ancestors (Conover and Decker 1991; Conover 1998).

Americans' perception of society also changed. People no longer spoke of the common man but instead embraced the notion of diversity. Citizens learned how to use the media and the political process to make their voices heard. This polarization of society made wildlife management decisions more controversial because no action could please everyone.

The changes in societal perceptions had both intended and inadvertent consequences. The passage of game laws, which protected wildlife from overexploitation by humans, and the adoption of science-based management practices had their intended result: populations of game species (e.g., deer, elk, turkey, and geese) and many fur-bearers (e.g., beaver) increased to levels not seen since colonial days. Likewise, predator populations, freed from unrestricted killing, recovered. These increasing wildlife populations produced some unforeseen negative consequences for society, particularly an increase in wildlife damage to crops and livestock (Conover and Decker 1991). Wildlife attacks on humans also increased as predator–human confrontations became more common, owing both to soaring predator populations and a growing enthusiasm for outdoor recreation.

Another new trend was the establishment of urban wildlife populations. Many wildlife species (e.g., white-tailed deer, Canada geese, red foxes, coyotes), which used to be found only in remote areas, moved into many U.S. metropolitan areas. Initially, these urban wildlife populations were encouraged by local residents. But as wildlife populations increased, some metropolitan residents became concerned with the negative consequences of large wildlife populations in their neighborhoods (Conover and Chasko 1985; Conover 1997). Deer–car collisions in the U.S. became more common and, by the 1990s, their number exceeded 1 million annually (Conover et al. 1995). Other problems included an increase in wildlife-vectored diseases, such as rabies, hantavirus, and Lyme disease, which were virtually unknown in the U.S. a few decades earlier.

WILDLIFE MANAGEMENT IN THE 21ST CENTURY: WHAT NOW?

Will the pendulum continue to oscillate? Future Americans could have a sense of déjà vu with regard to their encounters with wildlife. From the days of the Puritans until today, Americans have encroached upon wildlife habitat. Such trends will continue in the future as human populations increase, although this movement is counterbalanced by a movement of wildlife into urban human habitats. In the words of Anthony Brandt (1997):

> By moving into their habitat, by eliminating their predators, we have caused the explosion of deer and geese and beavers and moose and coyotes on what we persist in thinking is our property. We are the stewards of the world; we hold it in sacred trust. But [the wild animals aren't] "out there" any longer, somewhere in Montana or the rain forest of the Amazon basin. They are staring at us with big soulful brown eyes where our azaleas used to be.

Future generations of Americans may experience wildlife threats to their property, health, and even lives, in ways that their colonial ancestors could appreciate (Kellert and Berry 1980; Kellert and Westervelt 1982; Kellert 1985). A 1997 survey indicated that 65% of the families in North Haven, NY, had experienced Lyme disease. Brandt (1997) believed that this level of infection can only be described as a plague.

As we have seen, "progress" has been made in terms of saving wildlife. Will this progress continue in the 21st century? History has demonstrated that society will sacrifice wildlife resources for food resources when its food supply is threatened. Hence, the future of wildlife will be tied to our ability to increase our food productivity faster than the increase in the human population. Will this happen? Time will tell, but I am an optimist: despite Malthus' grim predictions in the 1700s about increasing populations causing famines, civilization has been winning this race.

SUMMARY

Wildlife have posed a great threat to pastoralists and farmers ever since humans first domesticated animals or started to farm. Early attempts at reducing wildlife damage were labor intensive and often involved having someone guard the livestock or crops constantly. Even before the advent of writing, ancient civilizations were using traps and poisons to reduce wildlife damage.

The dominant American view of wildlife from the 1600s until the end of the 1800s was dualistic; wildlife species were divided into good animals (those which had commercial value or could be eaten) and bad animals (those which threatened the colonists' safety or food supply). Philosophically, early colonial Americans believed that the environment was to be manipulated for man's purposes. Under the impacts of modernization and overexploitation of natural resources, Americans in the late nineteenth century began to appreciate the recreational value of wildlife and to develop a more protective attitude towards it. Still, the dichotomy between good and bad wildlife prevailed, with "good" species now being those that could be hunted. The world wars and the Great Depression halted the tilt toward a more protective approach to wildlife as Americans became more concerned with economic matters and agricultural productivity. Only during the prosperous post-World War II era, did the "ecological" approach to wildlife seem to gain ascendancy over the traditional dualistic, consumptive views.

Implementation of protective game laws and science-based wildlife management had their intended result: wildlife populations soared to levels not seen since colonial times. These increasing wildlife populations, in turn, had unexpected consequences as a movement of wildlife into urban areas began and wildlife damage intensified. Since World War II, more Americans have shown a greater interest in, and concern about, their wildlife legacy than at any time previously. But this increasingly diverse clientele for wildlife has produced rising tensions and deepening divisions within society over how wildlife should be managed. The result of all of these contentious issues has been the polarization of American society into local vs. national interests, urban vs. rural residents, hunters vs. antihunters, and "ecologists" vs. "utilitarians."

LITERATURE CITED

Ambrose, S. E., *Crazy Horse and Custer: The Parallel Lives of Two American Warriors.* Doubleday, New York, 1975.

Anderson, S. H., *Managing Our Wildlife Resources.* Prentice Hall, Englewood Cliffs, NJ, 1991.

Anonymous, Man versus mouse in 2500 B.C., *Sci. Am.,* 216: 60, 1967.

Belanger, D. O., *Managing American Wildlife: A History of the International Association of Fish and Wildlife Agencies.* University of Massachusetts Press, Amherst, MA, 1988.

Birkedal, T., Ancient hunters in the Alaskan wilderness: human predators and their role and effect on wildlife populations and the implications for resource management. In *Conf. Res. Resour. Manage. Parks Public Lands,* 7: 228–234, 1993.

Brandt, A., Not in my backyard. *Audubon,* 99(1): 86–90, 120–125, 1997.

Cawley, R. M., *Federal Land, Western Anger: The Sagebrush Rebellion and Environmental Politics.* University of Kansas Press, Lawrence, 1993.

Clark, A., *Barbarossa.* Quill, New York, 1985.

Conover, D. O. and M. R. Conover, Wildlife management in colonial Connecticut and New Haven during their first century: 1636–1736. *Trans. Northeast Sect. Wildl. Soc.,* 44, 1–7, 1987.

Conover, D. O. and M. R. Conover, Wildlife management by the Puritans. *Mass. Wildl.,* 39, 2–8, 1989.

Conover, M. R., Monetary and intangible valuation of deer in the United States. *Wildl. Soc. Bull.,* 25, 298–305, 1997.

Conover, M. R., Perceptions of American agricultural producers about wildlife on their farms and ranches. *Wildl. Soc. Bull.,* 26, 597–604, 1998.

Conover, M. R. and G. G. Chasko, Nuisance Canada goose problems in the eastern United States. *Wildl. Soc. Bull.,* 13, 228–233, 1985.

Conover, M. R. and D. J. Decker, Wildlife damage to crops: perceptions of agricultural and wildlife professionals in 1957 and 1987. *Wildl. Soc. Bull.,* 19, 46–52, 1991.

Conover, M. R., W.C. Pitt, K. K. Kessler, T. J. DuBow, and W. A. Sanborn, Review of human injuries, illnesses, and economic losses caused by wildlife in the United States. *Wildl. Soc. Bull.,* 23, 407–414, 1995.

Cronin, W., *Changes in the Land.* Hill and Wang, New York, 1983.

Dexter, F. B., *New Haven Town Records, Volume One.* New Haven Historical Society, New Haven, CT, 1917.

Di Silvestro, R. L., The Federal Animal Damage Control program. In *Audubon Wildlife Report.* National Audubon Society, New York, 130–148, 1985.

Doughty, R. A., *American Military History and the Evolution of Warfare in the Western World.* D. C. Heath, Lexington, MA, 1996.

Drummond, D. C., Unmasking Mascall's mouse traps. In *Vertebrate Pest Conf.,* 15, 229–235, 1992.

Dunlap, T. R., *Saving America's Wildlife.* Princeton University Press, Princeton, NJ, 1988.

Feldman, J. W., The Politics of Predator Control, 1964–1985. M.A. thesis, Utah State University, Logan, 1996.

Fitzwater, F. D., Mythology of vertebrate pest control. In *Vertebrate Pest Conf.,* 14, 12–15, 1990.

Forest and Stream. 1, 3, 1873.

Geller, L. D., *Pilgrims in Eden, Conservation Policies at New Plymouth.* Pride Publications, Wakefield, MA, 1974.

Hawthorne, D. W., G. L. Nunley, and V. Prothro, A history of the Wildlife Services program. *Probe*, 197, 1–5, 1999.

Hoadly, C. J., *Records of the Colony and Plantation of New Haven from 1638 to 1649*. Case, Tiffany and Co., Hartford, CT, 1857.

Howarth, D., *Waterloo: Day of Battle*. Atheneum, New York, 1968.

Hygnstrom, S. E., The evolution of vertebrate pest management — the species versus systems approach. In *Vertebrate Pest Conf.*, 14, 20–24, 1990.

Kay, C. E., Are ecosystems structured from the top-down or bottom-up: a new look at an old debate. *Wildl. Soc. Bull.*, 26, 484–498, 1998.

Keeler, C. E., *The Laboratory Mouse: Its Origins, Heredity and Culture*. Harvard University Press, Cambridge, MA, 1931.

Kellert, S. R., Public perceptions of predators, particularly the wolf and coyote. *Biol. Conserv.*, 31, 167–189, 1985.

Kellert, S. R. and J. K. Berry, *Knowledge, Affection and Basic Attitudes towards Animals in American Society.* U.S. Fish and Wildlife Service, Washington, D.C., 1980.

Kellert, S. R. and M. O. Westervelt, Historical trends in animal use and perception. In *North Am. Wildl. Nat. Resourc. Conf.*, 47, 649–664, 1982.

Marcus, R. D. and D. Burner, *America Firsthand from Reconstruction to the Present*. St. Martin's Press, New York, 1991.

Mascall, L., *A Booke of Engines and Traps to Take Polcats, Buzardes, Rattes, Mice and All Other Kindes of Vermine and Beasts Whatsoever, Most Profitable for All Warriners, and Such as Delight in this Kinde of Sport and Pastime*. John Wolfe, London, 1590. Reprinted by De Capo Press, New York, 1973.

McCade, R. E. and T. R. McCade, Of slings and arrows: a historical retrospection. P. 19–72 in L. K. Halls, Ed., *White-Tailed Deer: Ecology and Management*. Stackpole Books, Harrisburg, PA, 1984.

Mood, F., John Winthrop, Jr., on Indian corn. *New England Q.*, 10, 21–133, 1937.

Nash, R., *Environment and Americans: The Problem of Priorities*. R. E. Krieger, Huntington, NY, 1979.

Nettles, K. C., The beginning of money in Connecticut. *Trans. Wis. Acad. Sci. Arts Lett.*, 23, 1–48, 1927.

Norton, M. B., D. M. Katzman, P. D. Escott, H. P. Chudacuff, T. G. Paterson, and W. M Tuttle, Jr., *A People and a Nation: A History of the United States*. Houghton Mifflin, Boston, 1996.

Reed, N. P. and D. Drabelle, *The United States Fish and Wildlife Service*. Westview Press, Boulder, CO, 1984.

Reiger, G., The predator problem. *Field and Stream,* 97, 16–17, 1992.

Reiger, J. F., *American Sportsmen and the Origins of Conservation*. Winchester Press, New York, 1975.

Satchell, M., Uncle Sam's war on wildlife. *U.S. News and World Report*, 108(2), 36–37, 1990.

Topsell, E., The historie of four-footed beasts, 1607. Reprinted in M. South, Ed., *Topsell's Histories of Beasts*. Nelson-Hall, Chicago, 1981.

Trefethen, J. B., *An American Crusade for Wildlife*. Winchester Press, New York, 1975.

Trumbull, J. H., *The Public Records of the Colony of Connecticut Prior to the Union with New Haven Colony, May, 1665*. Brown and Parsons, Hartford, CT, 1850.

U.S. Department of Agriculture, *Animal Damage Control Program: Final Environmental Impact Statement. Volume Two*. U.S. Department of Agriculture, Animal Plant Health Inspection Service, Washington, D.C., 1994.

Wade, D. A., Predator damage control: recent history and current status. In *Proc. Vert. Pest Conf.*, 9, 189–199, 1980.

Walcott, R. R., Husbandry in colonial New England. *New England Q.*, 9, 218–253, 1936.
West, E., *The Way to the West: Essays on the Central Plains*. University of New Mexico, Albuquerque, NM, 1995.
Wright, E. N., *Bird Problems in Agriculture*. BCPC Publication, Croydon, England, 1980.

CHAPTER 3

Threats to Human Safety

"These [grizzly] bears being so hard to die rather intimidate us all; I must confess that I do not like the gentlemen and had rather be attacked by two Indians than one bear...."

Meriwether Lewis, 11 May 1805,
while on the Lewis and Clark Expedition

"It seems that earlier grotesque beliefs that predators were evil and had to be eradicated to make the world safe for lambs were overthrown at the cost of creating a new and beguiling myth of the benevolence of nature. With each new report of a human injured or killed by a bear we question where management went wrong. We failed to recall that most problems with bears in parks stem not from human malevolence but from too much benevolence. As with most conflicts between powerful adversaries, it is dangerous to appear weak."

McCullough (1982)

"For safe hiking in bear country, travel with a companion whom you can outrun."

Tongue-in-cheek suggestion by Adolf Murie

Only a small fraction of human–wildlife interactions result in human injury or death, but the loss of even one person is a horrible tragedy, and we must do whatever we can to keep people safe. These tragedies also can attract a great deal of attention and media coverage and have a large impact on public perception of wildlife.

There are three ways people become injured or killed from human–wildlife interactions. A person can 1. be bitten, clawed, gored, or attacked by wildlife; 2. be injured in a collision between an animal and an automobile or airplane; or 3. become ill from a disease or parasite which was passed along by a wildlife species. This chapter will cover the first two topics; the next chapter will deal with human illnesses.

WHY DO ANIMALS ATTACK PEOPLE?

Predatory Attacks

There are many reasons why animals attack humans; such attacks can be motivated by predatory, territorial, or defensive instincts. Predatory attacks occur when the animal views the person as prey. Predatory attacks on adult humans are very rare, in part because adults are so large that only the largest predators (e.g., cougars, tigers, lions, wolves, bears, and crocodiles) would consider them prey. A person's large size poses two problems for predators. One is that our size suggests that we could pose a formidable threat to the predator in defending ourselves (the risk of injury to the predator is too great). The second is that some predators, such as snakes, do not have the ability to tear flesh into bite-size pieces; hence, there is no reason for these predators to attack something that cannot be swallowed whole. Alligators and crocodiles also lack the teeth to chew off pieces of a large carcass. For this reason, alligators rarely make predatory attacks on adult humans. Crocodiles, however, have learned how to tear apart large carcasses by holding onto part of the carcass and twisting rapidly in the water. When other crocodiles join in, a large carcass can easily be torn into bite-size pieces. This behavioral ability allows crocodiles to feed on large mammals, such as zebras, wildebeests, and, occasionally, humans.

Some predatory attacks on humans occur because the predators underestimate the size of the object they are attacking. People are bitten every year by bluefish, which weigh less than 10 kg. In these cases, the fish probably saw an object, perhaps a hand or finger, in turbid water and bit it before realizing that it was part of something much larger. Undoubtedly, some shark and alligator attacks result from this same mistake. If a person is wading in knee-deep water, a submerged alligator might see the legs but not the rest of the body above the water and not realize that it is a person.

Young children do not have the benefit of large size to scare away potential predators. For this reason, a high proportion of predatory attacks on humans are directed at children, who are especially at risk when alone and undefended (Beier 1991). Even relatively small predators, such as coyotes, occasionally attack young children (Howell 1982; Carbyn 1989).

Territorial Attacks

Most wildlife species only exhibit territorial behavior towards conspecifics (individuals of the same species), but this is not always the case. Some species are aggressive towards interspecifics (individuals of other species) and may attack them. Mute swans are an example with which I am familiar. Some mute swans exhibit high rates of interspecific aggression and drive other birds, especially Canada geese, out of their territories (Conover and Kania 1994). Aggressive mute swans often threaten humans during the nesting season by flying at them as if attacking but stopping before making physical contact (i.e., bluff attack). However, I made the acquaintance of one particularly aggressive swan when I was checking nests on the

Connecticut River. As I was going up one tributary, a boater frantically waved at me and warned me not to proceed further because "there was a mad swan around the bend that attacked people." I did not pay attention because I knew that swans only make bluff attacks. However, this particular swan was not bluffing; when I went around the bend, it flew into the boat and attacked me. The attack ended after I pushed it out of the boat but not before both my legs and ego were bruised.

Most birds do not have the size or physical ability to cause serious injuries to humans. Nevertheless, serious injuries or deaths can indirectly result from some bird attacks. For example, a swooping hawk or eagle might cause someone to fall out of a tree or down a cliff. In Indiana, a man fell out of his boat and drowned after he was attacked by a mute swan.

Many large herbivores, such as bison, travel in herds and have a strict social hierarchy; subordinate animals must keep their distance from dominant individuals. Sometimes, people unknowingly invade the personal space of a dominant animal and this can provoke the animal to attack. This is especially a problem in parks, where tourists and photographers try to get as close as possible to wildlife. Usually, animals respond to this intrusion by moving away, but sometimes they attack the person. Tourists in Yellowstone National Park are injured every year by bison, elk, and moose.

Animals that are hand-reared by humans often imprint on their handlers. When they reach maturity, these animals often view humans as conspecifics and will attempt to drive people from their territories or attack people in an attempt to establish dominance. For this reason, it is dangerous to release hand-reared animals into the wild.

Defensive Attacks

Usually the animal is acting defensively when it attacks a human. An animal's best defense against a predator is to hide, flee, or at least keep its distance. However, when an animal is cornered, trapped, or constrained by a predator, its best chance to survive is to attack in hopes that it can inflict so much pain that the predator will give up and seek easier prey. For this reason, even small animals will bite or attack humans when they are being held or feel trapped. Animals will use whatever mechanism is available to them in their attempts to inflict pain. Cats use their claws to scratch, coyotes use their teeth to bite, bison use their horns to gore, moose use their front legs to kick, porcupines use their quills, and we all know what a skunk will do.

HOW OFTEN ARE HUMANS INJURED OR KILLED BY WILDLIFE?

We know surprisingly little about the frequency of wildlife attacks on humans in the U.S. This is especially true for bites or attacks by species that rarely cause serious injuries (e.g., squirrels, skunks, etc.). Often, what we do know is limited to data collected in a small area (Table 3.1). For instance, there have been scientific reports about coyote attacks on people in California and some national parks but

Table 3.1 Studies of Nonfatal and Fatal Injuries to Humans by Wildlife in Different Parts of the U.S. and Canada

Species	Location	Years of study	Injuries/year		Reference
			Nonfatal	Fatal	
Rodents	14 states[a]	1971–1972	4302	—[b]	Moore et al. (1977)
Venomous snakes	U.S.	?	7000	—	Minton (1987)
	U.S.	?	8000	9-15	Weiss (1990)
	U.S.	?	8000	16	Grenard (2000)
Skunks	14 states[a]	1971–1972	113	—	Moore et al. (1977)
Foxes	14 states[a]	1971–1972	76	—	Moore et al. (1977)
Sharks	U.S.	1990–1999	287	5	Int. Shark Attack Files (unpublished)
Black bears	U.S.	1960–1980	25	0.3	Herrero (1985)
Grizzly bears	North America	1900–1980	3.7	0.5	Herrero (1985)
Polar bears	Canada	1965–1985	1.0	0.3	Herrero and Fleck (1989)
American alligators	Florida	1948–1992	3.5	0.1	Gutierrez-Sanders (1992)
	U.S.	1990–1995	18	0.3	Conover and DuBow (1997)
Coyotes	Yellowstone	1960–1988	0.1	0.0	Carbyn (1989)
	Los Angeles	1975–1981	1.3	0.2	Howell (1982)
	California	1988–1997	2.0	0.0	Baker and Timm (1998)
Cougars	U.S.	1890–1990	0.2	0.05	Beier (1991)
	California	1986–1995	0.9	0.2	Mansfield and Charlton (1998)
Bison	Yellowstone	1978–1993	3.7	0.1	Conrad and Balison (1994)

[a] The study area included New Hampshire, Massachusetts, New York, New Jersey, Washington D.C., South Carolina, Georgia, Florida, Kentucky, Indiana, Illinois, North Dakota, South Dakota, Texas, and Arizona.
[b] Data not reported.

from nowhere else. In other parts of the world, information on the frequency of wildlife attacks on humans is even more sketchy than in the U.S. even though some countries have much greater problems with wildlife attacks than those experienced in the U.S. For instance, during a five-year period in India's state of Madhya Pradesh, 1094 people were injured or killed by wildlife attacks, including 735 by sloth bears, 138 by leopards, 121 by tigers, 34 by elephants, 29 by wild boars, 21 by gaur, 13 by wolves, and 3 by hyenas (Rajpurohit and Krausman 2000).

Sometimes, data are limited geographically but can be extrapolated to the entire country. For instance, 14 states (New Hampshire, Massachusetts, New York, New Jersey, South Carolina, Georgia, Florida, Kentucky, Indiana, Illinois, North Dakota, South Dakota, Texas, and Arizona) require doctors to report to public health officials when someone has been bitten by an animal (Moore et al. 1977). Because these states are widely distributed across the country, we can extrapolate the rates of wildlife attacks to the other 35 continental states with some degree of confidence. However, these data are conservative because they only include serious bites for

Table 3.2 Number of Human Injuries and Fatalities Annually in the U.S. from Bites or Attacks by Wildlife

Species	Injuries/ year	Fatalities/year
Rodents	27,000	?
Venomous snakes	8,000	15
Skunks	750	0
Foxes	500	0
Bears	30	1
Sharks	28	0.5
Alligators	18	0.5
Coyotes	2	0.0
Cougars	2	0.4

Table 3.3 Probability of an Attack Resulting in a Human Fatality in the U.S. and Canada

Predator	Probability of an Attack Resulting in a Human Fatality	Reference
Rodents	<0.0001	Moore et al. (1977)
Venomous snakes	0.002	Weiss (1990), Grenard (2000)
Alligators	0.03	Conover and DeBow (1997)
Bison	0.04	Conrad and Balison (1994)
Black bears	0.05	Herrero (1985)
Sharks	0.09	Int. Shark Attack Files (unpubl.)
Grizzly bears	0.11	Herrero (1985)
Cougars	0.20	Beier (1991)
Polar bears	0.30	Herrero and Fleck (1989)

which the victim sought medical attention; they do not include bites for which the person did not see a doctor.

About 36,000 people are bitten annually by wildlife in the U.S., and 10 to 20 people are killed (Table 3.2). Most victims (27,000 per year) are bitten by rodents. Although they do not receive much attention from the media, the number of bites by skunks (750 per year) and foxes (500 per year) far exceeds the number of attacks by bears (30 per year) or cougars (2 per year). One reason why attacks by bear and cougar receive so much attention is because a higher proportion of them result in a human fatality (Table 3.3).

SNAKEBITES

Venomous snakes bite about 8000 people annually in the U.S., resulting in about 15 human fatalities each year (Table 3.2). I found it surprising that only about 1 out of 500 venomous snakebites is fatal (Table 3.3). Obviously, medical advances for the treatment of snakebites (e.g., the use of anti-venoms) have been great, and my boyhood fears about deadly snakes were overblown. Rattlesnakes are responsible for 65% of all venomous snake bites annually in the U.S. and account for nearly all fatalities (Weiss 1990). The World Health Organization estimates that 1 million people are bitten by snakes annually throughout the world and that 40,000 people are killed.

Contrary to popular opinion, most snakebites in the U.S. occur within city limits or in suburbs, and only rarely are victims bitten while on hiking, fishing, or hunting trips. As Minton (1987) points out, a snakebite victim is most likely to be bitten by the snake which lives in his garden or under his house. Children between the ages of 5 and 12 are especially at risk if they live in areas where venomous snakes are plentiful because they are more likely to be barefoot and are less careful about where they step or place their hands. Children also are very curious about snakes and may try to catch them. In fact, half of the snakebites in the U.S. resulted from people deliberately handling snakes (Minton 1987; Grenard 2000).

BEAR ATTACKS

Black bears, grizzly bears, and polar bears all pose threats to people (Figure 3.1). About 30 people are attacked in North America annually by bears, primarily by black bears (Table 3.1). In the Madhya Pradesh state of India, sloth bears killed 48 people and wounded 638 from 1989 to 1994 (Rajpurohit and Krausman 2000). Many bear attacks occur after the animals lose their fear of humans. Herrero (1985) concluded that 90% of the black bears that injured people had habituated to humans and had been conditioned to people's food. Many of

Figure 3.1 A black bear. (From Hygnstrom et al. 1994. With permission.)

these attacks are defensive in nature. Black bears seeking food want to be left alone while they find and eat it. If they feel threatened by people trying to retrieve their food, protect their property, or just watch or photograph them, the bears may lash out. Such attacks are normally of short duration and end when the bear perceives that the threat has ended (Herrero and Fleck 1989). In contrast, most of the attacks on humans by polar bears are predatory in nature, and a high proportion of them result in a human fatality (Herrero and Fleck 1989). Many attacks on humans by grizzly bears result from sudden encounters where a bear, often with cubs, is surprised by a person in its immediate vicinity and feels threatened by that person (Herrero 1985).

Human injuries from bear attacks are particularly a problem in national parks where large numbers of bears and humans coexist. In Yosemite National Park, some black bears have learned that cars often contain food and spend their nights ripping apart unoccupied cars. Individual bears have even developed preferences for certain types of cars based on how easy they are to open. In 1998, Yosemite's black bears broke into 1103 vehicles causing over $600,000 in damage. However, the real danger is that eventually someone will be inside one of these cars or will attempt to scare away the offending bear and that person will become injured. Still, bears injure only a small proportion of park visitors. The probability of a visitor to a national park being injured by a bear during the 1980s was

- 328,000 visitors per injury in Denali
- 629,000 visitors per injury in Yosemite
- 763,000 visitors per injury in Glacier
- 1,157,000 visitors per injury in Yellowstone
- 2,561,000 visitors per injury in Great Smoky Mountain
- 4,048,000 visitors per injury in Banff (Herrero and Fleck 1989)

Similar problems occur worldwide (see Sidebar 3.1).

Sidebar 3.1 Tiger Attacks on People

In Africa and Asia, tigers pose a much greater threat to human safety than do cougars in North America. Before World War II, tigers killed 1000 to 2000 people annually in India (Perry 1964; McDougal 1991). Fortunately, such attacks are much less common today. In India's Sundarbans Forest, an 8000 km^2 mangrove forest containing 500 to 600 tigers, forest workers are at risk of being attacked. Several steps have been taken to try to teach the tigers not to attack people. Mannequins are dressed in old clothes, wrapped in electric wires, and placed in the forest. The idea is that tigers will get shocked when they attack the human dummy and develop a fear of people. Workers also take a goat or pig on a leash with them when they go into the forest. These animals protect the workers because tigers are more apt to attack these animals than people. Forest workers also wear masks on the back of their heads. These discourage tigers from attacking because these predators normally attack from behind and are reluctant to do so if the person appears to be watching them (Jackson 1991).

SHARK ATTACKS

Conflicts between humans and sharks occur worldwide (Figure 3.2). Approximately 50 people are attacked and 7 are killed by sharks annually (International Shark Attack Files, unpublished). The frequency of these attacks has increased in recent decades as water sports, especially surfing and scuba diving, have become more popular across the globe. Most shark attack victims during 1999 were surfers or wind-surfers (43%) and swimmers or waders (38%). During the 1990s, an average of 28 people were attacked annually by sharks in the U.S. Five of these attacks resulted in a human fatality. Not surprising, most shark attacks occurred in Florida (19 per year), Hawaii (3 per year), and California (3 per year).

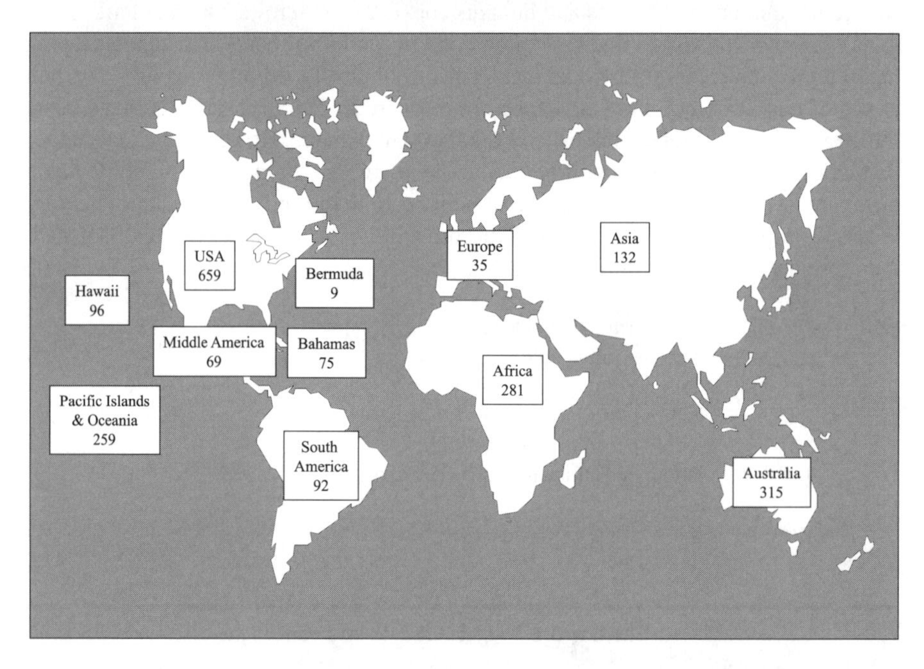

Figure 3.2 Locations of confirmed shark attacks through 1999. (Data from Int. Shark Files.)

ALLIGATOR ATTACKS

Conover and DuBow (1997) found 236 recorded cases of alligator attacks on humans during the 20th century (Figure 3.3). Of these cases, 218 (92%) occurred in Florida, 6 in South Carolina, 5 in Georgia, 4 in Alabama, 2 in Texas, and 1 in Louisiana. Most victims were males (84%) and ranged in age from 3 to 82 (Table 3.4). When attacked, 34% of the victims were totally immersed in the water (e.g., swimming, floating), 17% were partially in the water (e.g., wading), 38% were on shore, 2% in boats, and the location of 9% could not be determined (Conover and DuBow 1997). Victims were engaged in a variety of water sports when attacked. Alligators attacked people swimming, snorkeling, or scuba diving 76 times; fishing,

Figure 3.3 A nuisance alligator about to be translocated from a backyard pond. (Photo courtesy of Gary San Julian.)

clamming, or catching bait 21 times; playing golf or retrieving golf balls 24 times; and feeding animals 8 times. In 20% of the attacks, humans made the first contact (often inadvertently) by swimming into, jumping on, picking up, displaying, wrestling with, moving, or trying to harvest the alligator that attacked them.

Eight people (four males and four females) have died in the U.S. from alligator attacks (3% of all attacks) — seven in Florida and one in Georgia. Attacks on young children are 4 to 5 times more likely to be fatal than those on adults (half of the victims of fatal attacks were younger than 12). Four victims of fatal attacks were swimming or snorkeling prior to the attack; three were walking along the shore or wading (these were the three youngest victims); and the activity of one victim is unknown (Conover and DuBow 1997). Crocodiles pose a much greater problem in other countries than alligators do in the U.S. Worldwide, as many as 3000 people annually are attacked by crocodiles (Guggisberg 1972).

Table 3.4 Characteristics of People Injured by Predators in North America (Beier 1991; Conover and DuBow 1997)

	Alligator		Cougar	
	Nonfatal (%)	**Fatal (%)**	**Nonfatal (%)**	**Fatal (%)**
		Sex		
Male	84	50	72	80
Female	16	50	28	20
		Age (years)		
<10	12	38	41	70
11–20	16	25	15	20
>20	72	37	43	10

One critical question is why has there been a recent increase in alligator attacks? After all, alligators have posed a threat to human safety ever since man first set foot on the North American continent. For instance, early explorers of Florida reported that both they and the natives considered alligators dangerous and took precautions to protect themselves from these animals (Le Moyne 1591, Van Doren 1955, as cited by Hines and Keenlyne 1977). Yet, the number of people attacked by alligators has increased dramatically in recent decades. For instance, only 1 person was known to have been attacked by an alligator from 1900 to 1950, but 78 were injured in alligator attacks in the 1980s and 110 from 1990 to 1995.

What has caused this increase in the frequency of alligator attacks? Part of it is due to a much better reporting system. Today, state officials are much more likely to learn about an alligator attack than in the past, especially if there were only minor injuries (Conover and DuBow 1997). The most important factor for this increase, however, has been the expanding alligator population during the late-1900s. During the 1800s and 1900s, alligator populations were suppressed due to hunting, and the remaining alligators were confined to remote swamps where few humans ventured. Alligator populations began to recover beginning in 1969 with the passage of the Lacey Act Amendment, which effectively curtailed interstate shipment of illegal alligator hides. The Endangered Species Act in 1973 provided additional protection for alligators (Conover and DuBow 1997). As alligator populations have recovered, these animals have moved into suburban and urban lakes and canals where they may come into contact with humans daily. At the same time, the human population in Florida and residential development of waterfront property have increased.

Another factor for the increasing frequency of attacks is that many alligators have lost their fear of humans. A century ago, alligators would hide or flee at the first sight of a man because that person was likely to be an alligator hunter. Today, alligators have less to fear from humans and have learned that most people will not hurt them; some will even feed them. People also have lost their fear of alligators. A century ago, most Floridians rarely saw an alligator, and when they did see one, they stayed at a healthy distance. Today, some alligator attacks occur when people misjudge the danger that alligators pose and put themselves in harm's way. Each year, people are injured when they try to pet an alligator or pick it up, or when they see an alligator sunning itself on a road and try to push it off the highway.

In the U.S., wildlife agencies have responded to the threat of alligator attacks by implementing alligator management plans (Hines and Woodward 1980; Taylor et al. 1991). These plans often call for the removal of nuisance alligators and improvement of public education about alligators and the dangers they pose. Some states have started harvesting alligators. These efforts have been successful in reducing attacks in some areas, but further attacks are inevitable as long as humans and alligators share the same space.

ATTACKS BY WOLVES AND COYOTES

Coyotes, 10 to 20 kg canids (Figure 3.4), normally prey upon small mammals or the young of large ungulates and livestock, which they usually kill by clamping

Figure 3.4 Two coyotes.

their jaws on the prey's throat and neck, preventing the victim from breathing. Occasionally, a child is attacked and killed by a coyote. These attacks often appear to be predatory in nature because the coyotes direct their bites to the head and neck (Carbyn 1989). They occur in national parks, such as Yellowstone, Banff, and Jasper (Carbyn 1989), and in urban areas, such as Los Angeles, that have large coyote populations (Howell 1982). In recent years, coyote attacks on humans have become more common, especially in suburban parts of southern California. Baker and Timm (1998) reported that there were 53 coyote attacks on humans in California from 1988 to 1997, resulting in 21 human injuries.

Wolves are much larger than coyotes, weighing 40 to 70 kg, and are efficient killers of large mammals (e.g., deer, elk, and moose). Wolf populations are expanding in North America, both naturally and with the help of reintroduction efforts. There have been two wolf attacks on people in British Columbia, Canada, during the 1990s. One was by a wolf accustomed to acquiring food from humans (Streetly 2000). Wolf attacks on people have been more common in Europe and Asia. For instance during 1996, a healthy wolf attacked 76 children in Uttar Pradesh, India, and killed 50 of them (Jhala, Y. and B. Jethva, paper given at the Second International Wildlife Management Congress, 1999, and abstract published in *The Probe*, Issue 205).

COUGAR ATTACKS

During the last century, there have been 53 confirmed attacks on humans by cougars, resulting in 11 human deaths (Seidensticker and Lumpkin 1992), and the frequency of these attacks has increased in recent years. For example, there were only three people injured by cougars in California from 1890 to 1985, but nine people were injured from 1986 to 1995 (Mansfield and Charlton 1998). There are several reasons why cougar attacks on people have become more frequent. Cougar populations collapsed in the 19th and 20th centuries due to human attempts to exterminate them. In California, where the state paid a bounty for dead cougars, 12,461 were killed from 1907 to 1963 (Torres et al. 1996). Any individual cougars that survived developed a fear of humans and lived in remote areas where they rarely came in contact with humans. Recently, hunting restrictions have provided more protection for cougars from humans. For instance, cougars have not been hunted in California since 1973 and a ballot initiative (Proposition 117) was passed into law by the California voters in 1990 which designated the cougar as a "specially protected mammal" and prohibited the hunting of cougars (Torres et al. 1996).

As a result of these protective laws, cougar populations have recovered in many parts of the U.S., and these predators are moving into populated areas (Mansfield and Charlton 1998). Human populations are also increasing, outdoor recreation has become more popular, and people are spending more time in remote wilderness areas. All of these changes produced the same result: increased contact between humans and cougars.

In many parts of the world, free-ranging cats attack people. In Africa, approximately 20 people per year are killed by African lions. In India, both lions and tigers pose a threat to human safety (Sidebar 3.2).

Sidebar 3.2 Lion Attacks on Villagers around the Gir Forest Reserve Threaten Both People and Lions

The Asiatic lion once ranged from Greece to Bengal, but now only about 250 remain in the wild in a single reserve: the Gir Forest in western India. Their continued existence is threatened by their propensity to attack humans. From 1978 to 1991, an average of 15 people were attacked and 2 people killed annually by these lions. Most of the lion attacks on humans (82%) took placed outside the boundaries of the Gir Forest Reserve (Figure 3.5). A disproportionate number of these attacks are caused by subadults. This may be because these young lions are unable to establish a territory within the reserve due to the high density of lions already there.

The number of attacks on humans increased substantially after 1987 to 40 per year. Two things happened in 1987; the first was a record drought and the second was that the deliberate feeding of lions was curtailed. After this drought, lions began to feed on the bodies of their human victims. Prior to that year, lions were drawn into baiting stations where water buffalo calves were tethered for them to feed on. These baiting stations were created to provide viewing opportunities to tourists. They also may have helped the lions acquire a taste for livestock and become used to being around humans (Saberwal et al. 1994).

Villagers around the reserve were afraid of the lions because of their attacks on both humans and livestock. Most villagers (61%) said that the lions limited their ability to do things at night, and most people kept their livestock within walled compounds after dark. To protect their animals, some people (22%) even kept livestock in their

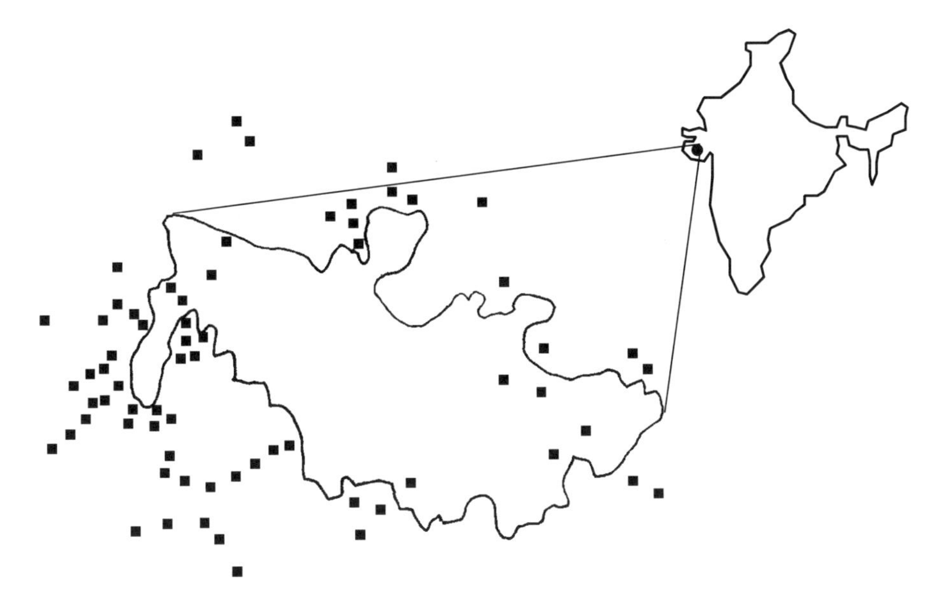

Figure 3.5 Locations of lion attacks on humans from 1988-1991 around the Gir Forest Reserve in India. (Adapted from Saberwal et al. 1994.)

houses at night. Due to these problems, the people living around the Gir Forest Reserve were hostile toward the lions and called upon the local government to take actions to end these attacks. Saberwal et al. (1994) recommended that problem lions be culled by either allowing sport-hunters to shoot them outside the reserve (with the money going to local villagers), or trapping them so that they could be relocated in distant parts of their former range or maintained in zoos.

ATTACKS BY LARGE HERBIVORES

Worldwide, more people are injured by large herbivores than by large predators. In Yellowstone National Park, bison injure more people than cougars and bears (Conrad and Balison 1994). From 1978 to 1992, bison injured 56 park visitors resulting in 2 human fatalities (Table 3.1). Of the 56 injured people, 36 were gored, usually in the buttock, thigh, or hip. Fourteen victims were tossed in the air after being hooked by the horns. In 19 cases, bison charged the victim (Conrad and Balison 1994). Moose also attack people when they feel threatened. During the 1990s, two people were killed by moose in Anchorage, AK. One man was stomped to death on the University of Alaska campus, and a woman was trampled to death in her yard. Hippopotamuses may pose a greater danger to people in Africa than predators, especially by turning over boats (Fetner 1987). Another danger is elephants; they kill 100 to 200 people annually in India (Veeramani et al. 1996) and an unknown number in Africa.

WHY HAS THERE BEEN A RECENT INCREASE IN WILDLIFE ATTACKS ON HUMANS IN NORTH AMERICA?

In North America, attacks by alligators, cougars, bears, coyotes, and probably bison and moose have been increasing in recent decades (Figure 3.6). Although these species span the taxonomic spectrum and live in different parts of the continent, the same factors are responsible for the recent increase. All of these animals are large and dangerous, and their populations have been rebounding from the early part of the 20th century when humans persecuted them. These animals, which currently enjoy either complete or partial legal protection, certainly have less reason to fear humans than they did previously, and some individual animals have habituated to humans and moved into urban areas. At the same time, human populations have increased and people are spending more time in remote areas frequented by these large animals. The result is that there is much more contact between these animals and humans, and occasionally these contacts have tragic consequences. For instance, a backcountry visitor in Yellowstone National Park is 38 times more likely to be injured by a bear than a "normal" visitor (Herrero and Fleck 1989). The increased risk for backcountry visitors is 84 times in Glacier National Park (Herrero and Fleck 1989).

Considering the large populations of humans and predators coexisting in the same areas, it is not surprising that large predators injure some people. What is really amazing is that so few people are attacked. Why don't hungry cougars and bears

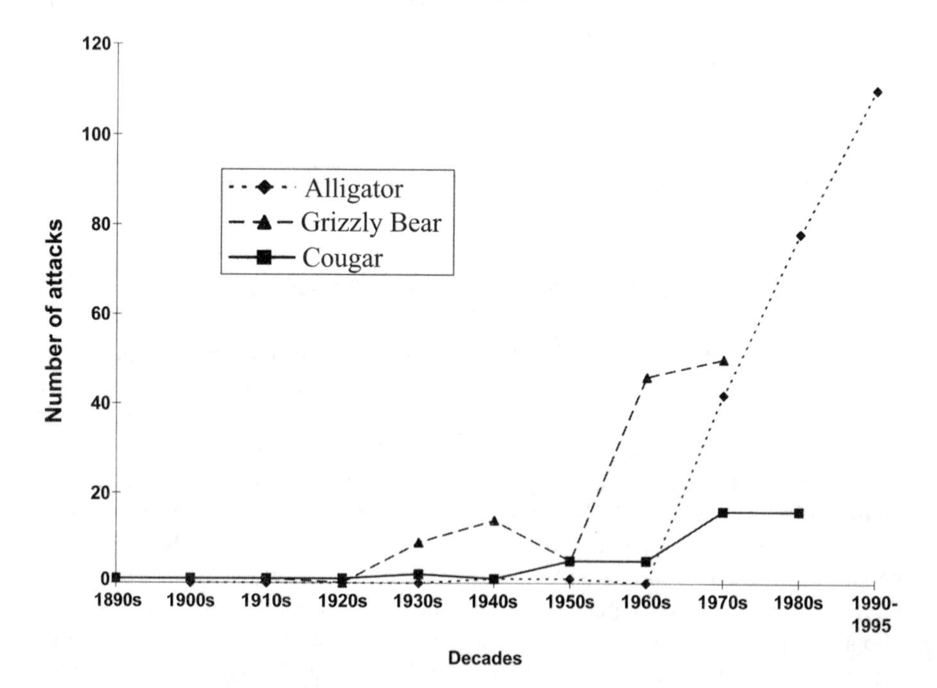

Figure 3.6 Number of humans attacked by alligators, cougars, and bears in the U.S. during recent decades.

commonly prey upon humans, especially children? People are more abundant and much easier to catch and kill than deer or elk. I suspect that part of the answer can be attributed to a type of historic memory passed from mothers to their offspring about how dangerous humans are. But as hunting of cougars and bears becomes less common due to a lack of interest by hunters or to the passage of laws that restrict the hunting of large predators, I believe that predator attacks on humans will become more common.

Seidensticker and Lumpkin (1992) noted that native bears in both India and North America stand up on their back legs when threatened by a big cat. The authors speculated that when erect these animals do not resemble the cat's normal prey (ungulates) and their heads and necks (where the cat makes its killing bite) are in the wrong place and too high to be attacked efficiently, especially when facing the lion. Perhaps the upright posture of humans inhibits the normal predator attack of lions. Seidensticker and Lumpkin (1992) also speculated that cougar attacks on humans occur most often when people lean over or squat, and hence, appear more similar to a lion's normal prey.

WHAT CAN BE DONE TO REDUCE THE FREQUENCY OF WILDLIFE ATTACKS ON HUMANS?

Although some people will be injured or killed by wildlife attacks as long as humans and animals coexist, a number of steps can be taken to reduce the frequency of such attacks. In some cases, we can identify the characteristics of animals that have a high probability of attacking people and remove the animals before they injure someone. Some examples of high threat animals include bears that raid campgrounds for food, alligators that are fed by people, mute swans that attack geese, and unusually belligerent bison. Mansfield and Charlton (1998) observed that pet depredations by cougars were useful indicators of lion activity in proximity to humans and may provide an early warning of potentially dangerous situations.

Another possible solution is to change animal behavior to ensure that large predators maintain a fear of man. If hunting becomes less common, new approaches must be discovered to discourage dangerous animals from becoming habituated to humans. Methods to teach nuisance bears to stay away from people have involved shooting them with rubber bullets or spraying them with pepper spray. Such methods, however, do not always work. Beier (1991) reported that aversive conditioning was tried with one cougar, which was shot with rock salt at close range after a near-attack. Only two weeks later, this animal repeated its aggressive behavior and had to be killed.

We also can change human behavior through education. We need to make sure that people have a healthy fear of dangerous animals. Most of the people injured by bison in Yellowstone National Park approached close to the bison while attempting to photograph it, pose with it for a photo, pet it, or feed it (Conrad and Balison 1994). We must teach people how to recognize a dangerous situation and how to avoid it. For instance, many grizzly bear attacks occur when a bear, often with cubs, is surprised by the sudden appearance of a person near it. Hikers could minimize the risk of such an attack by making loud noises to warn bears of their approach and by avoiding going through areas with dense vegetation or limited visibility (Herrero 1985). To minimize bear attacks, campers are urged to avoid camping near bear trails, seasonal feeding areas used by bears, or places where food or garbage has been left. Campers should also locate their cooking areas 100 m downwind of the tent, store food away from the camp in bear-proof containers, and locate their camp near a possible escape tree in grizzly bear country (Herrero 1985). Rajpurohit and Krausman (2000) believed that bear attacks in India could be reduced if people avoided bear habitat when bears are foraging during crepuscular periods and did not camp near water sources.

We also should let people know what to do if they are attacked by wildlife to minimize their risk of serious injury. In his analysis of cougar attacks on people, Beier (1991) found that aggressive responses by people were effective in averting a cougar from making an imminent attack. Running away was particularly futile unless the person was only a few meters away from safety. Once an attack had begun, playing dead was not an effective defense, but several attacks terminated after the victims screamed; fought back with their bare hands, sticks, rocks, or knives; or someone else at the scene yelled or made an aggressive response (Beier 1991).

How a person should respond to a bear attack depends on whether the attack is defensive or predatory in nature. For the latter, which is a common reason polar bears attack people, the victim should fight back. For black bear attacks in campgrounds, along roads, or other places where bears have become habituated to people, the bear is probably attacking because the person has gotten too close and the bear wants more space or is trying to get at the food the person might have (Herrero 1985). In such situations, the victim should back away and give up the food. When confronted by a grizzly bear, Herrero (1985) recommended climbing at least 5 m up a tree if there is enough time or to watch the bear closely and respond to its actions but not to stare at it because the bear may interpret this as threatening. Dropping something while fleeing may distract the bear and provide enough time for escape.

HUMAN INJURIES AND FATALITIES FROM UNGULATE–AUTOMOBILE COLLISIONS

Each year, thousands of people are injured when their vehicles collide with a free-ranging animal (Figure 3.8). In North America, most of these collisions involve white-tailed deer or mule deer. In Europe, automobiles hit roe deer, red deer, wild boar, chamois, moose, reindeer, white-tailed deer, and fallow deer (Table 3.5). Roe deer, however, are involved in most deer–automobile collisions in Europe.

Conover et al. (1995) estimated that the number of confirmed deer–vehicle collisions in the U.S. totaled 726,000 annually, based on dead deer found on the road or collisions reported to state authorities. This figure excludes any deer–vehicle collisions where deer died away from the highway or that were not reported to state authorities (Romin 1994). Because only about half of the deer–vehicle collisions are reported (Decker et al. 1990; Romin 1994), the actual number of

Figure 3.7 Elk stopping traffic as they cross a road in Colorado.

deer–vehicle collisions occurring annually in the U.S. is closer to 1.5 million. There also are an unknown number of collisions with moose and elk in the U.S., but we know there are about 600 moose–vehicle collisions yearly in Alaska alone (Romin 1994). Approximately 29,000 people are injured yearly in deer–automobile accidents, and 200 people lose their lives (Conover et al. 1995). Being hit by a vehicle is fatal to deer 92% of the time (Allen and McCullough 1976), which means that over 1.3 million deer die each year on U.S. highways.

Roads do not serve as natural barriers to deer and a deer's home range may include both sides of a road. Many deer–automobile collisions occur when deer try to cross the road. Erecting deer-proof fences along roads has reduced the movement of deer across roads (Falk et al. 1978; Feldhamer et al. 1986). However, fencing roads is not a panacea. One problem is that deer can often find ways to get around, over, or through fences. This creates a difficult problem because once deer are inside the fence, they may not be able to find a way out, and a deer trapped inside a highway fence faces a high risk of getting hit by an automobile. For these reasons, one-way gates (Figure 3.8) are built into highway fences so that these trapped deer have a way of getting off the road (Reed et al. 1974). Unfortunately, some deer do not recognize these gates as escape routes and ignore them. More recently, earthen ramps have been erected inside highway fences allowing deer inside a fence a method to jump over it (because the ramp is only built inside the fence, deer outside the fence cannot use it to jump in). There has not been enough research conducted on earthen ramps to know if they are used more by deer than the one-way gates.

Another problem with fencing roads is that many elk and mule deer herds are migratory, and if a fence blocks their ability to migrate between winter and summer ranges, the local deer or elk herd may not survive. For this reason, overpasses or underpasses (Figure 3.9) have been built to provide a way for animals to safely cross fenced roads (Reed et al. 1975; Ward 1982). These passes, however, are very expensive to build. For instance, a wildlife underpass system

Table 3.5 Annual Number of Automobile–Wildlife Collisions in European Countries where Data Are Available (Groot Bruinderink and Hazebrook 1996)

Country	Roe deer	Red deer	White-tailed deer	Fallow deer	Chamois	Moose	Reindeer	Wild boar
Austria	35,000	400			30			140
Denmark	10,000	90						
Finland			700			150	350	
Germany	12,000	950		1800	220			6900
Ireland		2		70				
Netherlands	2500	10						100
Norway	3200	400				1500	3	
Sweden	50,000	30						
						4000		30

Figure 3.8 A deer-proof fence built along a highway to keep deer from crossing the road and a one-way gate which is designed to allow deer trapped inside the right-of-way to get back to the outside.

built along a 2.5-mile stretch of U.S. Highway 441 in Florida cost $3.7 million. Often, deer and elk are reluctant to use the overpasses or underpasses because they want to avoid areas where they would be vulnerable to predators. Both underpasses and overpasses provide potential ambush sites for predators and also restrict the movements of deer and elk (Reed 1981).

Deer–automobile collisions can be reduced by modifying the road right-of-way (ROW). For instance, ROW should be wide and kept clear so that motorists can see deer approaching the road from the side and have time to respond (Bashore

Figure 3.9 An underpass beneath a highway built in the hopes that deer would use it rather than trying to cross the road.

et al. 1985). Also, many deer–automobile collisions occur when deer are attracted to the ROW by some resource, become startled by an approaching car, and try to escape by darting across the road. There are several resources along ROW which deer might find attractive. One of the most important is the vegetation planted there. Grass and bushes growing along the ROW are especially attractive to deer in forested areas. In many forests, little vegetation is within the reach of deer, and those plants that are may be of poor nutritional quality because they receive only partial sunlight and compete with the trees for water and soil nutrients. ROW plants, on the other hand, grow in full sunlight and are more nutritious because competition with trees is reduced. In arid regions, runoff from the road provides nearby vegetation with more water than vegetation growing further from the road. Because of this, vegetation along the road's margin will stay green longer than the surrounding vegetation. When the only green vegetation is along roadsides, hungry deer will be attracted to these areas.

One method to reduce the frequency of deer–automobile collisions is to reduce nutritional qualities of ROW plants, which are attracting hungry deer. ROW could be landscaped with bushes that are unpalatable or toxic to deer and seeded with less palatable grass species. Some grass cultivars contain a fungus within their leaves that produces alkaloids and makes the grass unpalatable to many herbivores (Aldrich et al. 1993; Schmidt and Osborn 1993; Conover and Messmer 1996). It is not known whether deer find fungus-infected grass less palatable than fungus-free grass, but this seems likely. If so, highway departments may be able to gain two levels of protection against herbivores by planting ROW with fungus-infected cultivars of unpalatable grass species.

Mowing increases the palatability of grass and should be reduced along ROW where deer–automobile collisions occur. Grass becomes rank when unmowed; the new leaves are covered by the old ones. This makes it hard for a herbivore to find new leaves (which are the most palatable) or to bite them without also getting a mouthful of old leaves. However, mowing removes the old leaves and makes the new leaves conspicuous and accessible; it also concentrates nutrients in the new leaves.

In mature forests, little sunlight reaches the floor and brush may be scarce. When roads cross such forests, brush and small trees often spring up along the edge of the ROW and the forest because more sunlight is reaching the forest floor. This is called the "edge effect." Deer like to rest and bed down in brush where there is some concealment. Hence, brush along the edge of the ROW may attract deer seeking cover. It should be removed if there is a threat of deer–automobile collisions.

In some areas, deer are attracted to ROW, especially in the spring, to lick the salt which was spread on the road in the winter to melt snow (Feldhamer et al. 1986). Deer could be discouraged from doing this by not using salt during the winter or by distributing salt blocks away from the road in areas where salt is spread on roads.

Deer–automobile collisions can also be avoided by changing driver behavior, especially by getting people to drive slower. Deer are most active during the two-hour period around sunrise and sunset; drivers should be particularly watchful for deer during these hours. Deer are most vulnerable to automobile collisions during the rut each fall when does are often chased by bucks. During this period, neither does nor bucks are as cautious as they are during the rest of the year.

Mule deer and elk are also vulnerable during the few weeks each spring and fall when they are migrating between their summer and winter ranges. Usually, migrating animals use the same paths or corridors to cross roads every year. In one innovative program, the Utah Division of Wildlife Resources placed special flashing signs along roads during the migratory period. Because these signs were only up for a few weeks, people did not habituate to them the way they do to permanent deer-crossing signs and actually slowed down.

Many deer travel in small groups, and does often are accompanied by their fawns (Figure 3.10). When these groups cross a road, the first deer often makes it safely to the other side, but the following deer (usually a fawn) is often the one that is killed. Too often, the fawn does not cross the road immediately, but once it realizes that an approaching car is going to come between it and its mother or the rest of the herd, it tries to dart across the road before the car reaches it. Deer–automobile collisions could be reduced by better informing drivers that deer often travel in groups and that if they see one deer crossing the road, there may be more crossing very soon. Public service announcements could communicate the message that it is the deer which the driver does not see that often causes the accident.

HUMAN INJURIES AND FATALITIES FROM BIRD–AIRCRAFT COLLISIONS

Humanity first flew in a "heavier-than-air" plane in 1903, and only a few years later (1912) suffered its first fatality from a bird–aircraft collision (Solman 1973).

Figure 3.10 Deer often travel is small groups; when they do, it is often the last deer that is struck by an automobile.

Since that time, planes have become more numerous, larger, and faster, and bird–aircraft collisions have increased exponentially.

In 1991, there were 2059 reported bird strikes by civilian aircraft in the U.S. (Conover et al. 1995). Because pilots are unaware of some strikes and too busy to report others, only about 25% of bird strikes are reported (Linnell et al. 1996, 1999; Cleary et al. 1997). If we extrapolate this reporting rate (25%) to the 2059 reported bird strikes, actual bird strikes by civilian aircraft would be approximately 8000 per year. From 1960 to 1988, 104 human fatalities and 34 serious injuries resulted from bird strikes with civilian planes in the U.S. (E. A. LaBoeuf, Fed. Aviation Admin., Washington, D.C., unpublished data); this averages 3.7 fatalities and 1.2 injuries per year.

Military aircraft are more vulnerable to bird strikes than commercial aircraft because they are designed more for speed and agility than for safety. Because of these design differences, military aircraft are less able to survive a bird strike. For example, many fighters have only one engine and will probably crash if the engine ingests a bird and shuts down, while commercial jets can still fly with the loss of an engine. Military aircraft also practice flying low to the ground to avoid detection by radar, but this is the same airspace used by birds.

The U.S. Air Force maintains the most accurate records on bird hazards to military planes. It reported an average of 3200 bird–aircraft collisions yearly (Conover et al. 1995). From 1987 to 1993, seven deaths were attributed to bird strikes with U.S. Air Force planes, but in 1995, 24 people lost their lives in a single crash when an AWAC aircraft collided with a flock of Canada geese during takeoff from Elmendorf Air Force Base in Alaska. This amounts to an average of 3.1 fatalities per year from 1987 to 1997. U.S. Navy officials estimate that the number and cost of bird–aircraft collisions for the Navy are similar to those for the Air Force, because the Navy flies along the coast where bird strikes are more likely. The U.S. Army has more aircraft than the U.S. Air Force but they are considered less vulnerable due to lower airspeeds and the location of training areas. Using U.S. Air Force fatality rates for all branches of the armed forces yields an estimate of approximately nine deaths per year due to bird strikes.

Birds pose a worldwide threat to aircraft, although the species that are involved vary among countries, depending upon the local bird populations (Table 3.6). For instance, vultures pose the greatest threat to aircraft in India, gulls in Great Britain, and raptors in Australia. Unfortunately, data are unavailable on the number of bird–aircraft collisions for most countries.

Efforts to design aircraft that can withstand a bird strike without sustaining damage have led to safer aircraft but have not eliminated the risk of a plane crash. The amount of damage caused by a bird strike depends upon several factors, such as the bird's mass (big birds generally cause more damage than small birds), aircraft speed, and where on the plane the strike occurs. Most dangerous are birds that strike the windshield or are ingested by an engine; the latter may cause the engine to fail. Bird strikes are more dangerous on takeoffs than on landings because a plane can still land if an engine fails.

Damage from a bird strike may go undetected for days or years until there is an equipment failure. An example of this problem was a DC-8 that struck a bird with one engine during takeoff from Rome. About 50 hours later, this engine failed during a flight over Alberta. Inspection in the repair shop showed blade failure and damage to the compressor (Blokpoel 1976).

Few birds fly more than 500 m above the ground; for this reason, 75% of all bird strikes experienced by civilian aircraft occur on or in the immediate vicinity of an airfield (Solman 1973; Blokpoel 1976). Therefore, efforts to reduce the frequency of bird–aircraft collisions are concentrated at airports and are designed to keep birds away from them. The underlying assumption behind these bird control efforts at airports is that a reduction in the localized avian population will result in fewer bird–aircraft collisions; this assumption appears to be accurate (Dolbeer et al. 1993). Bird control efforts at airports can involve removing items that attract birds. One example is draining surface water where birds drink; another is removing food

Table 3.6 Bird Collisions with Civilian Aircraft in Different Countries (Blokpoel 1976)

	Proportion of collisions involving different avian groups					
Bird group	**Great Britain (n = 406)**	**Germany (n = 81)**	**Russia (n = 234)**	**Canada (n = 81)**	**Australia (n = 423)**	**India (n = 90)**
Gulls	64	29	36	36	19	0
Waders/ shorebirds	16	11	4	5	17	0
Pigeons	3	3	7	2	5	7
Waterfowl	1	0	11	18	0	0
Corvids	3	6	7	2	0	0
Raptors	3	14	11	0	41	7
Vultures	1	0	0	0	0	86
Other birds	9	37	24	37	18	0

sources by closing adjacent landfills. Vegetation at airports can be managed to discourage birds as well (Brough and Bridgman 1980). For instance, at airports where geese gather to forage on the grass, the grass can be allowed to grow long by not mowing it. At tropical airports, wedelia appears to be an excellent ground cover because birds find it unpalatable as do rodents and insects (Linnell et al., in press). This helps reduce numbers of raptors and insectivorous birds that feed upon the rodents and insects.

When habitat management fails, bird numbers at airports have been reduced by using propane cannons, harassment techniques, or lethal means. John F. Kennedy International Airport in New York had a problem of jets striking gulls, due to the presence of a large gull nesting colony just off its runways. This problem was alleviated after gulls that crossed the runways were shot (Dolbeer et al. 1993).

SUMMARY

The three main types of wildlife attacks on humans are predatory, territorial, and defensive. Predatory attacks on adult humans are rare, for few predators are large enough to risk attacking such large and dangerous prey. Young children, however, are more at risk. Territorial attacks are often bluff attacks meant to scare away the intruder, but some individuals are more aggressive than others and may cause injury to humans. Defensive attacks are the most frequent and occur when an animal is unable to flee and attempts to inflict pain upon the person that it perceives to be a threat.

The frequency of wildlife attacks on humans in the U.S. has been increasing in the last few decades. Reasons for the increase include better reporting systems, increasing human and wildlife populations, humans moving into wildlife habitat, wildlife moving into suburban areas, and a change in behavior of both wildlife and humans toward each other. To reduce the frequency of wildlife attacks on humans, we can remove those individual animals which have a high probability of attacking people, teach dangerous animals to fear humans, and educate humans to recognize and avoid dangerous situations involving wildlife.

Collisions between vehicles and large wildlife species, especially deer, constitute a serious safety issue. Approximately 1.5 million deer–vehicle collisions occur annually in the U.S., resulting in 29,000 human injuries, 200 deaths, and over 1.3 million deer fatalities. Deer are often attracted to roadways by the presence of high-quality forage or salt. Methods to reduce deer–vehicle collisions include modifying the habitat along the roadway to make it less attractive to deer, developing methods that allow deer to cross the road safely, and erecting deer-proof fences to keep deer out. Another way to reduce collisions is by changing driver behavior.

Bird–aircraft collisions also result in human injuries and fatalities. Over 2000 bird strikes with civilian aircraft were reported in 1991 in the U.S., but the actual number of strikes may be closer to 8000. More than 9000 U.S. military planes may be involved in bird strikes annually. An average of four fatalities per year are caused by bird strikes with civilian planes, and an additional nine deaths are caused by bird strikes with U.S. military planes. Most bird strikes occur in the immediate vicinity of airports. To reduce bird strikes, vegetation at airports should be managed to make it as unattractive as possible to birds. Harassment of birds or lethal techniques can also be used at airports.

LITERATURE CITED

Aldrich, C. G., J. A. Paterson, J. L. Tate, and M. S. Kerley, The effects of endophyte-infected tall fescue on diet utilization and thermal regulation in cattle. *J. Anim. Sci.*, 71, 164–170, 1993.

Allen, R. E., and D. R. McCullough, Deer–car accidents in southern Michigan. *J. Wildl. Manage.*, 40, 317–325, 1976.

Baker, R. O. and R. M. Timm, Management of conflicts between urban coyotes and humans in southern California. In *Proc. Vert. Pest Conf.*, 18, 299–312, 1998.

Bashore, T. L., W. M. Tzilkowski, and E. D. Bellis, Analysis of deer–vehicle collision sites in Pennsylvania. *J. Wildl. Manage.*, 49, 769–774, 1985.

Beier, P., Cougar attacks on humans in the United States and Canada. *Wildl. Soc. Bull.*, 19, 403–412, 1991.

Blokpoel, H., *Bird Hazards to Aircraft.* Books Canada, Buffalo, NY, 1976.

Brough, T. and C. J. Bridgman, An evaluation of long grass as a bird deterrent on British airfields. *J. App. Ecol.*, 17: 243–253, 1980.

Carbyn, L., Coyote attacks on children in western North America. *Wildl. Soc. Bull.*, 17, 444–446, 1989.

Cleary, E. C., S. E. Wright, and R. A. Dolbeer, Wildlife strikes to civil aircraft in the United States, 1992–1996. Report DOT/FAA/AS/97-3, 1997.

Conover, M. R. and T. J. DuBow, Alligator attacks on humans in the United States. *Herpetol. Rev.*, 28, 120–124, 1997.

Conover, M. R. and G. S. Kania, Impacts of interspecific aggression and herbivory by mute swans on native waterfowl and aquatic vegetation in New England. *Auk,* 111, 744–748, 1994.

Conover, M. R., W. C. Pitt, K. K. Kessler, T. J. DuBow, and W. A. Sanborn, Review of human injuries, illnesses, and economic losses caused by wildlife in the United States. *Wildl. Soc. Bull.*, 23, 407–414, 1995.

Conover, M. R. and T. A. Messmer, Feeding preference and changes in mass of Canada geese grazing endophyte-infected tall fescue. *Condor*, 98, 859–862, 1996.

Conrad, L. and J. Balison, Bison goring injuries: penetrating and blunt trauma. *J. Wilderness Med.*, 5, 371–381, 1994.

Decker, D. J., K. M. Loconti-Lee, and N. A. Connelly, Incidence and costs of deer-related vehicular accidents in Tompkins County, New York. Human Dimensions Research Group 89-7. Cornell University, Ithaca, NY, 1990.

Dolbeer, R. A., J. L. Belant, and J. L. Sillings, Shooting gulls reduces strikes with aircraft at John F. Kennedy International Airport. *Wildl. Soc. Bull.*, 21, 442–450, 1993.

Falk, N. W., H. B. Graves, and E. D. Bellis, Highway right-of-way fences as deer deterrents. *J. Wildl. Manage.*, 42, 646–650, 1978.

Feldhamer, G. A., J. E. Gates, D. M. Harman, A. J. Loranger, and K. R. Dixon, Effects of interstate highway fencing on white-tailed deer activity. *J. Wildl. Manage.*, 50, 497–503, 1986.

Fetner, 1987. *The African Safari: The Ultimate Wildlife and Photography Adventure.* St. Martin's Press, New York, 1987.

Grenard, S., Is rattlesnake venom evolving? *Nat. Hist.*, 109(6), 45–49, 2000.
Groot Bruinderink, G. W. T. A. and E. Hazelbrook, Ungulate traffic collisions in Europe. *Conserv. Biol.*, 10, 1059–1067, 1996.
Guggisberg, C. A. W., *Crocodiles: Their Natural History, Folklore, and Conservation*. Stackpole Books, Harrisburg, PA, 1972.
Gutierrez-Sanders, C., Alligator attack statistics. Florida Game and Fresh Water Fish Commission, Office of Information Services, Tallahassee, FL, 1992.
Herrero, S., *Bear Attacks*. Nick Lyons, New York, 1985.
Herrero, S. and S. Fleck, Injury to people inflicted by black, grizzly or polar bears: recent trends and new insights. *Int. Conf. Bear Res. Manage.,* 8: 25–32, 1989.
Hines, T. C., and K. D. Keenlyne, Two incidents of alligator attacks on humans in Florida. *Copeia,* 1977(4), 735–738, 1977.
Hines, T. C. and A. R. Woodward, Nuisance alligator control in Florida. *Wildl. Soc. Bull.*, 8, 234–241, 1980.
Howell, R., The urban coyote problem in Los Angeles County. In *Proc. Vert. Pest Conf.* 10, 21–22, 1982.
Hygnstrom, S. E., R. M. Timm, and G. E. Larson, *Prevention and Control of Wildlife Damage*. University of Nebraska Cooperative Extension, Lincoln, 1994.
Jackson, P., *Tigers*. Chartwell Books, Secaucus, NJ, 1991.
Jhala, Y. and B. Jethra, abstract, Second Int. Wildl. Manage. Congr., *The Probe*, 205, 1999.
Le Moyne, J., Indorum Floridam proviniciam inhabitanitium elcones. Theordore Byr, Liege, 1591. In *Voyages en Virginie et en Floride*. Ducharte et Van Buggenhoudt, Paris, 1926.
Linnell, M. A., M. R. Conover, and T. J. Ohashi, Analysis of bird strikes at a tropical airport. *J. Wildl. Manage.*, 60, 935–945, 1996.
Linnell, M. A., M. R. Conover, and T. J. Ohashi, Biases in bird strike statistics based on pilot reports. *J. Wildl. Manage.*, 63, 997–1003, 1999.
Linnell, M. A., M. R. Conover, and T. J. Ohashi, Use of an alternative ground cover, wedelia, for reducing bird activity on tropical airfields. *J. Wildl. Res.*, 4, in press.
Mansfield, T. M. and K. G. Charlton. Trends in mountain lion depredation and public safety incidents in California. In *Proc. Vert. Pest Conf.*, 18, 118–121, 1998.
McCullough, D. R., Behavior, bears, and humans. *Wildl. Soc. Bull.*, 10, 27–33, 1982.
McDougal, C., Man-eaters. In J. Seidensticker and S. Lumpkin, Eds., *Great Cats: Majestic Creatures of the Wild*. Rodale Press, Emmaus, PA, p. 204–209, 1991.
Minton, S. A., Poisonous snakes and snakebite in the U.S. *Northwest Sci.,* 61, 130–137, 1987.
Moore, R. M., Jr., R. B. Zehmer, J. I. Moulthrop, and R. L Parker, Surveillance of animal-bite cases in the United States, 1971–1972. *Arch. Environ. Health*, 32, 267–270, 1977.
Perry, R., *The World of the Tiger*. Trinity Press, Worcester, England, 1964.
Rajpurohit, K. S. and P. R. Krausman, Human–sloth-bear conflicts in Madhya Pradesh, India. *Wildl. Soc. Bull.*, 28, 393–399, 2000.
Reed, D. F., Mule deer behavior at a highway underpass exit. *J. Wildl. Manage.*, 45, 542–543, 1981.
Reed, D. F., T. M. Pojar, and T. N. Woodard, Use of one-way gates by mule deer. *J. Wildl. Manage.*, 38, 9–15, 1974.
Reed, D. F., T. N. Woodard, and T. M. Pojar, Behavioral response of mule deer to a highway underpass. *J. Wildl. Manage.*, 39, 361–367, 1975.
Romin, L., Factors associated with mule deer highway mortality at Jordanell Reservoir, Utah. M.S. thesis, Utah State University, Logan, 1994.
Saberwal, V. K., J. P. Gibbs, R. Chellam, and A. J. T. Johnsingh, Lion–human conflict in the Gir Forest, India. *Conserv. Biol.,* 8, 501–507, 1994.

Schmidt, S. P. and T. G. Osborn, Effects of endophyte-infected tall fescue on animal performance. *Agric. Ecosystems Environ.*, 44, 233–262, 1993.

Seidensticker, J. and S. Lumpkin, Mountain lions don't stalk people? True or false. *Smithsonian,* 22(11), 113–122, 1992.

Solman, V. E. F., Birds and aircraft. *Biol. Conserv.*, 5, 79–86, 1973.

Streetly, J., Wolves between worlds. *BBC Wildl.,* 8(12), 59–60, 2000.

Taylor, D., N. Kinler, and G. Linscombe, Female alligator reproduction and associated population estimates. *J. Wildl. Manage.*, 55, 682–688, 1991.

Torres, S. G., T. M. Mansfield, J. E. Foley, T. Lupo, and A. Brinkhaus, Mountain lion and human activity in California. Testing speculations. *Wildl. Soc. Bull.*, 24, 451–460, 1996.

Van Doren, M., *Travels of William Bartram.* Dover Publications, New York, 1955.

Veeramani, A., E. A. Jayson, and P. S. Easa, Man–wildlife conflict: cattle lifting and human casualties in Kerala. *Indian For.*, 122, 897–902, 1996.

Ward, A. L., Mule deer behavior in relation to fencing and underpasses on Interstate 80 in Wyoming. *Transp. Res. Rec.,* 859, 8–13, 1982.

Weiss, R., Snakebite succor: researchers foresee antivenom improvements. *Sci. News.,* 38, 360–362, 1990.

CHAPTER 4

Zoonoses

"Ring around a rosy, pocket full of posey, ashes, ashes, all fall down."

A nursery rhyme from the Middle Ages which helped children face their fears about contracting the plague. The rhyme describes the plague's symptoms (ring of rosy), purported cures (pocket full of posey), and its consequences — death (all fall down).

Zoonoses literally means "diseases from animals." Zoonoses are human diseases for which vertebrates serve as reservoirs or hosts for the disease organism. Few disease organisms can exist outside a living body for very long. Without a wildlife species to serve as a reservoir, most zoonoses would cease to exist.

Each disease organism needs a way to spread from one host to another if it is to survive. The agents that diseases use to accomplish this are called vectors. For a human to become ill from a zoonotic disease, some vector must spread it from the wildlife reservoir to the human victim. Once infected, humans can also serve as vectors for some zoonoses that can be spread through human-to-human contact (i.e., they are contagious diseases). Most zoonoses, however, are not contagious.

Unfortunately, morbidity and mortality data (i.e., information on the number of people who become ill or die) from zoonoses are incomplete. The U.S. Department of Health and Human Services Centers for Disease Control and Prevention (CDC) keep track of "reportable diseases" in the United States, 15 of which are zoonoses. A "reportable disease" is one which physicians are required by law to report to public health officials. There are excellent data on reported cases of these diseases in the U.S., but they do not include any cases in which the infected person did not seek medical help or when the disease was misdiagnosed. Hence, CDC data are conservative (Table 4.1). The World Health Organization (WHO) also maintains morbidity and mortality data. Worldwide counts of disease cases are poor because many ill people do not seek medical attention; thus, many diseases are underreported.

The number of diseases that are classified as zoonoses varies from approximately 100 (out of 169 diseases listed in the 15th edition of the book, *Control of Communicable Diseases in Man*) to 3000 including over 2000 types of salmonella. A list

Table 4.1 Average Number of Cases per Year for 1991–1994 and 1995–1998 of 15 Notifiable Diseases in the U.S. for which Wildlife Species May Serve as a Vector or Reservoir[a]

Disease	Cases (mean per year)		Fatalities (mean per year)[b]	
	1991–1994	1995–1998	1991–1994	1995–1997
Brucellosis	112	97	0.2	0.7
Plague	13	7	1.2	1.3
Tularemia	145	—[c]	1.2	—[c]
Salmonellosis	43,507	44,259	50.2	58.3
Trichinosis	38	18	0.0	0.0
Leptospirosis	50	—[c]	1.0	—[c]
Lyme disease	10,165	14,439	—[d]	—[d]
Rocky Mountain spotted fever	513	549	10.0	8.7
Psittacosis	71	46	1.2	0.3
Encephalitis, eastern equine	—[e]	4[e]	0.7	1.3
Encephalitis, western equine	—[e]	0[e]	0.0	0.0
Encephalitis, California group	—[e]	97[e]	0.0	0.7
Encephalitis, St. Louis	—[e]	24[e]	3.7	2.3
Hantavirus	29[f]	25	14.3[f]	7.3
Rabies	3	3	2.0	3.0
Total	54,646	59,568	85.7	83.9

[a] Humans, pets, livestock, and other animals can also serve as the vector or reservoir to the infectious agent. Hence, wildlife are involved in an unknown proportion of these cases. Data were provided by U.S. Centers for Disease Control and Prevention.
[b] Fatalities are unavailable for 1998.
[c] Data unavailable: leptospirosis and tularemia were no longer notifiable after 1994.
[d] No fatality data are available for Lyme disease.
[e] Encephalitis morbidity data only available for 1998.
[f] Mean for 1992–1994 (Hantavirus data unavailable for 1991).

of zoonoses from Beran's (1994a) *Handbook of Zoonoses* is presented in Table 4.2. Obviously, all of them cannot be described here. Hence, this chapter will be limited to those diseases that pose serious threats to human health and cause public concern either in the U.S. or worldwide. For each disease, information is provided on its distribution, the infecting organism, the reservoir and vector species, how it is spread to humans, and its symptoms in humans.

BACTERIAL DISEASES

Plague

Plague is caused by the bacterium, *Yersinia pestis*. Humans contract this disease through contact with infected mammals (usually rodents), by being bitten by infected fleas, by eating infected meat, by direct contact with an infected carcass when field dressing an animal, or by inhaling the bacteria. There are three forms of plague: bubonic, septicemic, and pneumonic. Each form is caused by the same bacteria but differs in which organ system is attacked and by which method the bacteria are vectored to the infected person. Bubonic plague is transmitted through the bites of infected

Table 4.2 Review of Zoonoses from Beran (1994a)

Disease	Animal Reservoirs	How Vectored to Humans
	Bacteria	
Brucellosis	Livestock, bison, and elk	Ingestion, inhalation, or direct contact
Tuberculosis	Dairy cattle, and other captive and exotic animals	Inhalation
Tuberculoidoses	Chickens and free-ranging birds	Inhalation
Anthrax	Livestock	Inhalation, ingestion, or direct contact
Erysipelothrix infections	Hogs, rodents, and birds	Contact with skin wounds
Plague	Rodents	Flea bites or inhalation
Tularemia	Lagomorphs, rodents, and other free-ranging animals	Direct contact, ingestion, inhalation, or bites of ticks and biting flies
Tetanus	Livestock and wildlife	Punctures, cuts, and burns
Gas gangrene	Livestock and wildlife	Ingestion, inhalation, or wounds
Glanders	Horses, donkeys, and mules	Inhalation, wounds, and ingestion
Melioidosis	Hogs, sheep, goats, and rats	Exposure to organism via soil or water
Streptococcus suis infections	Hogs and livestock	Wounds, ingestion, or direct contact
Group C streptococcus	Horses and livestock	Direct contact or ingestion
Group L streptococcus	Hogs, dogs, and poultry	Wounds
Arcobacter infections	Livestock	Consumption or contact with contaminated water
Leprosy	Armadillos	Inhalation or direct contact
Pasteurellaceae	Mammals and birds	Bite or scratch
Capnocytophaga infections	Domestic animals	Bites, scratches, or direct exposure
Weeksella zoohelcum	Dogs and cats	?
M-5	Dogs	Bite wounds and infections
EF-4	Dogs and cats	Bite wounds and infections
Cat-scratch disease	Cats	Bites or scratches by cats
Rat-bite fever	Rats, mice, squirrels, and weasels	Bite or ingestion
Mammalian salmonellosis	Hogs, sheep, cattle, and rodents	Ingestion, fly bites, or inhalation
Avian salmonellosis	Birds, reptiles, and mammals	Ingestion
Campylobacteriosis	Poultry, swine, domestic animals, rodents, and primates	Ingestion or contact with infected animals
E. coli	Cattle	Ingestion
Yersinosis	Pigs, dogs, cats, and rats	Ingestion
Clostridium perfringens gastroenteritis	Livestock and wildlife	Ingestion
Botulism	Birds, fish, and mammals	Ingestion

continued

Table 4.2 (continued) Review of Zoonoses from Beran (1994a)

Disease	Animal Reservoirs	How Vectored to Humans
Staphylococcal food poisoning	Dairy cows, sheep, goats, and fish	Ingestion, direct contact, or inhalation
Vibrio parhaemolyticus infection	Crustaceans	Ingestion
Vibrio vulnificus infection	Oysters, shellfish, and mollusks	Ingestion
	Spirochetes	
Leptospirosis	Livestock, free-ranging and captive animals	Direct contact with leptospores
Lyme disease	Rodents and deer	Ticks
	Chlamydiosis	
Avian chlamydiosis	Waterfowl, pigeons, and psittacine birds	Direct contact or inhalation
Mammalian chlamydioses	Cats, sheep, and livestock	Direct contact or inhalation
	Rickettsia	
Rocky Mountain spotted fever	Rodents, lagomorphs, dogs, livestock, deer, and carnivores	Ticks
Q-fever	Arthropods, annelids, rodents, birds, and domestic animals	Inhalation, ticks, direct contact, or ingestion
Ehrlichiosis	Deer and rodents	Ticks
Murine typhus	Rodents and opossums	Inhalation, ticks, or fleas
Scrub typhus	Rodents	Chiggers
	Fungi	
Superficial mycoses	Variety of animal hosts	Infectious particle landing on skin
Aspergillosis	Soil or organic material	Inhalation
Blastomycosis	Dogs, wolves, and other mammals	Exogenous transmission
Thrush	Birds, mammals, and livestock	Endogenous source
Coccidioidomycosis	Wildlife	Inhalation
Cryptococcosis	Bird droppings, mammals, and reptiles	Direct contact or inhalation
Histoplasmosis	Soil enriched with bird droppings	Inhalation of airborne conidia
Protothecosis	Cats, cows, deer, and dogs	Direct contact
Pythiosis insidiosi	Horses, livestock, dogs, and soil	Wound contamination by flagellated zoospores
Sporotrichosis	Dogs, cats, horses, and armadillos	Subcutaneous inoculation by thorns or inhalation
Zygomycosis	Livestock, captive animals, and free-ranging animals	Inhalation, ingestion, or inoculation
Mycotoxicoses	Stored grains and feed	Ingestion

continued

Table 4.2 (continued) Review of Zoonoses from Beran (1994a)

Disease	Animal Reservoirs	How Vectored to Humans
	Virus	
Venezuelan equine encephalitis	Horses, donkeys, and rodents	Mosquitoes
Eastern encephalitis	Passerine and captive birds	Mosquitoes
Western equine encephalomyelitis	Free-ranging and captive birds	Mosquitoes
St. Louis encephalitis	Birds	Mosquitoes
Japanese encephalitis	Pigs, livestock, and Ardeid birds	Mosquitoes
California encephalitis	Rodents	Mosquitoes
Colorado tick fever	Squirrels and rodents	Ticks
Dengue and dengue hemorrhagic fever	Primates	Mosquitoes
Chikungunya fever	Primates	Mosquitoes
Yellow fever	Forest primates	Mosquitoes
Rift Valley fever	Livestock and other vertebrates	Bites of mosquitoes, flies, and midges or inhalation
West Nile fever	Birds and mammals	Mosquitoes
Crimean–Congo hemorrhagic fever	Small mammals, large ungulates, livestock, and birds	Tick bites or direct contact
Sindbis fever	Free-ranging and domestic birds	Mosquitoes and other insects
Vesicular stomatitis	Domestic and free-ranging mammals	Insect bite
Powassan encephalitis	Woodchucks, squirrels skunks, raccoons, voles, birds, and mice	Ticks
Tensaw virus	Free-ranging and domestic animals	Mosquitoes
Cache Valley virus	Free-ranging ungulates	Mosquitoes
Nothway virus	Horses and deer	Mosquitoes
Tlacotapan virus	Unknown	Mosquitoes
Main drain virus	Free-ranging leporid	Biting midge
Lokern virus	Jackrabbits	Biting midge
Mayaro fever	Monkeys and marmosets	Mosquitoes
Mucambo fever	Rodents	Mosquitoes
Rocio viral encephalitis	Free-ranging birds	Mosquitoes
Bussuquara fever	Rodents	Mosquitoes
Ilhéus fever	Free-ranging birds	Mosquitoes
Bunyaviral fever	Rodents and marsupials	Mosquitoes
Oropouche fever	Sloths and monkeys	Midges or mosquitoes
Guaroa fever	Free-ranging birds	Mosquitoes
Phlebotomus fever	*Phlebotomus* flies	*Phlebotomus* flies
Piry fever	Unknown	Unknown
Louping ill	Sheep and red grouse	Ticks
Tick-borne encephalitis	Hedgehogs, squirrels, mice, voles, and shrews	Ticks or ingestion

continued

Table 4.2 (continued) Review of Zoonoses from Beran (1994a)

Disease	Animal Reservoirs	How Vectored to Humans
Karelian fever	Passerine birds	Mosquitoes
Getah viral infections	Pigs and horses	Mosquitoes
Omsk hemorrhagic fever	Muskrats	Ticks or direct contact
Karshi	Rodents	Tick bites
California viral infections	Rabbits, hares, and hedgehogs	Mosquitoes
Batai viral infections	Cattle	Mosquitoes
Sandfly fever	Gerbils and squirrels	Sandflies
Bhanji fever	Bats	Inhalation, ingestion, ticks, or mosquitoes
Tamdy fever	Gerbils	Ticks
Kemerovo fever	?	Ticks
Syr–Darya Valley fever	?	Ticks
Dhori viral fever	?	Ticks
O'nyong-Nyong virus disease	?	Mosquitoes
Semliki forest disease	Monkeys and horses	Mosquitoes
Orungo virus	?	Mosquitoes
Wesselsbron viral disease	?	Mosquitoes, inhalation, or direct contact
Banzi fever	Rodents	Mosquitoes
Spondweni fever	?	Mosquitoes
Zika fever	Monkeys	Mosquitoes
Dugbe fever	?	Ticks
Nairobi sheep disease	Sheep and goats	Ticks
Quaranfil virus disease	Egrets and other birds	Ticks
Thogoto virus disease	Ruminants	Ticks
Kyasanur forest disease	Monkeys and small mammals	Ticks
Tembusu virus	Chickens	Mosquitoes
Jungra virus	Bats	Mosquitoes
Batai virus	Ungulates	Mosquitoes
Ingwavuma	Pigs and free-ranging birds	Mosquitoes
Chandipurid and Isfahan viruses	Hedgehogs	Phlebotomine flies
Murray Valley encephalitis	Water birds	Mosquitoes
Epidemic polyarthritis	Various animal reservoirs	Mosquitoes
Kunjin virus	Birds and cattle	Mosquitoes
Rabies	Free-ranging and captive mammals	Exposure to infectious saliva of rabid animals
Hemorrhagic fever with renal syndrome	Rodents	Direct contact or inhalation
Arenavirus infection	Rodents	Direct contact
Marburg virus	Primates	Direct contact
Ebola virus	Primates	Direct contact
Reston virus	Primates	Direct contact
Influenza A	Swine and horses	Inhalation or direct contact
Respiratory syncytial virus infection	Cattle and sheep	Inhalation
Parainfluenza viral infection	Hamsters, guinea pigs, and monkeys	Direct contact or inhalation

continued

Table 4.2 (continued) Review of Zoonoses from Beran (1994a)

Disease	Animal Reservoirs	How Vectored to Humans
Encephalomyocarditis	Rodents	Inhalation, intramuscular infections, or ingestion
Swine vesicular disease	Pigs	Inhalation or direct contact
Foot-and-mouth disease	Cattle, sheep, and goats	Inhalation
Lymphocytic choriomeningitis	Rodents	Inhalation and direct contact
Newcastle disease	Waterfowl and tropical birds	Ingestion, inhalation, or direct contact
Cowpox	Cows and cats	Direct contact
Monkeypox	Monkeys and squirrels	Ingestion or direct contact
Parapoxvirus	Sheep	Direct contact
Hantavirus	Rodents	Inhalation or direct contact
Tanapox	Monkeys	Arthropod bites
Herpesvirus simiae	Monkeys	Monkey bites
Human immunodeficiency virus (HIV)	Primates	Sexual contact or direct contact with blood or infected material

fleas, infecting the lymphatic system. Symptoms include swollen lymph nodes and fevers up to 40.5°C (106°F). Septicemic plague occurs when the bacterial strain invades the bloodstream. Symptoms appear in humans two to six days after infection and include fever and internal hemorrhaging. Pneumonic plague infects the lungs and results from inhalation of an airborne bacterium. Pneumonic plague progresses rapidly and is one of the few zoonotic diseases that is contagious. Signs of general infection (e.g., fever, flu-like symptoms) occur within one day of initial infection. By the second day, infected individuals experience respiratory distress and coughing of blood. Death usually occurs within four days of the initial infection. Fatality rates for untreated humans range from 50% for bubonic plague to 100% for septicemic and pneumonic plague (Gage et al. 1995). However, plague is treatable with prompt diagnosis. During an epidemic, *Yersinia pestis* may be spread by multiple vectors, and all three forms of plague (bubonic, pneumonic, and septicemic) can occur simultaneously (Figure 4.1).

There have been at least three major epidemics of plague in recorded history. The first began in Africa and swept through Constantinople in 542 A.D. (Gratz 1988). During the next 50 years, this epidemic killed millions of people as it spread throughout Europe. The second plague epidemic is known as the Black Death. It started in the 1300s in China and India and spread to Europe. Local outbreaks in Europe and Asia lasted for more than 400 years. One of these outbreaks, the Great Plague of London in 1664, killed over 100,000 people. In some areas, the Black Death killed a third of the human population. The third epidemic is called the Modern Epidemic; it started in Asia in the mid-1800s and killed 13 million people in India. In 1897, this epidemic reached North America. At the turn of the century in San Francisco there were 126 reported cases of plague and 122 of them were fatal. An outbreak in Los Angeles during 1924 resulted in 41 cases, 36 of which were fatal. The modern epidemic continues today in many parts of the world, including the western U.S. (Figure 4.2). In 1994, an outbreak in India resulted in many countries temporarily banning air travel to that country. Each year, at least 2000 worldwide contract plague (Poland et al. 1994).

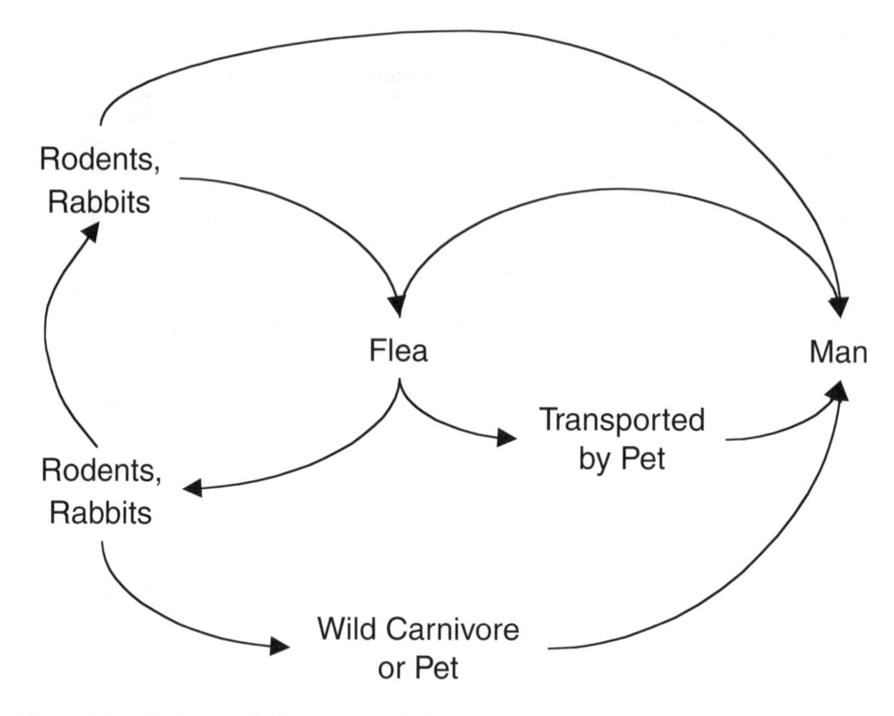

Figure 4.1 Pathways of plague transmission between animals and man.

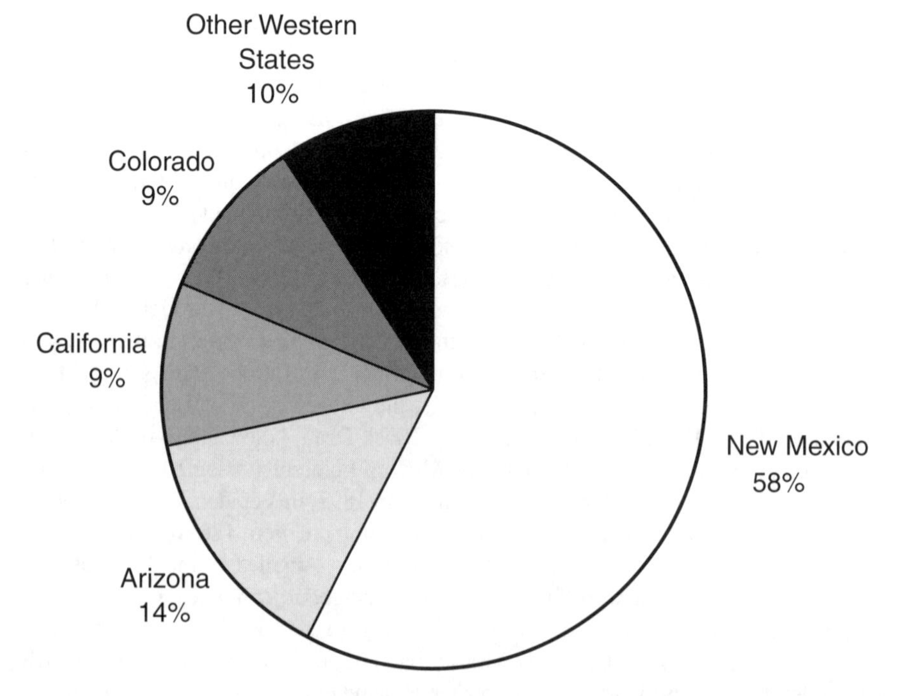

Figure 4.2 Distribution of human cases of plague among states in the U.S. from 1965 to 1990.

Rodents, especially rats, and other small mammals are the main reservoirs for *Y. pestis* in many parts of the world. A number of flea species serve to vector the disease between mammals and humans. In the western U.S., plague is endemic in prairie dog and ground squirrel populations in the four-corners region of Utah, Colorado, New Mexico, and Arizona and in ground squirrel populations in California (Figure 4.3).

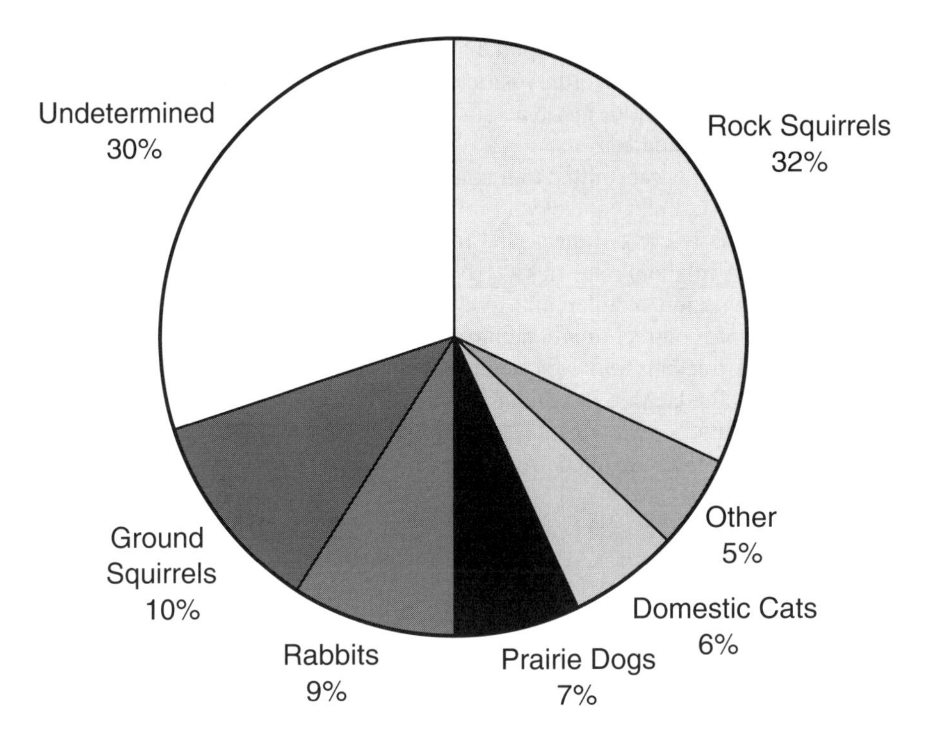

Figure 4.3 Animal sources of human plague infections in the U.S. during the 1980s.

A number of steps can help reduce the frequency of plague transmission from wildlife to humans. One of them is better public awareness of the disease. Many people are unaware that they can contract the disease from fleabites or from handling ground squirrels or prairie dogs. In areas where plague is endemic in mammal populations, habitats around homes and campgrounds should be modified to reduce ground squirrel and prairie dog populations. Small mammal populations should be monitored for plague outbreaks. When an outbreak occurs, insecticides can be used to kill fleas and to reduce the possibility of their spreading the disease to humans.

Tularemia

The bacterium *Francisella tularensis* causes tularemia, also known as rabbit fever. This disease can be difficult to diagnose in humans. Symptoms include sudden fever, chills, and fatigue and are usually followed by the appearance of skin ulcers and painful swelling of lymph nodes. Complications of this disease include respiratory distress, pneumonia, and/or meningitis. Before antibiotics were used, mortality in humans was between 5 and 15%, but it has now dropped to 1 to 3%. During the 1990s, there were an average of 145 cases of tularemia reported annually in the U.S. (Table 4.1).

Tularemia is found in Europe, Asia, and North America, including all U.S. states except Hawaii. It has a wider range of hosts than most zoonotic diseases, including more than 100 species of mammals and birds (Hopla and Hopla 1994). Rabbits, hares, and rodents are the most important reservoirs and sources of transmission of this disease to humans. In North America, these include eastern cottontails, desert cottontails, snowshoe hares, and black-tailed jackrabbits. Rodent hosts in the U.S. are voles, muskrats, and beavers.

Tularemia can be transmitted to humans in a variety of ways that include bites by infected insects, direct contact with infected animals, ingestion of contaminated water or improperly cooked meat, and inhalation of contaminated particles. The main vectors of tularemia are ticks (Ixodidae) and biting flies (Tabanidae). The mode of transmission of tularemia tends to vary by region. In the midwestern U.S., most people contract tularemia through direct contact with infected animals. In contrast, tick transmission is most common in the Southwest and Central States. Mosquitoes (Culicidae) are important vectors of tularemia in Russia and Sweden. On the other hand, a review of 1372 cases in Japan from 1924 to 1994 showed that 93% of the cases resulted from contact with infected free-ranging rabbits (Ohara et al. 1996).

Salmonellosis

Salmonellosis is generally referred to as food poisoning and results from an infection by one of the more than 2000 varieties (serotypes) of salmonella bacteria (Wray 1994). Symptoms include nausea, vomiting, abdominal pain, diarrhea, and high fever. Victims, especially children and the elderly, can become so dehydrated from diarrhea and vomiting that, without medical care, they can die.

Humans are highly susceptible to salmonella infections. About 44,000 human cases are reported yearly in the U.S. (Table 4.1), with millions more occurring worldwide (Gratz, 1988). The economic costs of the disease are high, both in prevention and treatment of the disease in humans. In Canada alone, these costs have been estimated at hundreds of millions of dollars annually (Todd 1989).

The disease is spread through the ingestion of salmonella bacteria. The ingested bacterium can come from eating inadequately cooked meat or food contaminated with feces from infected animals (Wray 1994). While a staggering number of species can serve as reservoirs for different strains of salmonella, commensal rodents are important both as reservoirs and vectors of the disease to humans (Meehan 1984).

Leptospirosis

Spirochetal bacteria of the genus *Leptospira* cause leptospirosis. Over 200 varieties of pathogenic leptospires have been identified. The disease occurs throughout the world and is known by several names that illustrate the symptoms and/or conditions under which it is transmitted to humans: seven-day fever, cane cutter's disease, Japanese autumnal fever, rice field worker's disease, swineherd's disease, canicola fever, marsh fever, swamp fever, mud fever, field fever, harvest fever, mouse fever, enzootic jaundice, and hemorrhagic jaundice.

Leptospirosis is one of the most widespread of all zoonoses. Pathogenic leptospires have been isolated in at least 82 different countries and evidence for the presence of this disease has been found in almost every country in the world. Endemic zones for this disease include areas characterized by wet climates. Monsoons create favorable conditions for leptospirosis and result in large endemic zones in China and Southeast Asia. About 50 cases of leptospirosis are reported annually in the U.S. (Table 4.1).

Leptospirosis can be transmitted between animals and humans through the urine of infected animals, especially commensal rodents (Meehan et al. 1984). Humans can be infected by skin contact (especially through cuts and scratches) with contaminated soil, vegetation, and water, and by the ingestion of contaminated food or water. Because of how the disease is transmitted, leptospirosis is common in farmers, fishermen, miners, and sewer workers. Human-to-human transmission is extremely rare, as only a single case has actually been recorded.

Virtually all mammalian species can be infected by the *Leptospira* bacterium, and leptospires have also been found in a few birds, reptiles, and invertebrates. Rodents and livestock are both important reservoirs of the disease, but their significance varies greatly among localities and over time (Torten and Marshall 1994). In a given region, the disease may be transmitted from rodents to humans in one year, and from cattle to humans in the next. The severity of the disease can also fluctuate; in some years most cases of leptospirosis in humans will be mild, while in other years a large number of severe cases may be observed. The cause of such shifts in virulence is unknown.

Leptospirosis is difficult to diagnose because its symptoms are extremely variable and often resemble many other diseases. In mild cases, this disease is usually misdiagnosed as influenza and the patient recovers after a few days. In intermediate cases, the symptoms include fever, malaise, neck aches, extreme weakness of muscles, bronchitis, pneumonia, anemia, skin lesions, vomiting, and muscular pains. In severe cases, the disease may result in jaundice, kidney failure, brain inflammation, and a mortality rate of 10 to 20%. Secondary complications of leptospirosis include eye infections that can appear up to a year after the original infection (regardless of its severity) and may cause blindness if left untreated. Antibiotics do not have a marked effect on the course of the disease unless taken during the first 24 hours after infection, but they can prevent secondary complications.

Because of its large number of host species, the existence of over 200 types of *Leptospira*, and its constantly shifting epidemiology, leptospirosis cannot be eradicated. However, rodent control in endemic areas and vaccination of livestock can greatly reduce the risk of transmission of this disease to humans. Other control measures include strict control of cattle importations across borders and the requirement of health certificates for imported livestock. A highly virulent variety of leptospirosis was accidentally introduced into Japan via a shipment of infected cattle.

Vaccines are available for humans and are widely used in endemic areas around the world. Domestic dogs are an important source of this disease in humans and vaccination of dogs is important for reducing risk. Wearing protective clothing, including gloves and high rubber boots, may work in laboratory conditions but is rarely practical for any extended period in the field. Whenever possible, fieldwork in infected areas should be avoided during the early morning, because dew can be contaminated by rodent urine.

Lyme Disease

The spirochetal bacterium that cause Lyme disease (*Borrelia burgdorferi*) was isolated for the first time in the U.S. in 1981. However, the disease itself had been described in Europe as early as 1909. The name was assigned to this disease after a number of cases were observed in the town of Old Lyme, CT, during 1975. The yearly number of reported cases of Lyme disease has increased rapidly in the U.S., rising from 497 during 1982 to over 14,000 during 1998 (Table 4.1).

Lyme disease occurs in North America, Europe, and Asia (mainly in Russia, northern China, and Japan). In the U.S., three endemic areas have been identified: the northeastern region, the north-central region (mainly Wisconsin and Minnesota), and the Pacific coastal region (primarily California, see Figure 4.4). The distribution of Lyme disease corresponds roughly to that of its main vectors, ticks of the *Ixodes ricinus* complex, including *I. dammini* in the eastern U.S., *I. pacificus* on the West Coast, *I. ricinus* in Europe, and *I. persulcatus* in Asia (Dennis and Lance 1994).

The life cycles of these vector ticks typically include three stages over two years: larval, nymph, and adult. Larvae feed on small mammals, birds, and lizards; nymphs feed on the blood of various vertebrates; and adults feed mostly on the blood of medium- to large-sized mammals (e.g., raccoons, canids, and bears). However, two

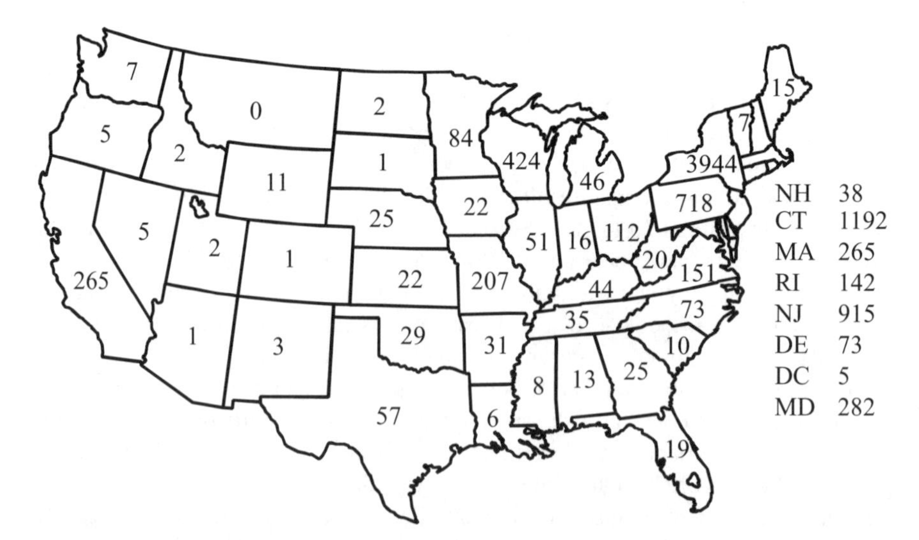

Figure 4.4 Geographic distribution of human cases of Lyme disease in the U.S. based on location of cases during 1991.

host species are particularly important to the survival of *I. dammini*: the white-footed mouse for immature stages and the white-tailed deer for adult ticks. Hundreds of adult *I. dammini* can live on a single deer, and all of them can become infected with *B. burgdorferi*.

The increase in Lyme disease during the last few decades in the eastern U.S. is attributed to an increase in *I. dammini* populations, which, in turn, has been linked to the rise in white-tailed deer populations in that part of the country. In some areas where deer were removed, tick populations decreased markedly but were not eliminated, due to the presence of other mammalian hosts. In California, the western fence lizard is the major host for immature *I. pacificus*, and mule deer is the primary host for adult ticks. In Europe, the bank vole and several species of *Apodemus* are the main reservoirs of Lyme disease.

Lyme disease is transmitted to humans primarily through the bite of nymphal ticks. One reason for this is because ticks have to remain attached for at least 24 hours before they start to transmit *B. burgdorferi*, and it is much easier for a person to overlook a small nymphal tick than a larger adult.

A characteristic early symptom of Lyme disease is an expanding ring-shaped rash from the site of the tick bite. Other early symptoms include low fever, fatigue, pain in muscles and joints, headache, swelling of lymph nodes, and conjunctivitis. Early symptoms may vanish, but weeks or months later, the disease produces more serious and persisting consequences, such as arthritis and cardiac problems. Because the disease organism infects nerves, neurological problems can also occur, including brain inflammation, partial paralysis, and some loss of motor control.

During the incubation period, Lyme disease can be easily treated with antibiotics. If left untreated in this early stage, the disease becomes more difficult to cure. In some cases, repeated and prolonged intravenous antibiotic treatment may be required and can cost more than $100,000 for a single patient. Although Lyme disease can be severe and debilitating, it is almost never fatal. A new vaccine is now available for humans.

The best preventive measure against the transmission of Lyme disease is personal protection from tick bites. Applying insect repellents containing DEET to clothing and skin and careful body inspection for ticks are both recommended when in tick-infested areas. Other preventive measures include tick control and habitat modifications. Pesticide applications have been successful in reducing tick populations near homes. Another approach is to distribute cotton balls impregnated with a pesticide in these areas. Mice use these cotton balls for nest material, and this eliminates ticks from their nests.

Control of Lyme disease by reducing deer populations is usually impractical except on isolated islands or inside deer-proof exclosures. Moderate reductions of deer populations have little effect on the density of ticks due to the presence of other reservoirs. Deer populations must be virtually eradicated before tick populations decline substantially.

Rocky Mountain Spotted Fever

The bacterium causing Rocky Mountain Spotted Fever (RMSF) is *Rickettsia rickettsii*. RMSF is only found in the Western Hemisphere. The disease received its name from a number of cases described in the late 19th century in the Rocky Mountains, but currently this disease is more prevalent in the eastern half of the U.S. RMSF is also called tick-borne typhus or tick typhus (Dumler 1994).

High fever and severe headaches characterize the early symptoms of RMSF. These are usually followed by a rash within three to five days. Other symptoms that can develop include muscle pain, nausea, vomiting, and abdominal pain. *R. rickettsii* increases the permeability of blood vessels by damaging the cells lining them; this can lead to hemorrhaging, edema, decreased blood volume, and shock. RMSF can be fatal because of complications involving heart problems and/or kidney failure.

Approximately 500 to 600 cases were reported annually in the U.S. during the 1990s (Table 4.1). Many more cases probably occurred but were misdiagnosed. Historically, the mortality rate for RMSF was 25 to 75%; with the advent of antibiotics, the rate has been reduced to less than 5% of reported cases in the U.S.

RMSF is transmitted to humans primarily through tick bites. The most important vectors are the ticks: *Dermacentor variabilis* in the eastern U.S., *D. andersoni* in the Rocky Mountains region, *Rhipicephalus sanguinaeus* in western and central Mexico, and *Amblyomma cajense* in southeastern Mexico and Central and South America. Ticks can be infected maternally or by feeding on an infected mammalian host.

Several mammalian species can serve as reservoirs for RMSF and as hosts for the vector ticks. Hosts for larval and nymphal ticks include voles, mice, rats, and other small mammals. Hosts for adult ticks include carnivores, deer, and domestic animals, especially dogs.

Preventive measures are similar to those recommended for Lyme disease and involve minimizing the frequency of tick bites by applying insect repellents when venturing into tick-infested areas. Applications of pesticides around buildings can reduce local tick densities temporarily but have to be repeated for long-term control. Dogs should wear tick-repellent collars and be inspected daily for ticks.

Scrub Typhus

Scrub typhus is caused by *Rickettsia tsutsugamushi* and is also called Tsutsugamushi disease or chigger-borne typhus. It is a serious disease with human mortality rates in untreated cases ranging from 1 to 60% depending on the disease strain. However, the disease responds well to antibiotics and therefore is very treatable. Symptoms include a lesion at the site of the infection, fever, headaches, and a rash.

The disease occurs mainly in Asia and the Pacific Islands. It has not been reported in Europe or the Americas. The disease is widespread in the endemic areas and can have heavy economic and social impacts in rural areas. For example, there are approximately 500,000 cases annually in Malaysia (Saunders et al. 1980).

The primary reservoirs of the disease are rats, but opossums and mice can also serve as hosts (Morlan et al. 1950). Mites or chiggers are the main vectors. The name scrub typhus originated with the observation that this disease is prevalent in areas where forests have been replaced by agriculture and pastures. The vegetational changes resulting from the creation of these "scrub" clearings allow rat populations to flourish.

Murine Typhus

This disease, also known as endemic typhus or flea-borne typhus, is caused by the bacteria *Rickettsia typhi*. Symptoms can include sudden high fever, chills, headache, malaise, muscle pain, and a rash that appears on the trunk several days after the onset of the disease (in 60% of cases). The fever and rash can last up to two weeks. Other symptoms include nausea and vomiting, abdominal pain, lung inflammation, and an inflammation of the spleen or liver. The fatality rate was approximately 5% before broad-spectrum antibiotics were available, but has now dropped below 1% (Gratz 1988). These antibiotics do not kill *R. typhi* but they prevent it from multiplying within infected cells while the immune system of the patient builds up a response against the disease. For this reason, the treatment must be continued for at least 14 days, even though symptoms usually disappear much sooner.

Murine typhus occurs on every continent except Antarctica, but is most prevalent in tropical and subtropical regions, especially where rats live in close proximity with human populations (Traub et al. 1978). During the first half of the 20th century, thousands of cases of murine typhus were reported annually in the U.S. Presently, this number has dropped to 60 to 80 cases per year, mostly from Texas, California, and Hawaii (Gratz 1988).

The main vector of murine typhus throughout the world is the oriental rat flea (*Xenopsylla cheopis*), although other flea species are involved in some areas. When a flea ingests infected blood, the disease organisms enter the cells lining the flea's mid-gut where they proliferate. These cells are then released and excreted with the flea's feces. *R. typhi* can stay viable for months, or even years, in flea feces. In most cases, murine typhus is transmitted to humans by contact with infected flea feces through a fleabite, scratch, wound, or through the eye membrane.

Preventive measures include the exclusion of rats from dwellings and the control of fleas and commensal rat populations. Flea control measures should be started prior to rat control, otherwise, fleas will leave dead rats and search for alternate hosts, such as humans. This can actually increase the risk of disease transmission (Gage et al. 1995).

Psittacosis (Ornithosis)

Psittacosis is also called avian chlamydiosis or ornithosis and is caused by *Chlamydia psittaci*, a bacteria of the order Chlamydiales. This disease occurs worldwide wherever birds are present. During the 1990s, the annual number of reported cases of psittacosis in the U.S. ranged from 40 to 100 (Table 4.1).

Transmission to humans usually occurs through inhalation of infectious feces or urine droplets, which originate from infected birds. Symptoms of psittacosis are fever, a dry cough, and an elevated respiration rate. While this disease is usually mild, it is insidious and can be fatal. The disease organism normally infects the lungs, but it may also enter the bloodstream. Psittacosis causes pneumonia, congestion of the spleen and liver, and sometimes congestion and edema of the brain and spinal cord. This disease is treated with antibiotics (Grimes 1994).

Reservoirs of psittacosis are birds, including waterfowl, seabirds, shorebirds, pigeons, doves, parrots, and poultry. Birds may become lifelong carriers, continuously shedding the disease organism in their feces and other secretions.

VIRAL DISEASES

Encephalitis

Insects and arachnids are the primary vectors of arboviruses — viral organisms that cause encephalitis, an inflammation of the brain or central nervous system (Beran and Steele 1994). For this reason, these diseases often are fatal. In most years, these viral diseases are rare in humans, but periodic outbreaks occur. Many different viruses are included among the arboviruses.

Eastern equine encephalitis (EEE), also called eastern encephalitis or sleeping sickness, is a serious disease in humans and fatal 30% of the time. Initial symptoms of the disease include fever, headache, and nausea. As the disease progresses, weakness, paralysis, seizures, coma, and death can result (Gibbs and Tsai 1994).

Periodic outbreaks or epidemics of this disease occur among both horses and humans in the eastern U.S. (Gibbs and Tsai 1994). During an outbreak, equine cases of EEE usually precede human cases, providing time for health authorities to react (Gibbs and Tsai 1994). Songbirds are the main reservoirs of this virus, and mosquitoes are responsible for vectoring it to mammals. Many urban areas in the U.S. have established mosquito control districts that use insecticides and water management to control mosquito populations. These efforts are intensified during an EEE outbreak to stop the spread of the disease. Horses can also be vaccinated against EEE.

Western equine encephalitis (WEE) is similar to EEE, in that it is a disease of horses and humans. Periodic outbreaks occur in the western sections of the U.S. and Canada. During a 1941 epidemic, over 2500 people were infected and 8 to 15% of the victims died (Iversen 1994). Free-ranging birds are the primary host, although the virus also occurs in mammals. The mosquito, *Culex tarsalis*, is the main vector (Iversen 1994). WEE is a reportable disease; surveillance programs have been established to detect the onset of a potential epidemic. Mosquito control is the primary approach to controlling WEE.

Venezuelan equine encephalitis (VEE) is another arbovirus that can cause serious illness in both horses and humans (Gratz 1988). Symptoms in humans range from an influenza-like fever to encephalitis. Fatalities are rare in humans (1% of cases) and most often occur in children. VEE is most prevalent in South America, but also occurs in Florida and Texas. During an epidemic from 1962 to 1964, over 50,000 human cases were reported in Ecuador and Venezuela (Gratz 1988).

Rodents serve as the primary reservoir for VEE and mosquitoes are the main vector. During epidemics, many different mammals, including horses, may contract this disease and transmission from horses to humans via mosquitoes can occur. Fortunately, an effective vaccine has been developed against this virus and millions of horses have been vaccinated (Gratz 1988).

California group viruses are a group of arboviruses including California encephalitis and La Crosse (LAC) virus. In most years, LAC is the most common cause of arboviral encephalitis in the U.S. and is endemic in Indiana, Illinois, Ohio, and Wisconsin (Gratz 1988). Small mammals (e.g., chipmunks and gray squirrels) are important reservoirs for LAC (Grimstad 1994), and mosquitoes are the vectors. Prevention usually involves avoidance of mosquito bites and control of local mosquito populations.

St. Louis encephalitis (SLE) was detected during a 1933 epidemic in St. Louis, MO, that resulted in at least 1000 human cases and 200 fatalities (Luby 1994). There are six known types of SLE in the Western Hemisphere: four occur in the U.S., one in Central America, and one in South America (Luby 1994). Free-ranging birds are the main reservoirs for this virus, and mosquitoes are the vectors. Prevention is aimed at mosquito abatement and disease surveillance.

Japanese encephalitis occurs across most of Asia (Hoke and Gingrich 1994) and is the most common cause of epidemic encephalitis in the world. Prevention involves avoiding mosquito bites and suppressing mosquito populations. A vaccine has been developed (CDC 1993; Hoke et al. 1988).

West Nile fever (WNF) is a viral infection whose symptoms in humans include fever, headache, muscle pains, and a rash. Occasionally, the disease can progress to encephalitis and can be fatal (Peiris and Amerasinghe 1994). Epidemics of WNF occur in Africa, Europe, and Asia. This disease has recently spread to the eastern U.S. Birds are the primary host of WNF, but it has also been detected in many free-ranging mammals. Mosquitoes are the main vectors for WNF, but mites and ticks may also be carriers (Peiris and Amerasinghe 1994). Prevention usually involves surveillance of free-ranging birds for WNR followed by mosquito control when the virus is detected.

Hantavirus

Hantavirus is a genus of viruses of the family Bunyaviridae. These viruses cause two major types of diseases: hemorrhagic fever with renal syndrome (HFRS) and hantavirus pulmonary syndrome (HPS).

HFRS is found worldwide, with the most virulent forms occurring in Eurasia. It was first reported in 1913 in the former Soviet Union (Yamanouchi 1994). HFRS is caused by a number of different viral strains including Hantaan, Dobrava/Belgrade, Seoul, and Puumala viruses. Approximately 100,000 to 200,000 cases of HFRS occur annually, mostly in Eurasia (Hart and Bennett 1999; Peters et al. 1999). China reported 90,000 human cases and 3000 deaths in 1984. Infections by Hantaan and Dobrava viruses are usually severe and result in a 5 to 15% mortality rate, whereas infections by Seoul and Puumala viruses are milder with a mortality rate of less than 1% (Peters et al. 1999).

HPS is found only in the Americas and was first recognized in 1993 following an outbreak of a "mystery illness" in the southwestern U.S. (Figure 4.5). The Sin Nombre virus, which was first isolated in 1993, causes most of this disease in the U.S. Other hantaviruses causing HPS have been discovered since 1993, including the Black Creek Canal, Bayou, New York, and Monongahela viruses in North

Figure 4.5 Geographic distribution of hantavirus in North America based on the first 100 human cases.

America, and the Andes, Laguna Negra, and Juquitiba viruses in South America (Hart and Bennett 1999). As of the end of 1998, 202 cases of HPS had been reported in the U.S. with a fatality rate of 41% (Khan et al. 1996; CDC 1999).

Both HFRS and HPS are characterized by an increase in the permeability of blood vessels, causing fever, muscle pain, malaise, and various gastrointestinal symptoms (e.g., nausea, vomiting, abdominal pain, and diarrhea). Additional symptoms for HFRS include blurred vision, lacrimation (watering of the eyes), and red eyes. The condition of the patient can rapidly degenerate into hypotension and shock. Other HFRS symptoms at this later stage are back pain and tenderness, hemorrhages, and kidney failure, whereas with HPS, the patients have trouble breathing as their lungs fill with fluid (pulmonary edema). Intensive care of HFRS and HPS patients is essential for survival in severe cases. Managing shock and adjusting blood volume are important parts of the treatment of both diseases.

Each strain of hantavirus is associated with one rodent species that serves as the primary reservoir. The distribution of each disease corresponds with the distribution of the reservoir rodent species. Important reservoir species include the striped field mouse for the Hantaan virus, the yellow-necked field mouse for the Dobrova/Belgrade virus, the Norway rat for the Seoul virus, the bank vole for the Puumala virus, the deer mouse for the Sin Nombre virus, the cotton rat for the Black Creek Canal virus, the rice rat for the Bayou virus, the white-footed mouse for the New York virus, *Oligoryzomys longicaudatus* for the Andes virus, and *Calomys laucha* for the Laguna Negra virus (Hart and Bennett 1999).

Infected rodents excrete viruses in their urine, feces, and saliva. Transmission of hantavirus to humans occurs mainly through inhalation of dried rodent excreta but can also occur through bites. Transmission between humans has never been reported except for the Andes virus in Argentina (Padula et al. 1998). Most reservoir species (except for the Norway rat) are found mainly in rural areas. For this reason, farmers, forestry workers, field biologists, and outdoor enthusiasts are most at risk (Childs et al. 1995). Some people have contracted hantavirus after cleaning areas that were contaminated with rodent feces (Zeitz et al. 1995).

Disease prevention involves minimizing contacts between humans and infected rodents and rodent droppings. This can be accomplished by reducing rodent populations within or near human dwellings or by rodent-proofing homes.

Rabies

Rabies viruses are part of the family of rhabdoviruses (Beran 1994b). Rabies is a terrifying disease with distinctive symptoms and is described in some of man's earliest writings. For example, laws for the Middle Eastern city of Eshnunna in 1885 B.C. required that mad dogs be caged, presumably as a measure to prevent the spread of rabies. Around 500 B.C., this disease was described by the Greek philosophers Democritus, Aristotle, and Hippocrates.

Rabies epidemics occurred in Europe throughout the Middle Ages. In Central America, many Spanish conquistadors died following the bite of vampire bats, according to a 1514 report (Beran 1994b). During 1989, there were 3200 human cases worldwide, with over 2000 of these occurring in Asia (Beran 1994b). During 2000, rabies killed five people in the U.S., the first such deaths since 1998. All but one of the deaths occurred when the person was bitten by a bat. Deaths were recorded in California, Georgia, Minnesota, New York, and Wisconsin (CDC unpublished report).

Rabies is transmitted to humans through exposure to saliva of rabid animals, typically through bites (Beran 1994b). Initial symptoms of rabies in humans include a tingling sensation and pain around the bite, and weakness or trembling developing in the affected limb. Patients progress into a hyperactive phase with a great sensitivity to visual, auditory, or tactile stimuli. Excessive salivation, perspiration, and lacrimation often occur. When patients attempt to swallow, fluids are expelled forcefully through the mouth and nasal passages due to an involuntary reflex action. This is both painful and terrifying to the patients, who often develop hydrophobia (a fear of fluids). This hyperactive phase is followed by a paralytic phase with the patient gradually developing stupor, paralysis, delirium, and coma (Beran 1994b). Once a rabies infection has progressed to the point where symptoms appear, it is almost always fatal to humans (only two people in recorded history have survived symptomatic rabies).

At present, rabies is endemic throughout the Americas, Europe, Africa, and Asia. It is absent from New Zealand and the Pacific Islands. Rabies is usually maintained through intraspecific contact, rather than from one species to another. Because of this, different strains of rabies evolved for different species or groups of species. For instance, canine-strain rabies can only maintain itself by spreading among dogs. Cats, cattle, sheep, goats, horses, and swine can all contract canine-type rabies from the bite of rabid dogs and die from the disease; all of these animals can also transmit the disease to humans. However, canine-type rabies virus cannot persist in any of these species. Canine-strain rabies was eliminated in the U.S. and Canada in the

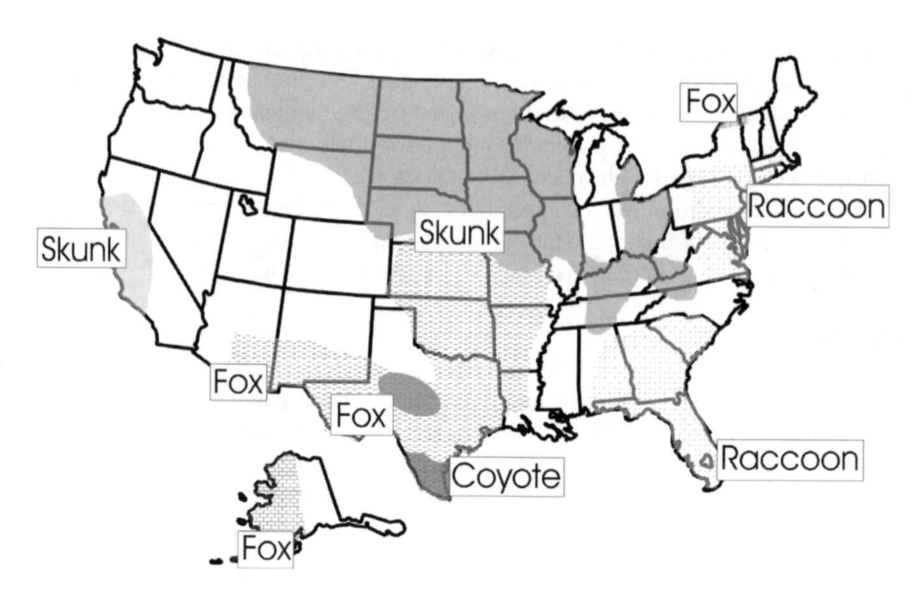

Figure 4.6 Geographic distribution of different strains of rabies in the U.S. in 1995.

1950s; unfortunately, it reappeared during the 1980s in Texas among dogs and coyotes (Beran 1994b).

The striped skunk is the predominant reservoir for rabies in the central part of the U.S. and Canada (skunk-strain rabies, Figure 4.6), and the raccoon serves as the reservoir in the eastern U.S. (raccoon-strain rabies). Arctic foxes range through the polar regions of North America, Europe, and Asia and serve as a major reservoir for a fox-strain rabies (fox-strain rabies also extends throughout the Intermountain West area). This rabies strain has moved south across Canada and the northeastern U.S. by infecting red and grey foxes (Beran 1994b). Rabies is common in many insectivorous, frugivorous, and vampire bats (bat-strain rabies). In some bat colonies, most of the bats have antibodies to rabies, indicating prior exposure to the disease. Transmission of rabies from bats to humans is rare although it has occurred and resulted from both bat bites and inhalation of aerosolized bat saliva in caves (Beran 1994b). In other parts of the world, jackals and mongooses are important reservoirs for rabies.

Stopping the spread of rabies in animals is the best means of protecting humans from this disease, but this requires the surveillance and accurate reporting of rabid animals. Prevention of rabies in cats and dogs can be accomplished through vaccination. Rabies spreads through mammal populations in a wave-like fashion with raccoon rabies spreading at a rate of about 40 km per year. Because of this slow rate of spread, it is possible to stop the spread of the disease in some free-ranging mammal populations; oral rabies vaccines have been placed in baits and dropped from airplanes in areas where the disease is spreading. Rabies control efforts in bats have been limited to vampire bats and have focused on destroying infected bat colonies, usually through the use of a toxin.

Pre-exposure immunization has been highly effective in protecting people from subsequent exposure to rabies. Vaccination is recommended for people who might be exposed to the virus in their everyday activities, such as wildlife biologists. After possible exposure to rabies, people can be treated by repeated injections of rabies immunoglobulin, providing that treatment is started immediately after infection, followed by a series of rabies vaccinations.

FUNGAL DISEASE

Histoplasmosis

Histoplasmosis is caused by the fungus, *Histoplasma capsulatum*, var. *capsulatum*. Humans usually contract histoplasmosis by inhaling airborne spores of the fungus. This fungus grows in soil enriched with feces of birds (e.g., blackbirds, starlings, and chickens). Areas with large accumulations of bat guano are also frequently infested. The source of histoplasmosis outbreaks can include old chicken coops, bell towers, and bird roosts. When the soil at these sites is disturbed, the spores become airborne and can be spread by air currents, endangering people downwind (Ajello and Padhye 1994).

Most infected people (90%) do not become ill or mistake histoplasmosis for a cold. The unfortunate 10% of humans who experience more serious side effects suffer through a wide variety of symptoms including coughing, chest pains, and labored breathing. In more severe forms, these symptoms may be accompanied by fever, night sweats, and weight loss. Most victims recover spontaneously from the acute form of the disease. However, later in life, radiographs may reveal calcifications in the lungs. In a few cases, the disease will spread from the lungs to other parts of the body. In these disseminated cases, symptoms include anemia, a decrease in the number of white blood cells, and weight loss. The disease may prove fatal if not diagnosed and treated promptly.

H. capsulatum is found throughout the world, but some regions are more heavily infested than others. In Latin America, areas with high infestation occur in Mexico, Guatemala, Venezuela, and Peru. In the U.S., histoplasmosis occurs primarily in the eastern states, especially in the Ohio River valley.

One prevention measure is to avoid activities in bird roosts, which might stir up clouds of dust. It is also recommended that work areas be wetted down and that protective masks be worn when working in high-risk conditions. Infested areas can also be decontaminated with sodium hypochlorite.

SUMMARY

Medical advances have greatly reduced the threat of zoonotic diseases from the days when a plague epidemic reduced human populations by a third. But even today, the hundreds of zoonotic diseases cause untold hardship and misery to people who are unfortunate enough to contract one. These diseases occur worldwide but are most common in developing countries where large commensal rodent populations and unhygienic conditions occur. One consequence of zoonoses is that they reduce people's positive outlook towards animals and their willingness to tolerate wildlife. For instance, people's tolerance of deer in their neighborhoods dropped after Lyme disease became prevalent.

LITERATURE CITED

Ajello, L., and A. A. Padhye, Systemic mycoses. Histoplasmosis. P. 493–495 in G. W. Beran, Ed., *Handbook of Zoonoses, Section A*. 2nd ed., CRC Press, Boca Raton, FL, 1994.

Beran, G. W., Ed., *Handbook of Zoonoses*. 2nd ed., CRC Press, Boca Raton, FL, 1994a.

Beran, G. W., Rabies and infections by rabies-related viruses. P. 307–357 in G. W. Beran, Ed., *Handbook of Zoonoses, Section B*. 2nd ed., CRC Press, Boca Raton, FL, 1994b.

Beran, G. W., and J. H. Steele, Concepts in bacterial, rickettsial, chlamydial, and mycotic zoonoses. P. 1–5 in G. W. Beran, Ed., *Handbook of Zoonoses, Section A*. 2nd ed., CRC Press, Boca Raton, FL, 1994.

Centers for Disease Control and Prevention, Inactivated Japanese encephalitis virus vaccine. Recommendations of the advisory committee on immunization practices (ACIP). *Morbidity Mortality Wkly. Rep.*, 42(RR-1), 1–15, 1993.

Centers for Disease Control and Prevention, Summary of notifiable diseases, United States 1998. *Morbidity Mortality Wkly. Rep.*, 47(53), 1–93, 1999.

Childs, J. E., J. N. Mills, and G. E. Glass, Rodent-borne hemorrhagic fever viruses: a special risk for mammalogists? *J. Mammal.*, 76, 664–680, 1995.

Dennis, D. T., and S. E. Lance, Lyme borreliosis. P. 265–280 in G. W. Beran, Ed., *Handbook of Zoonoses, Section A*. 2nd ed., CRC Press, Boca Raton, FL, 1994.

Dumler, J. E., Rocky Mountain spotted fever. P. 417–427 in G. W. Beran, Ed., *Handbook of Zoonoses, Section A*. 2nd ed., CRC Press, Boca Raton, FL, 1994.

Gage, K. L., R. S. Ostfeld, and J. G. Olson, Nonviral vector-borne zoonoses associated with mammals in the United States. *J. Mammal.*, 76, 695–715, 1995.

Gibbs, E. P. J. and T. R. Tsai, Eastern encephalitis. P. 11–24 in G. W. Beran, Ed., *Handbook of Zoonoses, Section B*. 2nd ed., CRC Press, Boca Raton, FL, 1994.

Gratz, N. G., Rodents and human diseases: a global appreciation. P. 101–169 in I. Prakash, Ed., *Rodent Pest Management*. CRC Press, Boca Raton, FL, 1988.

Grimes, J. E. 1994. Avian chlamydiosis. P. 389–402 in G. W. Beran, Ed., *Handbook of Zoonoses, Section A*. 2nd ed., CRC Press, Boca Raton, FL, 1994.

Grimstad, P. R., California group viral infections. P. 71–79 in G. W. Beran, Ed., *Handbook of Zoonoses, Section B*. 2nd ed., CRC Press, Boca Raton, FL, 1994.

Hart, C. A., and M. Bennett, Hantavirus infections: epidemiology and pathogenesis. *Microb. Infect.*, 1, 1229–1237, 1999.

Hoke, C. H., Jr. and J. B. Gingrich, Japanese encephalitis. P. 59–69 in G. W. Beran, Ed., *Handbook of Zoonoses, Section B*. 2nd ed., CRC Press, Boca Raton, FL, 1994.

Hoke, C. H., A. Nisalak, N. Sangawhipa, S. Jatanasen, T. Laorakapongse, B. L. Innis, S. Kotchasenee, J. B. Gingrich, J. Latendresse, K. Fukai, and D. S. Burke, Protection against Japanese encephalitis by inactivated vaccines. *New England J. Med.*, 319, 608–614, 1988.

Hopla, C. E. and A. K. Hopla, Tularemia. P. 113–126 in G. W. Beran, Ed., *Handbook of Zoonoses, Section A*. 2nd ed., CRC Press, Boca Raton, FL, 1994.

Iversen, J. O., Western equine encephalomyelitis. P. 25–31 in G. W. Beran, Ed., *Handbook of Zoonoses, Section B*. 2nd ed., CRC Press, Boca Raton, FL, 1994.

Khan, A. S., R. F. Khabbaz, L. R. Armstrong, R. C. Holman, S. P. Bauer, J. Graber, T. Strine, G. Miller, S. Reef, J. Tappero, P. E. Rollin, S. T. Nichol, S. R. Zaki, R. T. Bryan, L. E. Chapman, C. J. Peters, and T. G. Ksiazek, Hantavirus pulmonary syndrome: the first 100 U.S. cases. *J. Infect. Dis*., 173, 1297–1303, 1996.

Luby, J. P., St. Louis encephalitis. P. 47–58 in G. W. Beran, Ed., *Handbook of Zoonoses, Section B*. 2nd ed., CRC Press, Boca Raton, FL, 1994.

Meehan, A. P., *Rats and Mice*. Rentokil Limited, East Grinstead, U.K., 1984.

Morlan, H. B., E. L. Hill, and J. H. Schubert, Serological survey for murine typhus infection in southwest Georgia animals. *Public Health Rep.,* 65, 57–63, 1950.

Ohara, Y., T. Sato, and M. Homma, Epidemiological analysis of tularemia in Japan (*yato-byo*). *FEMS Immunol. Med. Microbiol.,* 13, 185–189, 1996.

Padula, P. J., A. Edelstein, S. D. Michael, N. M. Lopez, C. M. Rossi, and R. D. Rabinovich, Hantavirus pulmonary syndrome outbreak in Argentina: molecular evidence for person-to-person transmission of Andes virus. *Virology,* 241, 323–330, 1998.

Peiris, J. S. M. and F. P. Amerasinghe, West Nile fever. P. 139–148 in G. W. Beran, Ed., *Handbook of Zoonoses, Section B*. 2nd ed., CRC Press, Boca Raton, FL, 1994.

Peters, C. J., G. L. Simpson, and H. Levy, Spectrum of hantavirus infection: hemorrhagic fever with renal syndrome and hantavirus pulmonary syndrome. *Ann. Rev. Med.,* 50, 531–545, 1999.

Poland, J. D., T. J. Quan, and A. M. Barnes, Plague. P. 93–112 in G. W. Beran, Ed., *Handbook of Zoonoses, Section A*. 2nd ed., CRC Press, Boca Raton, FL, 1994.

Saunders, J. P., G. W. Brown, A. Shirai, and D. L. Huxsoll, The longevity of antibodies to *Rickettsia tsutsugamushi* in patients with confirmed scrub typhus. *Trans. R. Soc. Trop. Med. Hyg.,* 74, 253–257, 1980.

Todd, E. C., Preliminary estimates of costs of food borne disease in the United States. *J. Food Prod.,* 52, 595–601, 1989.

Torten, M. and R. B. Marshall, Leptospirosis. P. 245–264 in G. W. Began, Ed., *Handbook of Zoonoses, Section A*. 2nd ed., CRC Press, Boca Raton, FL, 1994.

Traub, R., C. L. Wisseman, and A. Farhang Azad, The ecology of murine typhus: a critical review. *Trop. Dis. Bull.,* 75, 237–317, 1978.

Wray, C., Mammalian salmonellosis. P. 289–302 in G. W. Began, Ed., *Handbook of Zoonoses, Section A*. 2nd ed., CRC Press, Boca Raton, FL, 1994.

Yamanouchi, K., Hemorrhagic fever with renal syndrome. P. 359–364 in G. W. Beran, Ed., *Handbook of Zoonoses, Section B*. 2nd ed., CRC Press, Boca Raton, FL, 1994.

Zeitz, P. S., J. C. Butler, J. E. Cheek, M. C. Samuel, J. E. Childs, L. A. Shands, R. E. Turner, R. E. Voorhees, J. Sarisky, P. E. Rollin, T. G. Ksiazek, L. Chapman, S. E. Reef, K. K. Komatsu, C. Dalton, J. W. Krebs, G. O. Maupin, K. Gage, C. M. Sewell, R. F. Breiman, and C. J. Peters, A case-control study of hantavirus pulmonary syndrome during an outbreak in the southwestern United States. *J. Infect. Dis*., 171, 864–870, 1995.

CHAPTER 5

Economics

"We haven't the money, so we've got to think."

Lord Rutherford

"If all of the economists were laid end to end, they still wouldn't reach a conclusion."

George Bernard Shaw

"An economist is an expert who will know tomorrow why the things he predicted yesterday didn't happen today."

Laurence Peter

ECONOMIC ANALYSES OF WILDLIFE VALUES

In Chapter 4, I explained that wildlife provides many different types of benefits to people, some of which are intangible and unquantifiable but are real nevertheless. Because so many wildlife benefits are intangible, wildlife biologists are reluctant to discuss the economics of wildlife. Such an attitude, however, is often counterproductive because it conveys the impression that wildlife has no economic value. A better approach is to try to assess the economic value of wildlife openly while keeping in mind that, in addition to monetary value, wildlife provides many intangible benefits of great importance.

Monetary benefits of wildlife are those that can be expressed or defined in dollars. Monetary values for an item are obtained in a free market system, where the price is set through open negotiations between willing sellers, who want to obtain as high a price as possible, and willing buyers, who want to pay as little as possible. If an item is both uncommon and demanded by many people, there will be many potential buyers who will compete against each other and the price will increase. If the item is easily obtained, then there will be many sellers who will compete against each other and the price will decrease.

The economic value of wildlife differs for everyone. Consider, for example, a local deer herd. Its economic value to a hunter is reflected by how much money this person is willing to pay for the right to hunt deer. For a landowner, the economic value of deer might best be determined by how much income he or she could obtain by leasing the hunting rights to the land. For a local community, it could be reflected by the amount of money spent in local businesses by tourists and hunters for lodging, meals, hunting equipment, entertainment, and supplies. For a state wildlife agency, it might be reflected by the amount of money it collects from hunting licenses. Obviously, when speaking of the economic importance of wildlife, we must always specify whose perspective we are using. Likewise, the costs of wildlife damage do not fall evenly on everyone. For instance, only a small proportion of automobile drivers will ever strike a deer. Also, deer herbivory will be a much greater problem for farmers and timber producers than for other people.

When calculating economic values based on purchases, it is necessary to consider how the money will likely be spent and how it might flow through the economy. To assess how hunting economically benefits a state economy, it may be necessary to exclude, or at least discount, the dollars spent by residents who hunt. Indeed, if these individuals could not hunt, they would most likely still spend money in the state but in some other manner than the purchase of hunting supplies and license fees. However, nonresident hunters would probably not come to the state if they could not hunt there, so their dollars would be lost to the state's economy.

There is also a multiplier effect of expenditures as money flows through a local economy. Let us consider a hypothetical example by tracking the movements of a $10 bill that is brought into a local community by a nonresident hunter eating breakfast at a local diner. The owner may pay this extra $10 to his cook to work overtime. The cook may then use the $10 to go to the local movie theater and the owner of the theater may use the $10 to purchase some items from a catalog, and the $10 leaves the local community. In this example, $10 brought into the community by the hunter produced $30 of economic activity in the local community. This is what is known as the multiplier effect.

Economic analyses also vary depending upon the question being addressed. If we are assessing the value of waterfowl to hunters in the U.S., we might randomly sample duck hunters and ask them how much money they would be willing to spend rather than give up the right to hunt waterfowl for a year. By multiplying the mean value that hunters would be willing to spend by the number of waterfowl hunters in the U.S., we could gain an approximation of the value of waterfowl hunting. However, if we are debating whether it is worthwhile to spend millions of dollars for habitat improvements to increase duck populations and hunter success, then the appropriate question is how much money would a duck hunter be willing to spend to harvest one more duck during a day of hunting? Indeed, if the money is not spent, the hunter can still go hunting and have success, but perhaps with slightly less success than if the habitat improvements were made. The monetary value of shooting one extra duck could be multiplied by the expected increase in duck numbers that would result from the habitat improvements. The resulting value can be considered the rate of return or the interest earned annually on the money invested in habitat improvements. We can then decide whether making the habitat improvements would be a wise economic investment. This type of economic approach is referred to as a marginal value analysis.

WHAT ARE THE SOURCES OF ECONOMIC DATA ABOUT THE POSITIVE VALUES OF WILDLIFE?

There are several ways to assess the economic values of wildlife. Listed below are six methods commonly employed by economists. This list is not complete but rather provides a sample of the range of techniques that can be used to measure economic values.

1. Money

People spend money when they take a trip to hunt or view wildlife or to fish. For a fishing trip, the costs might include fishing license fees, fishing equipment and supplies, clothing, meals, lodging, gas, wear on the car, etc. If you add up all of the money that someone spends while fishing, then the value or enjoyment that this individual derives from fishing must be higher than his or her costs, otherwise he or she would not do it. The actual value, or enjoyment, may be several times higher than the amount expended. Therefore, this method underestimates the true value of wildlife, but it does provide a solid, conservative estimate of value.

2. Time Expended

Time is a precious commodity for everyone. One way to value wildlife is to determine the total number of hours people spend in using the wildlife resource and then multiply this by how much money they would be earning if they were working. This method is based on the realization that most people could work overtime or obtain a second job if they wanted to do so. Their decision to enjoy wildlife rather than to work means that they value their time enjoying wildlife more than the extra income they are missing by not working. Time expenditures are often combined with actual monetary expenditures to obtain more realistic values.

3. Income-Producing Ability

Landowners can generate income by leasing the hunting rights to their property. The value of a hunting lease is set by the traditional balance of supply and demand. Hence, hunting leases near cities bring much more than those in remote parts of the country. Likewise, scarce hunting opportunities, such as a hunting lease on a prime waterfowl marsh in an arid region, would bring much more than one in the middle of the Prairie Potholes, where marshes are abundant. The wildlife resource can be viewed like a bond or bank account, and the money obtained from the hunting lease can be viewed as the dividend or interest. If a farmer is receiving $1000 annually for hunting rights, and banks are paying 5% interest on savings accounts or bonds, then the wildlife resource of the farm has a value of $20,000 as an income-producing investment, because 5% of $20,000 is $1000.

4. Increase in Property Values

Throughout much of the U.S., advertisements for farmland real estate iterate phrases such as "hunter's paradise," "good deer hunting," and "abundant wildlife." Real estate agents add these phrases because they know that land which has good wildlife populations will attract buyers and a higher bid. This price premium for land with wildlife can be used to determine the value of wildlife from the standpoint of the landowner (hedonic pricing method).

5. Willingness to Pay

This analysis, called the contingent valuation method, is based on asking the person interviewed (i.e., the respondent) how much money he or she would be willing to pay for some opportunity or item. For instance, we might ask tourists visiting Yellowstone National Park how much money they would be willing to pay for the opportunity to see a wolf during their visit to the park. These values could then be summed to determine the value of wolves to park visitors.

6. Willingness to Do Without

This method of evaluation explores how much a person would have to be paid to give something away or to forego the opportunity to engage in some activity. For instance, to judge the value of beaver ponds to landowners in Toombs County, Georgia, we might go to the County Courthouse and obtain a list of all landowners in the county and ask them if there are beaver ponds on their land. If so, we could ask them how much money we would have to pay them to let us remove the ponds. By adding up all the answers, we could then determine the value of all beaver ponds in the county from the perspective of local landowners. This is another aspect of the contingent valuation method; it generally results in higher values than the willingness to pay method.

WHY IS IT IMPORTANT TO HAVE ACCURATE ECONOMIC DATA ABOUT LOSSES FROM WILDLIFE DAMAGE?

Wildlife causes a myriad of problems in the U.S. (e.g., deer–automobile collisions, diseases, reduced agricultural productivity, and nuisances). Unfortunately, little information is available on the extent of human lives lost, damage to property, and losses of economic and agricultural productivity caused by wildlife. Because of this, making rational decisions about how to manage wildlife is difficult. For instance, if we do not know how much damage deer are causing, how can we determine the optimal deer population in terms of maximizing its value to society? Lack of data also makes it difficult to prioritize projects or to effectively allocate resources to solve wildlife problems (Conover and Decker 1991). Are we better off spending money trying to reduce deer–automobile collisions or bird–airplane collisions? Should the U.S. Public Health Service use its funds to try to control Lyme disease or rabies? If a chemical company developed an effective deer repellent, how much money might the repellent generate? Do coyotes kill enough sheep to justify the U.S. government spending $10 million a year to kill coyotes? These questions can only be answered by knowing the extent of losses caused by wildlife.

MEASURING WILDLIFE DAMAGE BY MAKING A DIRECT ASSESSMENT OF LOSSES

There are two major methods used to obtain information about the extent of losses from wildlife damage. The first is to make visits to sites where damage is occurring and directly measure the extent of the losses. The second is to identify the people who are suffering from the damage and ask them about their losses. There are advantages and disadvantages to each approach. The direct assessment of damage is the most accurate way to collect data about economic losses but is labor intensive and, for that reason, is most practical for small-scale or localized problems (Table 5.1). Let us consider the steps we would need to follow to directly assess losses due to wildlife.

The first step is to determine exactly what we want to assess. This is a critical question because we will want to be able to extrapolate our findings to a particular group. We can only do so if the group is precisely defined in terms of location (e.g., county, state, country, world) and class membership (e.g., a soybean field, all soybean fields, all crops). The question of what we want to survey is harder to answer than it initially seems. For instance, if we want to assess cattle losses to predators and need to contact cattle ranchers, we broach the question of how to define a cattle rancher. If someone has one steer as a pet, would that person be a cattle rancher? Are dairy farmers or veal producers (whose calves are kept inside) cattle ranchers?

Next, we must determine how we are going to collect our samples. No lists of corn fields or wheat fields or cattle herds are available, so we usually have to locate these fields or herds through their owners. Normally, we cannot survey every field or herd. Instead, we will survey a sample of them. Thus, we will sample corn farmers, wheat farmers, or cattle ranchers in order to sample their fields or herds. To produce accurate results, all members in the group must have an equal chance of being sampled. This means that the investigator must have some way to identify all members in the group or at least a random sample of them. For instance, to survey all U.S. cattle ranches, we might obtain a membership list from the U.S. Cattlemen's Association or from the American Farm Bureau Federation, a list of cattle ranchers from a company that maintains and sells such lists or from the U.S. Bureau of the Census, or a list of subscribers to the magazine *Beef Today*. We might also obtain such a list through the U.S. Department of Agriculture's Farm Services Agency, which has offices in most counties in the country. Obviously, one question with the use of each of these methods is whether the list is

Table 5.1 Examples of Studies Which Have Directly Measured Economic Losses Due to Wildlife Damage (dollar values are adjusted for inflation and expressed as the value of a dollar in the year 2000).

Commodity	Wildlife species	Location	Extent of wildlife damage: % of production	Extent of wildlife damage: $/ha	Reference
Alfalfa	Belding's ground squirrels	SE Oregon	44	?	Kalinowski and deCalestra (1981)
		NE California	25	?	Sauer (1984)
		NE California	35–45	$409	Whisson et al. (1999)
Field corn	Birds (mostly blackbirds)	Pennsylvania	0.6	$5.27	Wakeley and Mitchell (1981)
	All wildlife species	Midwest U.S.	0.7	$4.80	Wywialowski (1996)
	Birds		0.2	$1.32	
	Deer		0.2	$1.60	
	Unidentified		0.3	$1.87	
	Birds (mostly weavers)	Somalia	0–2	?	Bruggers (1980)
Rice	Birds (mostly quelea)	Somalia	0–33	?	Bruggers (1980)
Millet	Birds	Chad	12	?	Manikowski and Da Camara-Smeets (1979)
Wheat	Lesser bandicoot rats	Bangladesh	12	?	Sultana et al. (1983)
	Canada geese	U.S.	16–30	?	Flegler et al. (1987)
	Brant geese	Britain	6–10	?	Summers (1990)
	Pink-footed geese	Scotland	15	?	Patterson et al. (1989)
Barley	Pink-footed geese	Scotland	7	?	Patterson et al. (1989)
	Pink-footed geese	Denmark	7–20	?	Lorenzen and Madsen (1986)

Stored grain at markets	Rats	Pakistan, Punjab Province	0.3	?	Ahmad et al. (1995)
Sunflowers	Birds	Sacramento Valley, CA	0–5	$0–169 (<$16 for 68% of fields)	Avery and DeHaven (1982)
	Birds	Sacramento Valley, CA	2	?	Avery and DeHaven (1984)
Peanuts	Birds	Central Oklahoma		<$2	Mott et al. (1972)
	All wildlife species	Punjab, Pakistan	5	$54	Brooks et al. (1988)
Coconuts	Black rats	Columbia, San Andres Island	34	?	Valencia (1980)
		Columbia, Gorgona Island	60	?	
		Columbia	20	?	
Blueberries	Birds	Florida (1988)	15–19	?	Nelms et al. (1990)
		Florida (1989)	17–75	$800–16,000	
Wine grapes	Birds	Central California	2	$324	DeHaven (1974)

representative of all cattle ranchers or whether it is biased in some way. For instance, the subscribers to *Beef Today* may be better educated and more willing to try new ideas than the average cattle rancher.

As most farmers own more than one field, we will also randomly select fields on each farm. Once we know which individual fields to sample, we need some way to estimate losses. This will involve one or more site visits and a measure of the damage. If the field is too large to assess in its entirety, we must divide it into units small enough to be measured and use some random method to determine which units should be surveyed.

The fourth step involves determining how best to assess damage, which will vary with each crop and also with the intended use of the crop. Consider the problem of trying to determine the amount of damage birds have caused in a corn field. If it is a sweet corn field and the ears are intended for the fresh food market, we could simply count the number of corn ears which have been damaged by birds and determined to be unfit for sale (Figure 5.1). On the other hand, if the grain will be harvested for animal feed, it will be necessary to determine how much of the grain biomass has been lost due to bird damage. We might be able to do this by determining what proportion of the kernels on an ear of corn have been eaten by birds. However, one problem with this approach is that not all kernels are the same size; those toward the end of an ear are smaller. As bird damage occurs at the ear's end, damage will be concordantly overestimated by this approach, although only slightly. Alternatively,

Figure 5.1 For sweet corn, an ear damaged by blackbirds is worthless even when only a few kernels are eaten because no one will buy a damaged ear.

if bird damage makes the corn plants more susceptible to disease, the actual level of damage caused by birds might be much higher than just the number of corn kernel that they consumed (Woronecki et al. 1980). Some diseases produce alfatoxin, and if levels of this toxin in the harvested grain exceed permissible levels, the entire crop may become almost worthless.

Another problem with trying to estimate damage is that plants will try to compensate for damage (Belsky 1986). For instance, if an ear of corn is damaged early, the plant will divert its nutrients to the second ear and to the remaining kernels on the damaged ear, and these will grow bigger than would otherwise be the case. For instance, winter grazing in British wheat fields by Brant geese reduced the leaf and shoot biomass by 75%, but by the time the fields were harvested, there was only a 6 to 10% reduction in grain yields (Summers 1990). However, the compensatory ability of a plant is limited; if only a few kernels remain, they cannot grow to massive size. A plant's ability to compensate for damage also depends on how much time the plant has to make this compensation. Poche et al. (1982) reported that wheat plants could compensate for damage caused by lesser bandicoot rats cutting young shoots but not after the wheat plants reached the booting stage. Sunflowers could compensate for seeds removed by blackbirds but only if damage occurred early (within two weeks of when flowering ended) and if less than 15% of the seeds were removed (Figure 5.2; Cummings et al. 1989). In the case of bird damage to ripening corn ears, the damage occurs so late in the growing season (when the seeds are close to being ripe) that the plant lacks time to compensate for their loss (Woronecki et al. 1980).

Figure 5.2 Sunflowers cannot compensate for seeds removed by blackbirds after the seeds have matured or if more than 15% of the seeds are removed. (Photo courtesy of George Linz.)

ASSESSING THE EXTENT OF WILDLIFE DAMAGE BY SURVEYING PEOPLE

Although wildlife damage is best assessed through direct measurement, this method is usually so labor intensive and costly that it can only be used for wildlife problems that are limited in scope. It is simply too costly to survey an adequate number of randomly selected fields across a large geographic area. Hence, this method has been used almost exclusively to measure losses to a single crop or item and in a limited area (Table 5.1). There are hundreds of different grains, forage species, vegetables, trees, landscaping plants, etc. grown in the U.S. Given the diversity of birds and mammals that also occur in the U.S., almost all of these crops are damaged by some wildlife species. It is therefore unlikely that we will be able to obtain an accurate accounting of wildlife damage in the U.S. by direct measurement any time soon. Thus, most economists are forced to rely upon the estimates of the people who are suffering the damage (e.g., farmers and ranchers).

One way to sample people is to conduct a systematic survey method. The technique is based on deciding whom we wish to question (defining the target audience), determining a method to randomly select the members of the target audience to survey, and determining how to question these people. People might be questioned during a face-to-face meeting or a telephone call, by sending them a questionnaire through the mail, or by e-mailing them. However, it is often difficult to get a high percentage of people to respond to the survey or answer the questions. Some people will never respond to a survey, and the scientist must then determine if the potential answers from those who did not respond differ from those individuals who did (i.e., the respondents). For instance, people who are interested in the subject matter are more likely to respond than those who are uninterested, and this could bias the results if not corrected. This is called a nonresponse bias. For example, people who have contracted Lyme disease or are afraid of it would be more likely to respond to a survey about this disease than someone who has never heard of it. Hence, the incidence of Lyme disease among respondents might be higher than it is in the general population. A false conclusion might be drawn from the survey unless the scientist checks for a nonresponse bias.

Another difficulty with using survey data is that the perceptions of the respondents may not be accurate. They may be either overestimating or underestimating the actual level of damage. There are two reasons why perceptions may be inaccurate. When wildlife damage is very conspicuous, damage may be overestimated. Alternatively, farmers may underestimate deer damage in an alfalfa field because deer are likely to come into the fields at night and be undetected. Also, if deer remove a small part of the crop every night, the farmer may not even be aware of their damage, unless there is some part of the field where deer have been excluded. Otherwise, it can be difficult for the farmer to judge how big a plant might have been if it had not been browsed. However, most farmers and ranchers can accurately estimate wildlife damage to their production.

LOST OPPORTUNITY COSTS

One economic cost of wildlife damage that is difficult to quantify is that of lost opportunity. That is, individuals may forego an opportunity to make money because they know that wildlife damage would be too high. For instance, a farmer growing crops near a large bird roost might not plant sunflowers, because he knows that the birds would eat all the sunflower seeds. In this case, there is no actual damage to measure because sunflowers were never planted, but the farmer still suffered a loss. However, it was a lost opportunity rather than a physical loss. Another example is that ranchers around Yellowstone National Park may go out of business because they can no longer raise livestock profitably due to predation by the reintroduced wolves. If this happens, it may appear that the wolves have little impact on ranchers because there are no longer any livestock upon which to prey, but such would not be true. While wildlife biologists may acknowledge a lost opportunity cost associated with wildlife damage, I am unaware of any studies which have tried to measure it.

Another cost of wildlife damage that is difficult to quantify is a reduction in a person's quality of life due to wildlife problems or the fear of problems. For example, few people are attacked by sharks, yet the fear of such an attack prevents many people from swimming in saltwater and reduces the enjoyment of a day on the beach for many others.

ECONOMIC ASSESSMENT OF WILDLIFE DAMAGE IN THE U.S. AND WORLDWIDE

Several studies have examined the economic losses caused by wildlife damage. I have provided mean values whenever a source gave a range of values or when more than one reference had data. All dollar figures in the chapter are U.S. dollars and have been standardized to the value of a dollar in the year 2000 using the composite deflator provided by the U.S. Office of Management and Budget (2000). Hence, dollar figures reported here may not be the same as they were in the original citations because I adjusted them for inflation.

Deer–Automobile Collisions

Conover (1997a) estimated that the number of deer–automobile collisions in the U.S. amounted to 726,000 annually, based on collisions reported to state authorities or dead deer found on roads (Figure 5.3). These figures exclude deer that die off the highway from collisions and deer–vehicle collisions that are not reported to state authorities (Romin 1994). Because only about half of the deer–vehicle collisions are reported (Decker et al. 1990; Romin 1994), approximately 1.5 million deer–vehicle collisions occur annually in the U.S., as well as 600 moose–vehicle collisions (Romin 1994).

Figure 5.3 Over a million deer–vehicle collisions occur yearly. (Photo courtesy of Mark McClure.)

The average cost to repair a vehicle after a reported deer collision is $1644 (Conover et al. 1995; Conover 1997a). When multiplied by the estimated 726,000 reported collisions, this equals $1.2 billion. An estimate of repair costs for the unreported deer–vehicle collisions is not available, but it is probably much less than that for reported accidents mainly because people are more likely to report an accident when there is a higher level of damage to their vehicle. If we assume that the average repair cost for unreported deer collisions is at least $500 (repairing automobiles is so expensive that it takes very little physical damage before the bill exceeds this amount), then total damage from unreported collisions is about $360 million and damage from all collisions (both reported and unreported) is approximately $1.6 billion annually.

In addition to vehicle damage, approximately 29,000 people are injured and 200 people lose their lives yearly in deer–automobile accidents in the U.S. (Conover et al. 1995). I do not have enough information to estimate what the economic consequences of these human injuries and fatalities might be. Likewise, 1.4 million deer die annually from automobile collisions. These deaths are also an economic loss because of the economic benefits that could have been provided by deer, had they not been killed by automobiles. However, I lack the information necessary to quantify what this loss might be in economic terms.

Bird–Aircraft Collisions

Approximately 8000 civilian aircraft and a similar number of military aircraft collide with birds or other wildlife annually (Chapter 4). Annual costs of wildlife strikes to U.S. civilian and military aircraft exceeded $300 million during the 1990s (Dolbeer et al. 2000). When a bird strike results in a crash, the cost from that single bird strike can be catastrophic (Figure 5.4). For instance, in 1995, an AWACs military plane crashed after colliding with a flock of geese on takeoff, killing all 24 members of the crew and destroying the $200 million aircraft.

Figure 5.4 When birds and airplanes both share the same airspace at the same time, the result can be a collision that causes the plane to crash. (Photo courtesy of C. R. Madsen and U.S.D.A. Wildlife Services.)

Wildlife Damage to Households

Conover (1997b) surveyed 1000 households, randomly selected from the 100 largest metropolitan centers in the U.S. Over half of the respondents (61%) reported that they or their household had a problem with one or more wildlife species during the prior year and suffered a mean loss of $73 in damage. Mice were the most common culprits, followed by squirrels, raccoons, moles, pigeons, starlings, and skunks (Table 5.2).

Almost half (42%) of all urban households reported that they tried to solve a wildlife damage problem in the prior year (Figure 5.5) and spent an average of $38 in the attempt. Unfortunately, 52% reported that their efforts to solve the problem were unsuccessful. When these results are extrapolated to the 60 million metropolitan households in the U.S. (160 million residents), metropolitan households suffered $4.4 billion a year in wildlife damage despite spending $2.3 billion and 268 million hours trying to prevent these problems (Conover 1997b). If we conservatively value people's time at the minimum wage ($6.15 in 2000), the total labor cost would be $1.6 billion. Hence, the total cost of wildlife damage (actual damage plus money and time spent to prevent the problem) to metropolitan residents equals approximately $8.3 billion.

An additional 34 million households (92 million residents) live in smaller cities, towns, and rural areas (U.S. Bureau of the Census 1992). Because wildlife populations should be higher in rural areas, I assume that these households suffer at least as much from wildlife problems as do people living in large metropolitan areas. This would mean that the total annual cost of wildlife damage to rural

Table 5.2 Percentage and Number of U.S. Agricultural Producers (2 Million Total) and Metropolitan Households (60 Million) Who Reported That the Following Wildlife Species Had Caused Damage on Their Property in the Prior Year (Conover 1997b, 1998)

	Agricultural producers		Metropolitan households	
Wildlife species	**%**	**Number (in thousands)**	**%**	**Number (in thousands)**
Deer	51	1086	4	2,400
Raccoons	25	522	16	9,600
Coyotes	24	501	?	?
Ground hogs	21	438	4	2,400
Beavers	19	396	?	?
Blackbirds	18	376	6	3,600
Mice	18	376	22	13,200
Starlings	16	334	11	6,600
Rabbits	16	334	7	4,200
Foxes	12	250	?	?
Skunks	9	19	11	6,600
Squirrels	?	?	18	10,800
Moles	?	?	14	8,400
Pigeons	?	?	12	7,200

Figure 5.5 This metropolitan homeowner has covered his bushes to protect them from deer.

households (damage plus money and time spent to prevent the problem) would conservatively amount to $4.2 billion.

Wildlife Damage to the Timber Industry

The U.S. timber industry is centered in the Southeast, the Northeast, and the Pacific Northwest. Each area has it own distinctive wildlife problems.

Southeast. Beaver is the primary wildlife species that causes damage to southern timber. Beaver impoundments have flooded more than 288,000 ha in 6 of the 13 southern states (Arner and Hepp 1989). In Mississippi, Arner and DuBose (1980) estimated an annual loss of $3.6 million in timber production on 23,000 ha of inundated lands (Figure 5.6). If extrapolated to 288,000 ha of flooded land in the Southeast, this would equal an annual loss of $43 million.

Figure 5.6 Beaver impoundments have flooded hundreds of thousands of hectares of timberland in the U.S. (Photo courtesy of Gary San Julian.)

Beavers also damage nonimpounded timber by felling and gnawing on trees (Figure 5.7). Bullock and Arner (1985) reported that in Mississippi, damage of this type to merchantable timber averaged $893 per ha in bottomland forests. This damage was believed to have accumulated over a decade and represented an annual loss of $90 per ha. There are 13 million ha of bottomland forests in the Southeast, and beaver damage to timber occurs in all of them (McKnight and Johnson 1981). If we assume that beaver losses to bottomland forests in Mississippi are similar to those which occur in bottomland forests in other southeastern states, then beaver damage to southern timber is $1.1 billion annually. Furthermore, seedlings and young trees of all of the important timber species in the bottomland hardwoods are damaged by deer, rabbits, and rodents (McKnight and Johnson 1981), but the amount of damage by these species has not been quantified.

Figure 5.7 Beaver gnawing on trees causes losses to the timber industry. (Photos courtesy of Gary San Julian.)

Northeast. Although many wildlife species cause damage to forests in the Northeast, white-tailed deer cause most of the economic losses in this region. Experiments using deer exclosures have demonstrated that their browsing on tree seedlings and sprouts can reduce tree growth and result in understocked stands and longer rotations before trees are ready for harvest (Richards and Farnsworth 1971; Marquis 1974; Marquis and Brenneman 1981). Furthermore, deer browsing can change the species composition of forested stands (Marquis and Grisez 1978). Unfortunately, the tree species most desirable for timber production, because of their value as sawtimber, are also the species which deer prefer to browse (Marquis and Brenneman 1981).

Marquis (1981) estimated that annual timber losses across Pennsylvania's Allegheny hardwood forest amounted to over $102 per ha. This high-quality hardwood forest covers 6.5 million ha in Pennsylvania. Hence, deer-induced losses in this region may be $655 million per year. Pennsylvania's Allegheny hardwood forest represents 14% of the oak–hickory (*Quercus–Carya*) type hardwood that covers 45 million ha in the northeastern U.S. (Gedney and Van Sickle 1979). Commercial forests cover 72 million ha in the Northeast (Gedney and Van Sickle 1979). Unfortunately, losses to timber production from wildlife damage are unknown for most of this area but deer populations are high throughout the region. If we assume that deer damage in the oak–hickory forests in the rest of the Northeast (38 million ha) is at least 25% of the damage that occurs in the Alleghenies, or $25 per ha per year, then deer damage to the oak–hickory forest industry in the Northeast is approximately $1.6 billion annually ($655 million in the Alleghenies and $950 million in the rest of the Northeast).

Northwest. Animal damage to forests and forest regeneration in the Pacific Northwest is caused by numerous wildlife species. The principal damaging species are beaver, deer, elk, mountain beaver, pocket gophers, porcupine, and rabbits (Black et al. 1979; Borrecco and Black 1990). These species cause damage to Douglas fir, lodgepole pine, and ponderosa pine at all stages of stand growth, resulting in tree deaths or reduced growth rates (Figure 5.8).

Newly planted stands are especially vulnerable. In 1988, measures were taken to prevent wildlife damage on 57,500 ha of U.S. Department of Agriculture Forest Service land in the Pacific Northwest. These measures cost $8.4 million, not including the cost of reforesting the sites and any loss in timber value (Borrecco and Black 1990).

In a large-scale study using exclosures, Black et al. (1979) found that, after five years, wildlife damage in Washington and Oregon reduced the survival of newly planted Douglas fir by 20%, their height by 24%, and the survival and height of ponderosa pine by 31% and 22%, respectively. Based on these data, Brodie et al. (1979) calculated two estimates of wildlife damage to the 158,000 ha planted in 1976 with Douglas fir and ponderosa pine in Oregon and Washington. The two estimates yielded an annual loss of $320 million and $1.1 billion. These values exclude damage to established trees, which can be killed or have their growth rates reduced by being debarked or girdled by bears, porcupines, or squirrels (Sullivan et al. 1993). I know of no studies documenting the extent of these losses.

Wildlife Damage to Agricultural Production

Conover (1998) surveyed a random sample of 2000 farmers and ranchers and found that 80% had suffered wildlife damage in the prior year, and 53% reported that the damage exceeded their tolerance. Problems were caused most often by deer, raccoons, coyotes, and ground hogs (Table 5.2). When asked to categorize their losses

Figure 5.8 Deer herbivory on young tree seedlings can prevent some clear-cuts from being regenerated in trees. (Photo courtesy of Barrie Gilbert.)

from wildlife damage in the prior year, 22% reported losses of less than $100, 45% between $100 and $999, 23% between $1000 and $4999, and 6% between $5000 and $9999. Three percent said that their losses in the prior year exceeded $10,000. In addition, farmers and ranchers reported spending an average of 44 hours and $1000 in the prior year trying to prevent a wildlife problem. When extrapolated to the nation's 2,088,000 agricultural producers, Conover (1998) estimated that, despite spending 90 million hours and $2 billion trying to prevent wildlife problems, they still suffered $2 billion in damage annually. This equals $4.5 billion if we value their time at $6.15/hour.

Total Economic Losses Due to Wildlife Damage in the U.S.

Wildlife damage in the U.S. costs approximately $22 billion annually (Table 5.3). I believe this to be a conservative number, as I usually rounded down or based the figures on conservative assumptions. These losses include the cost of property damage caused by wildlife and the time and money people spent trying to prevent wildlife damage. Presumably, damage from wildlife would have been much higher if people had not spent their time and money to prevent a problem. That is, I assume that people were rational and spent time and money when they believed that their savings in damage outweighed the costs of the preventive steps.

These figures do not include economic losses from human illnesses or injuries caused by wildlife. For instance, when someone is injured or becomes ill, there is

Table 5.3 Summary of Wildlife Damage Occurring Annually in the U.S. (the values include both damage to property and money and time spent to prevent the problem)

Problem	Dollars (in billions)
Damage from deer–automobile collisions	1.6
Damage from bird–aircraft collisions	0.3
Damage to agricultural producers	4.5
Damage to timber industry	
Southeast	1.2
Northeast	1.6
Northwest	0.6
Damage to metropolitan households	8.3
Damage to rural households	4.2
Economic losses from human injuries, fatalities, and illnesses which result from wildlife-related incidents	Unknown, but in the billions
Total losses	22.3

both the cost for medical treatment and wages lost while the person is recuperating. I also have chosen not to put a value on a human life or on human pain and suffering. However, courts often do so during lawsuits. When this happens, the cost of a human life or injury can be substantial. For instance, the state of Arizona settled out of court with a person mauled by a bear for more than $4 million.

DO HIGH LEVELS OF WILDLIFE DAMAGE MEAN THAT WILDLIFE POPULATIONS ARE TOO HIGH?

Economic losses, injuries, and diseases caused by wildlife are a substantial problem, but this does not imply that wildlife populations are too high or should be controlled. All wildlife species or populations have both positive and negative values. To determine the net value of a wildlife species, all of the positive values must be summed up and all the negative values must be subtracted from them. I used such an approach to determine the positive and negative values of deer (Conover 1997a). I calculated that the amount spent by U.S. residents to hunt deer was $5 billion and the amount spent on deer-related nonconsumptive recreation was $2 billion. I estimated that the consumer surplus of these activities (value received by deer hunters and recreationists which they did not have to pay for) conservatively exceeded $7 billion. Hence the positive value of deer for recreation was $14 billion annually.

Negative values for deer were calculated as $1.6 billion in automobile damage from collisions with deer, $100 million in damage to agricultural productivity, $750 million in damage to the timber industry, and $376 million in damage to households (including money spent by households to prevent deer damage). Adding these up, the negative value of deer was about $3 billion annually, excluding the cost of illnesses due to Lyme disease and human injuries and deaths caused by deer–automobile collisions. Hence, the net value of deer was highly positive (approximately $11 billion annually).

WILDLIFE DAMAGE TO AGRICULTURAL PRODUCTION IN OTHER PARTS OF THE WORLD

Thus far, I have concentrated on economic losses in the U.S., but wildlife cause damage to agriculture in most countries. Hence it is worthwhile to consider wildlife problems in other parts of the world.

South America

Many North American birds that winter in South America cause agricultural damage; the dickcissel is probably the species that is most destructive to grain crops. During the winter, flocks numbering in the thousands forage in ripening grain fields, especially grain sorghum and rice, in Venezuela and Colombia (Besser et al. 1970). Farmers use lethal means to control these birds.

Several species of parakeet and doves cause significant damage to sunflowers in South America (Linz and Hanzel 1997). Some of the most destructive species in Argentina and Uruguay are the spot-winged pigeon, Picazuro pigeon, and the eared dove. All of these bird species forage in large flocks after the nesting season. Additionally, monk parakeets cause agricultural damage in Argentina, Bolivia, Brazil, Paraguay, and Uruguay.

Iceland

Due to its isolation in the far north of the Atlantic Ocean, Iceland has only one native mammal, the arctic fox, and a small human population. Still, Iceland has its share of wildlife damage, which include fox predation on lambs and predation by fox, raven, and mink (introduced to Iceland in the 20th century) on common eiders, which nest in large colonies and are captured annually for their down. Farmers also have to contend with goose damage to their fields and ranchers worry about feral reindeer competing with their livestock for forage. Methods to reduce wildlife damage include erecting predator-proof fences around eider colonies, allowing the hunting of reindeer, and killing fox and mink by trapping, denning, and shooting. As early as 1295 A.D., laws were passed requiring each farmer to kill at least one fox per year (P. Hersteinsson, article published in the TWS Wildlife Damage Management Working Group Newsletter, Summer 1998, 5(3):10–11).

Great Britain

British farmers consider the European rabbit to be their greatest wildlife pest due to its herbivory in grain fields, horticultural crops, and pastures (Wilson and McKillop 1986; Hardy 1990). Often, rabbits are controlled by fumigating their burrows and by trapping, but some Britons consider these methods to be inhumane. Fencing is often used to exclude rabbits from agricultural fields. Rats and house mice occupy many farm buildings and cause post-harvest losses to stored grain and feed. They also serve as a reservoir for diseases, such as leptospirosis, trichinosis, and salmonella (Hardy 1990).

Several bird species also cause agricultural damage. Sparrows and doves are pests in sunflower fields across Europe (Linz and Hanzel 1997). Wood-pigeons damage growing fields of rapeseed and bullfinches ruin fruit in orchards. During winter, large numbers of Brant geese graze in fields of winter grain, as do Canada geese, an exotic species introduced from North America (Patterson et al. 1989; Summers 1990; McKay et al. 1993). Starlings cause nuisance problems in urban areas and economic losses at livestock facilities (Hardy 1990).

Red deer, roe deer, fallow deer, and muntjac deer damage coppice woodlands and broadleaf plantations in Great Britain (Putman 1994; Moore et al. 1999). In Europe, roe deer and red deer damage young tree plantations of Norway spruce, Sitka spruce, Scots pine, and other coniferous species (Bergquist and Örlander 1998).

Ungulates cause over 500,000 automobile collisions in European countries, including England. Groot Bruinderink and Hazebroek (1996) estimate that these resulted in 300 human fatalities (including 50 in France, 10 to 20 in Sweden, 14 in Great Britain, and 25 in the former German Democratic Republic), 30,000 human injuries, and $1.1 billion (U.S. dollars) in property damage.

Sidebar 5.1 Problems Caused by Wildlife to Villagers Near India's Sariska Tiger Reserve

Predation on livestock and wildlife damage to crops are becoming serious problems around many of the world's wildlife reserves and national parks. Two thirds of the villagers near India's Sariska Tiger Reserve spent considerable time and money protecting crops and livestock from wildlife, primarily by guarding them or building fences. Despite these efforts, nearly half of the households living near the Sariska Tiger Reserve experienced losses to their livestock or crops due to wildlife depredation. Nilgai and wild boar caused most of the crop losses; tigers and leopards were the main predators of livestock. Despite their losses to wildlife, villagers had a positive attitude about the tiger reserve because it also provided them with fodder and firewood (Sekhar 1998).

Southeast Asia

Rice is the primary crop in this region, and rodents are often identified as a major problem limiting rice yields. In many countries, rodents consume up to 15% of the crop prior to harvest (Table 5.4). Many farmers lack rodent-proof storage facilities for grain. Because of this, additional losses occur after harvest. For instance, Ahmad et al. (1995) found that the average grain shop in Pakistan was home to 40 rats. Given the large number of poor subsistence farmers in Southeast Asia, a reduction of rice yields due to rodents is a serious problem that results in much human misery and hunger. Often, multiple crops of grain and rice are produced yearly in some parts of Southeast Asia, and this practice provides ample food for rodent populations to reach high levels. The high densities of people and rodents living in close association with each other result in serious problems with zoonoses (see Chapter 4 for a more thorough discussion of this topic).

Table 5.4 Estimates of Annual Rodent Damage, Adapted from Jackson (1977), with Some Additional References for Asia and Madagascar as Noted (monetary values have been adjusted for inflation and expressed in year 2000 U.S. dollars).

Location	Crops	Loss	Remarks and/or additional references
		Worldwide	
	Food grains or crops	2 million tons	Girish et al. (1974)
		200 million tons	Cotton (1963)
		33 million tons	Townes and Morales (1953)
		U.S.	
Nationwide	Food grains or crops	5 million tons	
		3–4%; $2–7 billion	
Florida	Sugar-cane	5–11%	
Hawaii	Macadamia nut	16%	
	Sugar-cane	$20 million	
		Latin America	
Argentina	Sugar-cane	1 ton sugar/ha	
Jamaica	Coconuts	5%	
Guyana	Sugar-cane	12,500 tons	During rodent outbreak
		Asia	
Bangladesh	Wheat	12%	Sultana et al. (1983)
	Grain	1–3%; 300,000 tons	Sultana and Jaeger (1992)
	Peanuts	3%; 3000 tons	Brooks et al. (1988)
Cambodia	Rice	0.1%; 3000 tons	Jahn et al. (1999)
	Oil palms	3–36%	
China	Grain	7,500,000 tons	Zhang et al. (1999)
India	Betel nut	Up to 20%	
	Coconuts	3–17%	Loss of nuts
	Peanuts	1–4%	

	Tea bushes	Up to 50%	Feeding on roots
	Barley	5–12%	
	Rice	6–9%	
	Rice	1400 kg	
	Sorghum	6%	
	Food grains or crops	2–4%, 2 million tons	
India (Punjab)	Peanuts	49 kg/ha	Data from 3 villages
	Sugar-cane	161–240 kg/ha	
	Wheat	52 kg/ha	Data from 3 villages
India (Tanjore district)	Rice	50,000 tons	
India (Uttar Pradesh)	Sugar-cane	$40 million	
Indonesia	Rice	40%	
		17%; 9 million tons	Leung et al. (1999)
		17%; 8 million tons	Singleton and Petch (1994)
Indonesia (Java)	Sugar-cane	30–100%	
Iraq	Sugar-cane	1.6%	
	Corn	68% of ears	
Korea	Soybean	2–12%	
	Sweet potatoes	3%	
	Rice	4%	
	Wheat and barley	1%	
Malaysia	Oil palm	164 kg oil/ha	
	Rice	3–5%	Singleton and Petch (1994); Singleton et al. (1999)
		87,000 tons	
Nepal	Food grains or crops	10–30%; 400,000 tons	
Pakistan	Food grains or crops	5%	
Philippines	Castor bean	20–80%	
	Coconuts	1600 nuts/ha	
	Sugar-cane	2–10%	
	Rice	5%	
		1–3%	Singleton and Petch (1994); Singleton et al. (1999)
		$67 million	

continued

Table 5.4 Estimates of Annual Rodent Damage, Adapted from Jackson (1977), with Some Additional References for Asia and Madagascar as Noted (monetary values have been adjusted for inflation and expressed in year 2000 U.S. dollars) (continued)

Location	Crops	Loss	Remarks and/or additional references
Thailand	Rice	1.5%; $35 million	Boonsong et al. (1999)
		$4.5 million	Singleton and Petch (1994)
Tibet (Qinghai Plateau)	Grass forage	15 million tons	Fan et al. (1999)
Vietnam	Rice	300,000 tons	Brown et al. (1999)
		Pacific Islands	
Fiji	Cacao	9–63%	
	Coconuts	7–28%	
French Polynesia	Coconuts	25–30%	
Gilbert and Ellis Islands	Coconuts	21–73%	
New Hebrides	Coconuts	20–50%	
Samoa	Cacao	25%	
Solomon Islands	Cacao	2%	
Thaiti	Coconuts	27–47%	
Tarawa	Coconuts	23%	
Taveuni	Coconuts	28%	
Tokelau Island	Coconuts	30–40%	
		Africa	
Comoro Islands	Coconuts	37%	
Egypt	Cotton	Up to 30%	
	Sugar-cane	5–20%	
Ethiopia	Food grains or crops	33%	
Ivory Coast	Coconuts	10–15%	
Kenya	Barley	23%	During outbreak of field rats
Madagascar	Rice	2–3%; 40,000 tons	Duplantier and Rakotondravony (1999)
Mali	Food grains or crops	30–35%	

Mauritius	Sugar-cane	Up to 15 tons/ha
Nigeria	Cacao	10%
Senegal	Peanuts	5–10%

Table 5.5 Examples of Reported Economic Losses to Agriculture Due to Bird Damage on an Annual Basis (Adapted from De Grazio 1978, monetary values have been adjusted for inflation and expressed in year 2000 U.S. dollars)

Location	Year	Crops	Loss	Major Species	Remarks
			Canada		
10 provinces	1976	Agriculture	$233 million	Seed-eating birds	
4 provinces	1976	Agriculture	$90 million	Waterfowl	
Ontario	1971	Sweet cherries	$173,000; 2.8%	Birds	Niagara Peninsula
			U.S.		
Nationwide	1971	Corn	6.8 million bushels	Blackbirds	24 states; 2500 fields examined
	1971	Emerging corn	32.5 million bushels	Blackbirds, pheasants	Questionnaire survey in 25 states
	1972	Blueberries	$6–7 million, 5%	Birds	14 states
	1972	Grapes	$12 million	Starlings, sparrows, finches	Questionnaire survey in 13 states
	1972	Cable damage	$700,000	Woodpeckers	Bell Telephone Co.
	1973	Utility pole damage	$600,000	Woodpeckers	Damage to one private company
Arkansas	1963	Oats	$600,000	Blackbirds	
	1963	Rice	$16 million	Blackbirds	22 counties
	1968	Wheat	$1 million	Blackbirds	
California	1972	Almonds	$300,000	Crows	Tulare County
	1973	Grapes	$14 million	Starlings, finches	9 counties
	1971	Rice	$300,000	Blackbirds	Sacramento Valley
	1976	Miscellaneous	$35 million	Birds	

continued

Table 5.5 Examples of Reported Economic Losses to Agriculture Due to Bird Damage on an Annual Basis (Adapted from De Grazio 1978, monetary values have been adjusted for inflation and expressed as year 2000 U.S. dollars) (continued)

Location	Year	Crops	Loss	Major Species	Remarks
Louisiana	1972	Pecans	$4 million	Crows	
Michigan	1974	Cherries	$6–15 million	Birds	
Minnesota, North Dakota	1972	Sunflowers	1.2%	Blackbirds	18 counties
New Jersey	1962	Miscellaneous	$5 million	Birds	
Ohio	1975	Corn	$6 million	Blackbirds	
Oklahoma	1970	Peanuts	≤ $150/acre	Blackbirds	
	1939	Sorghum, corn, pecans	$100 million	Crows	Questionnaire survey
Virginia	1970	Miscellaneous	$21 million	Birds	
Wyoming	1977	Sheep (lambs)	9%	Golden eagles	2 counties
Latin America					
Dominican Republic	1971	Rice	$27 million	Black-headed weaver	
Hispaniola	1971	Grain, peas, tomatoes	$117 million	Black-headed weaver	
Uruguay	1975	Sunflowers	$6 million	Parakeets, doves	
	1974	Wheat	$800,000	Doves	
Europe					
England	1970	Cereal grains	$5–10 million	Wood-pigeon	
Germany	1968	Grapes, cherries	$11 million	European starling	
Asia					
Pakistan	1974	Rice	20%	Birds	
	1974	Sorghum	80%	Birds	
Thailand	1978	Rice	Up to 50%	Birds	

Africa				
Kenya	1952	Grain	$5 million	Birds
Madagascar	1967	Rice	1.2 million tons	Red fody
Morocco	1971	Wheat, rice, sunflowers	$16 million	House sparrow, Spanish sparrow
Nigeria	1971	All crops	$10 million	Birds
Senegal	1976	Rice, sorghum, other grains	100,000–200,000 tons	Birds
South Africa	1953	Sorghum	$10 million	Red-billed quelea
Somalia	1978	Rice, millet, sorghum	$350/acre	Birds
Sudan	1969	Sorghum	$900,000	Red-billed quelea
Tunisia		Olives	15,000 tons	European starling

Rodent control often involves using toxicants and traps to try to reduce their numbers. In many areas, rodents are consumed by humans, and this provides another incentive to trap them (Singleton et al. 1999). Rodent-proof fences are employed to keep rodents out of high-valued crops, but these fences are so expensive that they are not cost effective for most fields.

Africa

Africa is blessed with an abundance of wildlife, but they can pose problems for the continent's farmers and ranchers. Many primates, such as red-tailed monkeys, olive baboons, and chimpanzees, raid agricultural fields throughout Africa. More than 40 species of birds attack grain crops in northern Africa, with most of the damage caused by red-billed quelea, village weavers, yellow-backed weavers, and glossy starlings (Manikowski and Da Camara-Smeets 1979). Quelea and weaver birds live in semi-arid parts of Africa and breed in large colonies during the rainy season. During the rest of the year, they form large flocks that can number in the millions (Crook and Ward 1968). These birds feed on insects and grass seeds and cause extensive damage to ripening grain fields. Many scientists believe that they cause more agricultural damage than any other bird species in the world. It is not uncommon for these birds to reduce grain yields in sorghum and wheat fields by more than 50% (Table 5.5). Rock pigeons and red-eyed turtle doves are the most important avian species that damage sunflower fields in South Africa (Linz and Hanzel 1997).

Africa's large number of ungulate species often compete with livestock for forage and serve as reservoirs for certain livestock diseases, such as foot-and-mouth disease. In some parts of Africa, these losses are partially offset by the income from tourists who come to see and hunt the native ungulates (De Boer and Baquete 1998; Hill 1998).

African elephants pose a difficult problem for people who farm close to parks or preserves, owing to their propensity to forage in agricultural fields (Hill 1998). Their damage not only includes the large amounts of food they consume, but they also trample vegetation under foot and cause considerable damage to fences and structures. During a 14-month study, elephants outside Kenya's Amboseli National Park raided crops 457 times, resulting in $200,000 in crop damage and 12 human deaths and injuries (E. E. Esikuri and D. F. Stauffer, paper given in 1999 at the Second International Wildlife Management Congress and abstract published in *The Probe*, Issue 205). Over 300 elephants left Cameroon's Waza National Park during 1992 to search for food (Tchamba 1996). By the time they returned, they had killed two people and destroyed over 5000 ha of farmland. During 1993, about 400 elephants repeated the excursion, causing twice as much damage to crops as the year before and killing four people.

Australia

Wildlife damage in Australia is caused primarily by exotic species, which were brought to the continent by European settlers. Feral pigs, red foxes, European rabbits, goats, donkeys, horses, water buffalo, rats, and house mice all thrived in their new surroundings and have become serious environmental threats. Indeed, the native flora and marsupial fauna did not evolve with these mammalian herbivores and predators and too often cannot compete with them or defend against them. These exotic mammals also have become major agricultural pests in Australia. The loss to Australia's wool industry due to rabbits is over $110 million annually (Williams et al. 1995).

To control the populations of these exotic species, lethal means have been employed, including shooting, poisoning, and releasing exotic diseases. For instance, to control European rabbits, myxoma virus was released in 1950. It had a dramatic impact on rabbit populations initially, but its impact has waned over time: the virus has become less virulent, and the rabbits have evolved some resistance to the disease. In 1995, rabbit calicivirus accidently escaped and has reduced some rabbit populations by 90%, while having little impact on others. Current research is aimed at developing means to sterilize large numbers of rabbits.

Sidebar 5.2 Consequences of a Mouse Plague in Australia

The house mouse is an exotic species in Australia, where it arrived with European settlers. Most years, its numbers are low, but about once a decade, its populations erupt to incredible levels. Such an eruption, or mouse plague, occurred in 1993 in South Australia. Crop losses were estimated at $45 million and mostly resulted from destruction of newly planted legume and grain fields. Livestock producers also suffered from mice consuming and fouling livestock feed and gnawing on livestock. Problems were not limited to the agricultural community. Mice migrated into towns, where they caused considerable problems to households (mean loss $280), schools ($2700), hospitals ($2100), rural suppliers ($7400), food retailers ($2300), and hotels and motels ($2400). These losses included damage to supplies and equipment and hours spent trapping and cleaning up after the mice. Fires also became more common as mice gnawed through electrical wires. The estimated cost of the plague to businesses and community services was about $900,000 and damage to households was $2.2 million not counting labor costs for households (Caughley et al. 1994).

The quality of life was diminished for everyone living in the affected area. People reported that the stress caused by having large numbers of mice in their homes and work places and the constant need to clean up their mess was the worst aspect of the plague. Mice were everywhere and under everything. They invaded the walls, cupboards, wardrobes, and food supplies and built nests behind furniture and kitchen appliances. They ran over people as they lay in bed at night. Mice are scent markers and urinate constantly. This, combined with their droppings and decomposing mice located in inaccessible places, created a stench that lasted for months in many homes (Caughley et al. 1994).

Other problems are caused by the native wildlife. Livestock compete with kangaroos and wallabies for forage and are preyed upon by dingoes, a canid brought to Australia by the Aborigines before the arrival of Europeans. Native birds, such as silvereyes, galahs, and sulfur-crested cockatoos, cause losses to sunflower, grain, and fruit growers. These birds can destroy several times the amount of sunflower seed that they consume by decapitating the sunflower head (Linz and Hanzel 1977).

While losses caused by these species may be low nationwide, they cause significant problems in local areas (Bomford 1992).

LITERATURE CITED

Ahmad, E., I. Hussain, and J. E. Brooks, Losses of stored foods due to rats in grain markets in Pakistan. *Int. Biodeterioration Biodegradation*, 36, 125–133, 1995.

Arner, D. H. and J. S. DuBose, The impact of the beaver on the environment and economics in the southeastern United States. In *Int. Wildl. Conf.*, 14, 241–247, 1980.

Arner, D. H. and G. R. Hepp, Beaver pond wetlands: a southern perspective. P. 117–128 in L. M. Smith, R. L. Pederson, and R. M. Kaminski, Eds., *Habitat Management for Migrating and Wintering Waterfowl in North America.* Texas Tech University Press, Lubbock, 1989.

Avery, M. L. and R. DeHaven, Bird damage to sunflowers in the Sacramento Valley, California. In *Proc. Vert. Pest Conf.*, 10, 197–200, 1982.

Avery, M. L. and R. DeHaven, Bird damage chronology and feeding behavior in two sunflower fields, Sacramento, California, 1982. In *Proc. Vert. Pest Conf.*, 11, 223–228, 1984.

Belsky, A. J., Does herbivory benefit plants? A review of the evidence. *Am. Nat.*, 127, 870–892, 1986.

Bergquist, J. and G. Örlander, Browsing damage by roe deer on Norway spruce seedlings planted on clearcuts of different ages: 1. Effect of slash removal, vegetation development, and roe deer density. *For. Ecol. Manage.*, 105, 283–293, 1998.

Besser, J. F., J. W. De Grazio, and K. H. Larsen, The dickcissel – a problem in ripening grains in Latin America. In *Bird Control Sem.,* 5, 141–143, 1970.

Black, H. C., E. J. Dimock, II, J. Evans, and J. A. Rochelle, Animal damage to coniferous plantations in Oregon and Washington: Part 1. A survey, 1963–1975. *For. Res. Lab. Res. Bull. 25.* Oregon State University, Corvallis, OR, 1979.

Bomford, M., Review of research on control of bird pests in Australia. In *Proc. Vert. Pest Conf.*, 15: 93–96, 1992.

Boonsong, P., S. Hongnark, K. Suasa-ard, Y. Khoprasert, P. Promkerd, G. Hamarit, P. Nookam, and T. Jäkel, Rodent management in Thailand. P. 338–357 in G. R. Singleton, L. A. Hinds, H. Leirs, and Z. Zhang, Eds., *Ecologically-Based Management of Rodent Pests.* Australian Center for International Agricultural Research, Canberra, Australia, 1999.

Borrecco, J. E. and H. C. Black, Animal damage problems and control activities on national forest system lands. In *Proc. Vert. Pest Conf.*, 14, 192–198, 1990.

Brodie, D., H. C. Black, E. J. Dimock, II, J. Evans, C. Kao, and J. A. Rochelle, Animal damage to coniferous plantations in Oregon and Washington: Part II. An economic evaluation. *Res. Bull. 26.* Forest Research Laboratory, Oregon State University, Corvallis, 1979.

Brooks, J. E., E. Ahmad, and I. Hussain, Characteristics of damage by vertebrate pests to groundnuts in Pakistan. In *Proc. Vert. Pest Conf.*, 13, 129–133, 1988.

Brown, P. R., N. Q. Hung, N. M. Hung, and M. van Wensveen, Population ecology and management of rodent pests in the Mekong River Delta, Vietnam. P. 319–337 in G. R. Singleton, L. A. Hinds, H. Leirs, and Z. Zhang, Eds., *Ecologically-Based Management of Rodent Pests.* Australian Center for International Agricultural Research, Canberra, Australia, 1999.

Bruggers, R. L., The situation of grain-eating birds in Somalia. In *Proc. Vert. Pest Conf.*, 9, 5–16, 1980.

Bullock, J. F., and D. H. Arner, Beaver damage to non-impounded timber in Mississippi. *South. J. Appl. For.*, 9, 137–140, 1985.

Caughley, J, V. Monamy, and K. Heiden, Impact of the 1993 mouse plague. *Occasional Papers Series* 7. Grains Research and Development Corporation, Canberra, Australia, 1994.

Conover, M. R., Monetary and intangible valuation of deer in the United States. *Wildl. Soc. Bull.*, 25, 298–305, 1997a.

Conover, M. R., Wildlife management by metropolitan residents in the United States: practices, perceptions, costs, and values. *Wildl. Soc. Bull.*, 25, 306–311, 1997b.

Conover, M. R., Perceptions of American agricultural producers about wildlife on their farms and ranches. *Wildl. Soc. Bull.*, 26, 597–604, 1998.

Conover, M. R. and D. J. Decker, Wildlife damage to crops: perceptions of agricultural and wildlife professionals in 1957 and 1987. *Wildl. Soc. Bull.*, 19, 46–52, 1991.

Conover, M. R., W. C. Pitt, K. K. Kessler, T. J. DuBow, and W. A. Sanborn, Review of human injuries, illnesses, and economic losses caused by wildlife in the United States. *Wildl. Soc. Bull.*, 23, 407–414, 1995.

Cotton, T. R., *Pests of Stored Grain Products.* Burgess Publishing, Minneapolis, MN, 1963.

Crook, J. H. and P. Ward, The quelea problem in Africa. P. 211–229 in R. K. Murton and E. N. Wright, Eds., *The Problems of Birds as Pests. Institute of Biology Symposium 17.* Academic Press, London, 1968.

Cummings, J. L., J. L. Guarino, E. E. Knittle, Chronology of blackbird damage to sunflowers. *Wildl. Soc. Bull.*, 17, 50–52, 1989.

Decker, D. J., K. M. Loconti-Lee, and N. A. Connelly, Incidence and costs of deer-related vehicular accidents in Tompkins County, New York. *Human Dimensions Research Group 89-7.* Cornell University, Ithaca, New York, 1990.

De Boer, W. F. and D. S. Baquete, Natural resource use, crop damage and attitudes of rural people in the vicinity of the Maputo Elephant Reserve, Mozambique. *Environ. Conserv.*, 25, 208–218, 1998.

De Grazio, J. W., World bird damage problems. In *Proc. Vert. Pest Conf.*, 8, 9–24, 1978.

DeHaven, R. W., Bird damage to wine grapes in central California, 1973. In *Proc. Vert. Pest Conf.*, 6, 248–252, 1974.

Dolbeer, R. A., S. E. Wright, and E. C. Cleary, Ranking the hazard level of wildlife species to aviation. *Wildl. Soc. Bull.*, 28, 372–378, 2000.

Duplantier, J.-M. and R. Rakotondravony, The rodent problem in Madagascar: agricultural pest and threat to human health. P. 441–459 in G. R. Singleton, L. A. Hinds, H. Leirs, and Z. Zhang, Eds., *Ecologically-Based Management of Rodent Pests.* Australian Center for International Agricultural Research, Canberra, Australia, 1999.

Fan, N., W. Zhou, W. Wei, Q. Wang, and Y. Jiang, Rodent pest management in the Qinghai-Tibet alpine meadow ecosystem. P. 285–304 in G. R. Singleton, L. A. Hinds, H. Leirs, and Z. Zhang, Eds., *Ecologically-Based Management of Rodent Pests.* Australian Center for International Agricultural Research, Canberra, Australia, 1999.

Flegler, E. J., Jr., H. H. Prince, and W. C. Johnson, Effects of grazing by Canada geese on winter wheat yield. *Wildl. Soc. Bull.*, 15, 402–405, 1987.

Gedney, D. R. and C. Van Sickle, Geographic context of forestry. P. 301–318 in W. A. Duerr, D. E. Teeguarden, N. B. Christiansen, and S. Guttenberg, Eds., *Forest Resource Management.* Saunders Publishing, Philadelphia, 1979.

Girish, G. K., K. K. Arora, and K. Krishnamurthy, Studies on rodents and their control. Part 10. Storage losses in foodgrains by rats. *Bull. Grain Technol.*, 12, 189–192, 1974.

Groot Bruinderink, G. W. T. A. and E. Hazebroek, Ungulate traffic collisions in Europe. *Conserv. Biol.*, 10, 1059–1067, 1996.

Hardy, A. R., Vertebrate pests of U.K. agriculture: present problems and future solutions. In *Proc. Vert. Pest Conf.*, 14, 181–185, 1990.

Hill, C. M., Conflicting attitudes towards elephants around the Budongo Forest Reserve, Uganda. *Environ. Conserv.*, 25, 244–250, 1998.

Jackson, W. B., Evaluation of rodent depredations to crops and stored products. *EPPO Bull.*, 7, 439–458, 1977.

Jahn, G. C., M. Solieng, P. G. Cox, and C. Nel, Farmer participatory research on rat management in Cambodia. P. 358–371 in G. R. Singleton, L. A. Hinds, H. Leirs, and Z. Zhang, Eds., *Ecologically-Based Management of Rodent Pests*. Australian Center for International Agricultural Research, Canberra, Australia, 1999.

Kalinowski, S. A. and D. S. deCalestra, Baiting regimes for reducing ground squirrel damage to alfalfa. *Wildl. Soc. Bull.*, 9, 268–272, 1981.

Leung, L. K. P., G. R. Singleton, Sudarmaji, and Rahmini, Ecologically-based population management of the rice-field rat in Indonesia. P. 305–318 in G. R. Singleton, L. A. Hinds, H. Leirs, and Z. Zhang, Eds., *Ecologically-Based Management of Rodent Pests.* Australian Center for International Agricultural Research, Canberra, Australia, 1999.

Linz, G. M. and J. J. Hanzel, Birds and sunflowers. P. 381–394 in *Sunflower Technology and Production, Agricultural Monograph 35*, 1997.

Lorenzen, B. and J. Madsen. Feeding by geese on the Filso Farmland, Denmark, and the effect of grazing on yield structure of spring barley. *Holarctic Ecol.*, 9, 305–311. 1986.

Manikowski, S., and Da Camara-Smeets, M., Estimating bird damage to sorghum and millet in Chad. *J. Wildl. Manage.*, 43, 540–544, 1979.

Marquis, D. A., The impact of deer browsing on Allegheny hardwood regeneration. Forest Service Research Paper NE-308, Broomall, PA, 1974.

Marquis, D. A., The effect of deer browsing on timber production in Allegheny hardwood forests of Northwestern Pennsylvania. Northeast Forest Experiment Station, Broomall, PA, 1981.

Marquis, D. A. and R. Brenneman, The impact of deer on forest vegetation in Pennsylvania. Northeast Forest Experiment Station, Broomall, PA, 1981.

Marquis, D. A. and T. J. Grisez, The effect of deer exclosures on the recovery ofvegetation in failed clearcuts on the Allegheny Plateau. Forest Service Research Note NE-270, Broomall, PA, 1978.

McKay, H. V., J. D. Bishop, C. J. Feare, and M. C. Stevens, Feeding by brant geese can reduce yield of oilseed rape. *Crop Prot.,* 12, 101–105, 1993.

McKnight, J. S. and R. L. Johnson, Hardwood management in southern bottomlands. *For. Farmer,* 35(5), 31–39, 1981.

Moore, N. P., J. D. Hart, and S. D. Langton, Factors influencing browsing by fallow deer *Dama dama* in young broad-leaved plantations. *Biol. Conserv.,* 87, 255–260, 1999.

Mott, D. F., J. F. Besser, R. R. West, and J. W. DeGrazio, Bird damage to peanuts and methods for alleviating the problem. In *Proc. Vert. Pest Conf.*, 5, 118–120, 1972.

Nelms, C. O., M. L. Avery, and D. G. Decker, Assessment of bird damage to early-ripening blueberries in Florida. In *Proc. Vert. Pest Conf.*, 14, 302–306, 1990.

Patterson, I. J., S. Abdul Jalil, and M. L. East, Damage to winter cereals by greylag and pink-footed geese in northeast Scotland. *J. App. Ecol.*, 26, 879–895, 1989.

Poche, R. M., M. D. Mian, M. E. Haque, and P. Sultana, Rodent damage and burrowing characteristics in Bangladesh wheat fields. *J. Wildl. Manage.*, 46, 139–147, 1982.

Putman, R. J., Deer damage in coppice woodlands: an analysis of factors affecting the severity of damage and options for management. *Q. J. For.*, 88, 45–54, 1994.

Richards, N. A. and C. E. Farnsworth, Effect of cutting level on regeneration of northern hardwoods protected from deer. *J. For.*, 69, 230–233, 1971.

Romin, L., Factors associated with mule deer highway mortality at Jordanelle Reservoir, Utah. M.S. thesis, Utah State University, Logan, 1994.

Sauer, W. C., Impact of the Belding's ground squirrel, *Spermophilia beldingi*, on alfalfa production in northeastern California. In *Proc. Vert. Pest Conf.*, 11, 20–23, 1984.

Sekhar, N. U., Crop and livestock depredation caused by wild animals in protected areas: the case of Sariska Tiger Reserve, Rajasthan, India. *Environ. Conserv.*, 25, 160–171, 1998.

Singleton, G. R., L. A. Hinds, H. Leirs, and Z. Zhang, Eds., *Ecologically-Based Management of Rodent Pests*. Australian Center for International Agricultural Research, Canberra, Australia, 1999.

Singleton, G. R. and D. A. Petch, A review of the biology and management of rodent pests in Southeast Asia. *Technical Report 30,* Australian Center for International Agricultural Research, Canberra, Australia, 1994.

Sullivan, T. P., H. Coates, L. A. Jozsa, and P. K. Diggle, Influence of feeding damage by small mammals on tree growth and wood quality in young lodgepole pine. *Can. J. For. Res.*, 23, 799–809, 1993.

Sultana, P., J. E. Brooks, and R. M. Poché, Methods for assessing rat damage to growing wheat in Bangladesh, with examples of applications. In *Vert. Pest Control Manage. Mater. Symp.*, 4, 231–238, 1983.

Sultana, P. and M. M. Jaeger, Control strategies to reduce preharvest rat damage in Bangladesh. In *Proc. Vert. Pest Conf.*, 15, 261–267, 1992.

Summers, R. W., The effect on winter wheat of grazing by brant geese *Branta bernicla. J. App. Ecol.*, 27, 821–833, 1990.

Tchamba, M. N., History and present status of the human/elephant conflict in the Waza–Logone region, Cameroon, West Africa. *Biol. Conserv.*, 75, 35–41, 1996.

Townes, H. and J. Morales, Control of field rats in the Philippines with general reference to Kotabato. *Bull. Plant Ind. Dig.*, 16(12), 3–12, 1953.

U.S. Bureau of the Census, *Statistical Abstract of the United States. 111th ed.* U.S. Government Printing Office, Washington, D.C., 1992.

U.S. Office of Management and Budget, *Historical Tables, Budget of the United States Government, Fiscal Year 2001*. U.S. Government Printing Office, Washington, D.C., 2000.

Valencia, D. G., Rat control in coconut palms in Colombia. In *Proc. Vert. Pest Conf.*, 9, 110–113, 1980.

Wakeley, J. S. and R. C. Mitchell, Blackbird damage to ripening field corn in Pennsylvania. *Wildl. Soc. Bull.*, 9, 52–55, 1981.

Whisson, D. A., S. B. Orloff, and D. L. Lancaster, Alfalfa yield loss from Belding's ground squirrels in northeastern California. *Wildl. Soc. Bull.*, 27, 178–183, 1999.

Williams, K., I. Parer, B. Coman, J. Burley, and M. Braysher, *Managing Vertebrate Pests: Rabbits*. Bureau of Resource Sciences/CSIRO Division of Wildlife and Ecology, Australian Government Publishing Service, Canberra, Australia, 1995.

Wilson, C. J. and I. G. McKillop, An acoustic scaring device tested against European rabbits. *Wildl. Soc. Bull.*, 14, 409–411, 1986.

Woronecki, P. P., R. A. Stehn, and R. A. Dolbeer, Compensatory response of maturing corn kernels following simulated damage by birds. *J. App. Ecol.*, 17, 737–746, 1980.

Wywialowski, A. P., Wildlife damage to field corn in 1993. *Wildl. Soc. Bull.*, 24, 264–271, 1996.

Zhang, Z., A. Chen, Z. Ning, and X. Huang, Rodent pest management in agricultural ecosystems in China. P. 261–284 in G. R. Singleton, L. A. Hinds, H. Leirs, and Z. Zhang, Eds., *Ecologically-Based Management of Rodent Pests*. Australian Center for International Agricultural Research, Canberra, Australia, 1999.

CHAPTER 6

Environmental Damage and Exotic Species

"Intercontinental invasions of non-indigenous species are leading towards gradual homogenization of the earth's biota [and] a homogenized global biota would also be a substantially impoverished biota."

Cox (1999)

"The most common reason for a species to become extinct is biological pollution – the invasion of its habitat by an exotic species."

Anonymous

"Take heed the fate of flightless rails, the kiwi of the north, in nameless graves are moldering those, who numberless went forth. These tiny birds, by all beloved, despite peculiar habits, owe their demise to rodents, the productive race of rabbits."

Walker and Hudson (1945)

Some species have increased in abundance to where their current populations are much higher than they were historically. These population increases often are the result of human-induced environmental changes. There is no word or phrase to describe species whose current population exceeds historical levels due to human-caused environmental changes. Hence, I will refer to these species as being "anthropogenic abundant" or "AA." Many native birds we think of as "common" due to their current abundance, such as mourning doves, American robins, mockingbirds, cowbirds, and red-winged blackbirds, are AA species. These birds have benefitted from humans converting the vast forests of North America into farms, fields, pastures, and house lots. Populations of coyotes also became AA after humans decimated the populations of cougars and gray wolves which suppressed coyote numbers through competition and predation.

Often, AA species cause environmental changes. When these changes are not to society's liking, we call it environmental degradation or destruction. Because humans are nostalgic and do not like changes in the environment, most environmental changes are considered environmental destruction. For instance, many AA species have contributed to the decline of some endangered species through excessive predation, competition, or disease transmission (Goodrich and Buskirk 1995).

One example of an AA species in North America is the white-tailed deer. Within the last few decades, populations of this species have expanded greatly due to habitat changes caused by agriculture and silviculture, along with a discontinuance of over-harvesting of deer by humans. For some people, the recovery of white-tailed deer populations is one of the great accomplishments of modern game management, while others view it as one of its greatest disasters (Warren 1997). The high densities of white-tailed deer in many parts of the eastern U.S. have caused profound environmental changes. In those areas, deer have altered the species composition of forests. Palatable trees, such as aspen, sugar maple, and cherry, can no longer regenerate, owing to high rates of deer herbivory on their seedlings and saplings. This has lead to dominance of unpalatable plant species, such as American beech (Figure 6.1). Tree species which are both slow growing and palatable, such as eastern hemlock and northern white cedar, may be eliminated entirely from stands where they have existed for hundreds of years (Frelich and Lorimer 1985; Waller and Alverson 1997).

Figure 6.1 Deer browsing on tree seedlings have prevented some tree species from regenerating and eliminated understory vegetation in many forests. (Photo courtesy of Dennis Austin.)

Deer browsing also has a profound impact on understory vegetation and herbaceous plants because these plants never outgrow a deer's reach and are subject to repeated browsing (Warren 1991). Many rare plants, especially some lilies and orchids, cannot tolerate deer browsing. To save them, nature reserves now erect deer-proof fences around these plants (Miller et al. 1992). At even higher deer densities, seedlings and saplings of all tree species and understory vegetation can be eliminated by deer browsing, resulting in open, park-like forests where only ferns can survive in the understory (Waller and Alverson 1997). These vegetation changes alter ecosystem processes by changing nutrient cycling, quantity and quality of plant litter available to decomposers, and soil microenvironments (Augustine and McNaughton 1998).

Changes in understory vegetation impact the avian community in forests by causing a decline in bird species that depend on understory plants for nesting or foraging (deCalesta 1994; McShea 1997). Some birds, such as black-and-white warblers, black-throated green warblers, and wild turkeys, are uncommon in areas where understory vegetation has been stripped by browsing deer (Casey and Hein 1983). High deer densities also can suppress squirrel and rodent populations by competing with them for acorns (Ostfeld et al. 1996). In Massachusetts, areas with high deer densities had few forb species and more graminoid species than other areas. Because of these habitat changes, populations of southern red-backed voles and short-tailed shrews declined, while densities of white-footed mice increased (Brooks and Healy 1988). Thus, high deer densities can have a profound impact upon the avian and mammalian communities in deciduous forests (Warren 1991).

The field of wildlife damage management has the responsibility of trying to reduce or prevent environmental damage caused by wildlife. One way to prevent this damage is to rectify the original environmental changes that allowed the AA species to become so abundant. For example, Sargeant and Arnold (1984) noted that coyotes displaced foxes and suggested that the problem of excessive predation by red foxes on nesting birds could be reduced if coyote populations were allowed to increase in the Prairie Potholes. Competition between gulls and other colonially nesting birds for nesting sites could be reduced by closing human garbage dumps where gulls feed, thereby reducing gull populations. Dairy farms and livestock operations could be modified so that they provide less food for the brown-headed cowbird, an AA bird that parasitizes the nests of other bird species. Reducing cowbird populations has resulted in higher nesting success for many birds that are parasitized by it, including the endangered Kirtland's warbler (Sidebar 6.1).

Sidebar 6.1 Reducing Nest Parasitism by Brown-headed Cowbirds to Protect the Endangered Kirtland's Warbler

Kirtland's warblers nest in the pine–oak forests of Michigan. Their population had declined to a few individuals due to loss of suitable habitat and high rates of nest parasitism by brown-headed cowbirds, which lay their own eggs in the nests of Kirtland's warblers. The parents of parasitized nests are tricked into raising baby cowbirds rather than their own offspring. Brown-headed cowbirds are abundant along forest edges; thus, fragmentation of the pine–oak forests has made more Kirtland's warbler nests vulnerable to cowbird parasitism. From 1966 to 1971, 69% of all warbler nests

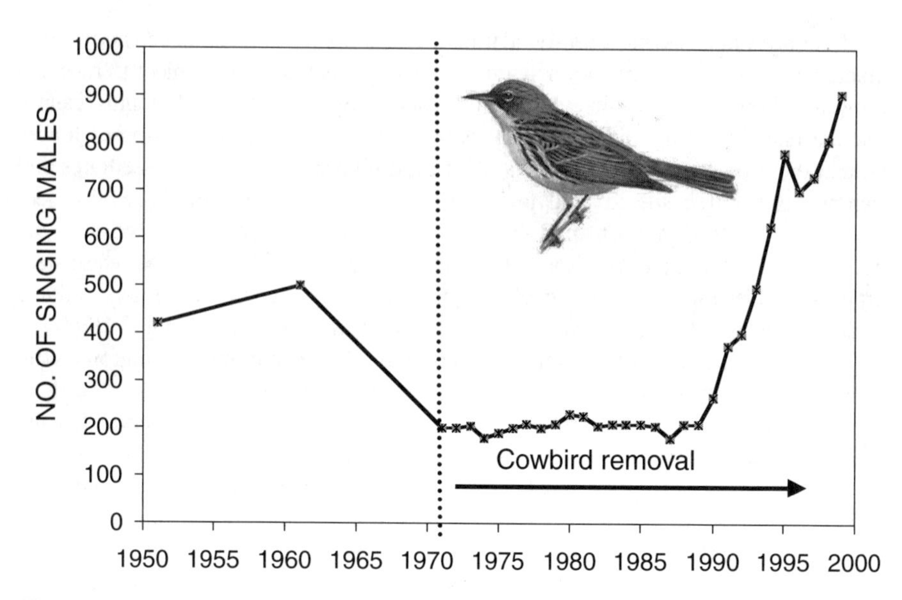

Figure 6.2 Changes in the population of Kirtland's warbler before and after brown-headed cowbirds were removed from the warbler's nesting area. (Figure courtesy of Richard Dolbeer.)

were parasitized. During this time, reproductive rates decreased to 0.8 fledged young per nest, and the total population of breeding males dropped to less than 200 individuals. To counter this, a trapping program was started in 1972 to remove cowbirds from the breeding area of the warbler (Kelly and DeCapita 1982). From 1972 to 2000, over 110,000 cowbirds were trapped. At the same time, controlled fires were used to produce large tracts of suitable warbler breeding habitat (Mayfield 1993). These combined efforts resulted in a fourfold increase in the Kirtland's warbler population by 1999 (Richard Dolbeer, personal communication). Although the trapping program had a great impact on the population of Kirtland's warbler (Figure 6.2), the annual loss of 4000 to 6000 cowbirds had no impact on the brown-headed cowbird population because approximately half of the 100 million cowbirds alive each spring die within a year due to natural causes (Richard Dolbeer, personal communication).

Many environmental changes caused by humans either simply cannot be reversed or the cost of doing so would be too high. For instance, our need for food and fiber will prevent us from allowing farms to reconvert to the climax forests that were originally there. In these cases, other approaches are needed to reduce the environmental harm caused by AA species. One such approach is to reduce populations of AA species when they threaten an endangered species or pose a danger to the environment. For example, some predator populations (e.g., coyotes, red foxes, striped skunks, American badgers, great horned owls, and American kestrels) have been controlled to protect endangered species, such as San Joaquin kit foxes, California least terns, black-footed ferrets, desert tortoises, peregrine falcons, and whooping cranes (Goodrich and Buskirk 1995).

Before the arrival of settlers, the northern Prairie Pothole region in North America used to be a safe area for waterfowl to nest because few mammalian predators could survive the long winters. However, things changed with the arrival of American settlers, who inadvertently provided food and shelter for predators during the winter and also eradicated wolves from the region. Hence, populations of skunks, foxes, and raccoons increased. Concomitantly, much of the prairie was converted to agricultural fields, forcing ducks to concentrate their nesting in the remaining patches of prairie. This made it easier for predators to find nesting ducks than when nests were spread across the entire landscape. Because of these changes, nesting success of ducks in the Prairie Potholes has declined due to high rates of predation. It is unlikely that mammalian predator populations will ever return to their low historic levels. Hence, scientists and biologists in the field of wildlife damage management are trying to develop and employ methods to increase the reproductive success of ducks in the presence of high densities of predators. Some of their techniques are discussed in Chapter 16.

Many of the most serious problems in the world in terms of environmental destruction and economic damage are not caused by abundant native species but rather by exotic species. The rest of this chapter will deal with exotic species — the problems they have caused, methods to stop their spread, and ways to eradicate or control their populations.

WHAT IS AN EXOTIC SPECIES?

Exotic, introduced, invasive, nonindigenous, alien, or nonnative species are terms used interchangeably to describe an animal living outside its normal range. They are even sometimes referred to as "biological pollution." Temple (1992) defined an exotic species as one that exists outside its natural range as a result of human activity. In contrast, a native or indigenous species is one that occurs within its "normal range" or the area that it occupied before the arrival of Europeans. Some people, however, argue that this definition should be changed to encompass the time before any humans arrived in the area, not just Europeans. For instance, the popular definition of a native species in Hawaii is one which existed there before the arrival of Captain James Cook in 1800, the first European explorer to visit Hawaii. However, the Hawaiian flora and fauna he observed had already been altered by Polynesians, who had earlier introduced several plants and animals into the Hawaiian Islands (Wilcove 1989; Temple 1992). Hence, the debate is whether the species the Polynesians brought with them should be considered exotic or native species of Hawaii.

One part of Temple's definition is "... with the help of man." In a few cases, a species arrived and occupied a new area following the arrival of Europeans but did so entirely by its own means and without help (either intentional or unintentional) from humans. These are considered native species, not exotic species. However, if a species expanded its geographic range in response to human-caused changes to the habitat, then the species would be considered an exotic species because its natural dispersion was made possible by unnatural events (Temple 1992). Examples would include mammals that expand their range northward in response to new sources of winter food provided by people.

The question becomes even more clouded when trying to define an exotic species in Europe, Asia, and Africa, where humans have existed for thousands of years. Often, there are no data to confirm whether or not a species was introduced by humans or was a native to the area (Carlton 1996).

IMPACTS OF EXOTIC SPECIES ON THE NATIVE BIOTA

Populations of exotic species often reach higher levels than in their native range because the diseases, parasites, competitors, or predators that limited their population in their native range may be absent in the new area they have colonized. For instance, starlings, which are an exotic species in North America, have become one of the most abundant birds on the continent. Exotic populations, especially when abundant, can cause several problems for native species. Exotic species may outcompete them for food or shelter. Normally, only one species can occupy the same niche at the same time; if the needs of the exotic species and the native species are similar enough, only one will survive. For instance, the establishment of gray squirrels into England caused a decline in the native red squirrel due to competition for food (Kenward and Holm 1989). Jones (1980) reported that the native Mauritius parakeet is threatened with extinction because the exotic rose-ringed parakeet can exclude it from nest cavities. In Australia, feral goats compete for food and water against two endangered native species: the yellow-footed rock wallaby and the brush-tailed rock wallaby (Bomford 1991).

The arrival of an exotic predator can have a great impact on the native fauna, especially on oceanic islands where native species evolved in the absence of predators. To illustrate this, over 90% of reptiles that have gone extinct since the 1600s are island species (Honegger 1981); 93% of avian extinctions (King 1980) and 81% of mammalian extinctions (Ceballos and Brown 1995) also occurred on islands. One reason why animals on islands that were free of predators are so vulnerable to exotic predators is that they lack defensive ability and wariness (Case and Bolger 1991). For instance, exotic rats have been responsible for the decline or extinction of birds from over 30 islands (Atkinson 1985), and exotic rats or cats have reduced seabird numbers by tens of millions (Olsen 1977; Moors and Atkinson 1984). The introduction of rats to predator-free islands caused more avian extinctions than their introduction to islands with native predators (Atkinson 1985). Likewise, the introduction of mammalian predators (e.g., rats, domestic cats, Indian mongooses, etc.) to predator-free islands caused a higher extirpation rate of endemic reptiles than when exotic predators reached islands which already contained predators (Case and Bolger 1991).

Another problem is that invasions of exotic species can cause habitat changes for which the endemic species are ill-suited. Introduced carp have been implicated in the decline of waterfowl in many wetlands. Not only do they directly compete with some waterfowl species for food, but their spawning and feeding activities also increase water turbidity, which decreases the populations of invertebrates used by many birds as food. Carp also alter wetland habitats by destroying aquatic vegetation used by birds for food, cover, and nesting. In just one year, carp caused an 80% decline in pondweed in an Oregon lake (Ivey et al. 1998).

Establishment of one exotic species can cause a cascade effect by creating conditions allowing other exotic species to successfully compete against the native species. An exotic predator may feed much more heavily on endemic animals, because they cannot defend themselves from predators as effectively as exotic animals that originated from areas with numerous predators (Case and Bolger 1991). Hence, the arrival of an exotic predator may give another exotic animal a strong competitive advantage over a native species that it might not have been able to outcompete otherwise. Similarly, exotic plants might initially have difficulty competing with native plants but introduction of an exotic herbivore could change this. Exotic herbivores often overgraze endemic plants which have not evolved mechanisms to protect themselves from herbivory. Absence of these endemic plants creates openings in which exotic plants can flourish.

On Hawaii, feral pigs had a hard time getting established due to a lack of protein in their diet. This changed after another exotic species was introduced — earthworms. After that, earthworms provided the protein needed by pigs and became a large part of the pigs' diet. This allowed pig populations to soar. The high pig densities, in turn, caused some native plants, such as ferns, to decline and exotic plants to increase in density (Stone and Loope 1987).

Another problem with the introduction of exotic species is that they might interbreed with genetically similar native species, causing a loss of genetic integrity in the native population. For example, elk hybridized with red deer in New Zealand, and sika deer and red deer interbred in Europe (Demarais et al. 1990). Native Iberian waterfrogs hybridized with introduced frogs (Arano et al. 1995). Mallards introduced in New Zealand, Hawaii, Australia, and Florida threatened the integrity of indigenous duck populations by interbreeding with them (Simberloff 1996).

Exotic species can also spread exotic diseases (O'Brien 1989). In Queensland, Australia, a rapid decline in frog populations may be attributed to a virus introduced by pet fish (McNeely 2000). In another example, harp seals carrying the distemper virus found their way from the Arctic Ocean to the North Sea. Seals native to the North Sea were not resistant to this virus and their populations crashed due to distemper, but later were able to recover (Boye 1996 as cited by Gebhardt 1996).

On a global basis, exotic species have had a profound impact. They have been the major cause for 42% of the extinctions in reptiles, 25% in fish, 22% in birds, and 20% in mammals (Cox 1999). To investigate exotic species in more detail, I want to examine the consequences of exotic species on native biota in different parts of the world. Exotic species that have caused great change include diseases, plants, insects, and vertebrates. However, we will limit our discussion to vertebrates.

HAWAII

The flora and fauna of the Hawaiian Islands evolved over a period of 40 million years. Due to their isolated location in the middle of the Pacific Ocean, a new species colonized them every 50,000 to 100,000 years on average prior to the arrival of humans. Hence, their 100 endemic land birds evolved from as few as 20 colonizing species and the islands' native flora of 1000 to 2000 species evolved from about

300 colonizing species (Stone and Loope 1987). Hawaii's native avifauna evolved in the absence of mammalian predators, humans, and most avian diseases. Hence, it was ill-prepared for the arrival of the Polynesians 1400 years ago along with their fellow travelers: rats, dogs, pigs, and jungle fowl. The Polynesians also introduced 32 plant species they used for food or fiber (Stone and Loope 1987). The end result was that at least 38 of Hawaii's native bird species including 18 species of honey creepers, 8 geese, 7 rails, 3 owls and 2 flightless ibises became extinct soon after the arrival of Polynesians (Wilcove 1989).

When Captain Cook first visited Hawaii in 1778, there were still at least 50 native land bird species remaining. However, a second wave of exterminations began after European and American settlers introduced livestock and other mammals. Even more devastating was that some of the introduced birds were infected with malaria and avian pox. Hawaii's endemic birds lacked resistance to these diseases, and most birds living below the elevation of 1000 m were eradicated. Birds above this elevation were spared because the exotic mosquito that served as the vector for these avian diseases could not survive at such high elevations (van Riper et al. 1982). The result was the loss of 13 more native species. Consequently, more than two thirds of Hawaii's native birds have become extinct since the arrival of humans (Wilcove 1989). Hawaii's avian community has not been silenced, but native birds have been replaced by over 45 exotic bird species from across the world, which have become established throughout Hawaii (Eldredge 1992). These exotic birds have been joined in Hawaii by 19 exotic fish species, 18 exotic mammal species, and 17 exotic reptile and amphibian species (Cox 1999).

GALAPAGOS ISLANDS

The native fauna of the Galapagos Islands was made famous by the writings of Charles Darwin and is composed mainly of endemic reptiles (e.g., iguanas and the giant tortoise) and birds (e.g., Darwin's finches). These native species, however, are threatened by exotic species, including feral cattle, donkeys, horses, goats, and pigs, introduced during the 1800s by the first settlers. Many native plants disappeared due to overgrazing by exotic herbivores, which also prompted the spread of exotic plants (Hamann 1979). Feral pigs preyed upon the eggs of native reptiles, including the giant tortoise and green turtle. Introduced black rats preyed upon dark-rumped petrels and baby tortoises, and caused the extermination of four endemic rice rat species. Feral dogs ate iguanas, fur seals, boobies, and penguins (Barnett 1986; Brockie et al. 1988).

In response, campaigns have been conducted to eradicate these exotic animals from some of the Galapagos Islands. Feral goats were eliminated from several islands by the Ecuador National Park Service. Feral cattle were removed from San Cristobal, Santa Cruz, and Santa Maria islands. Feral dogs were eliminated on San Cristobal and Santa Maria islands by shooting and poisoning. Exotic rat populations have also been reduced on a couple of the smaller islands (Brockie et al. 1988).

GUAM

Brown tree snakes accidentally reached the Pacific island of Guam, presumably sometime after World War II, when vehicles and equipment harboring some of those snakes were brought in from New Guinea (Figure 6.3; Rodda et al. 1997). Due to the inability of amphibians and reptiles to cross large stretches of open ocean, Guam's native fauna evolved in the absence of snakes and thus was unable to adapt to predation by the brown tree snake. Consequently, 10 of 13 species of Guam's native forest birds, 2 of its 3 native mammals, and 6 of 12 native lizard species were extirpated (Savidge 1987; Case and Bolger 1991; Fritts and Rodda 1998). Furthermore, all three remaining native birds and the one remaining native mammal are endangered. Shortly after the introduction of brown tree snakes, other species were introduced: musk shrews, green anoles, black drongos, and curious skinks. Brown tree snakes prey upon these exotic species and without them may not have been able to reach the high densities that lead to such a dramatic loss of biodiversity (Fritts and Rodda 1998).

Figure 6.3 The brown tree snake. (Photo courtesy of Joe Caudell.)

AUSTRALIA

Twenty-seven bird species have established exotic populations in Australia, along with two exotic reptiles and one amphibian (Bomford 1991). Mammals, however, have been the most successful in invading Australia and devastating the native biota. This is not surprising because all of the native mammals are marsupials, allowing exotic mammals to find open niches and outcompete native species. Twenty-six species of exotic mammals have become established in Australia since its colonization by Europeans. These include 12 species of domestic animals that established feral populations, 9 species purposely released for hunting, 4 com-

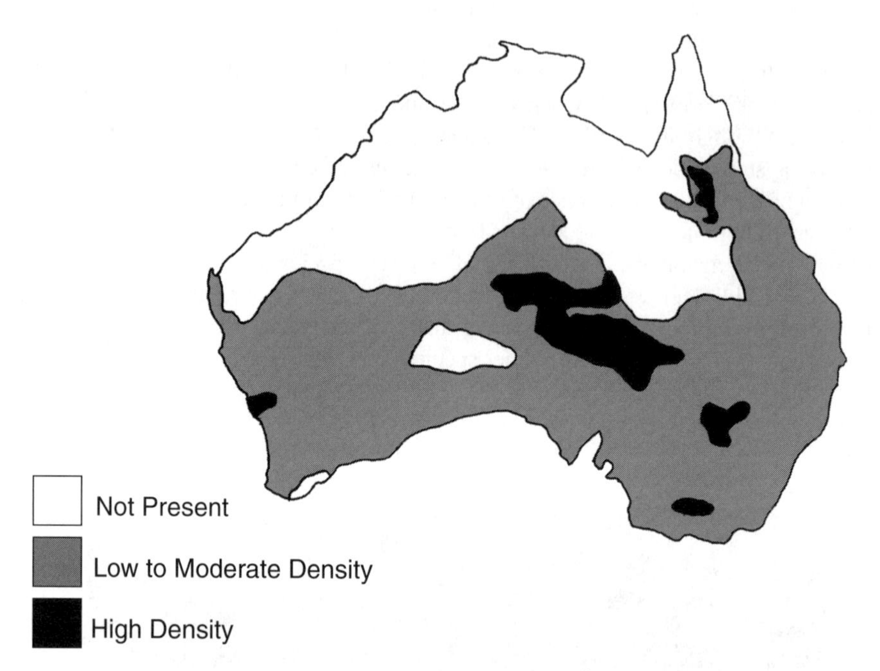

Figure 6.4 Distribution of European rabbits in Australia. (Adapted from Wilson et al. 1993.)

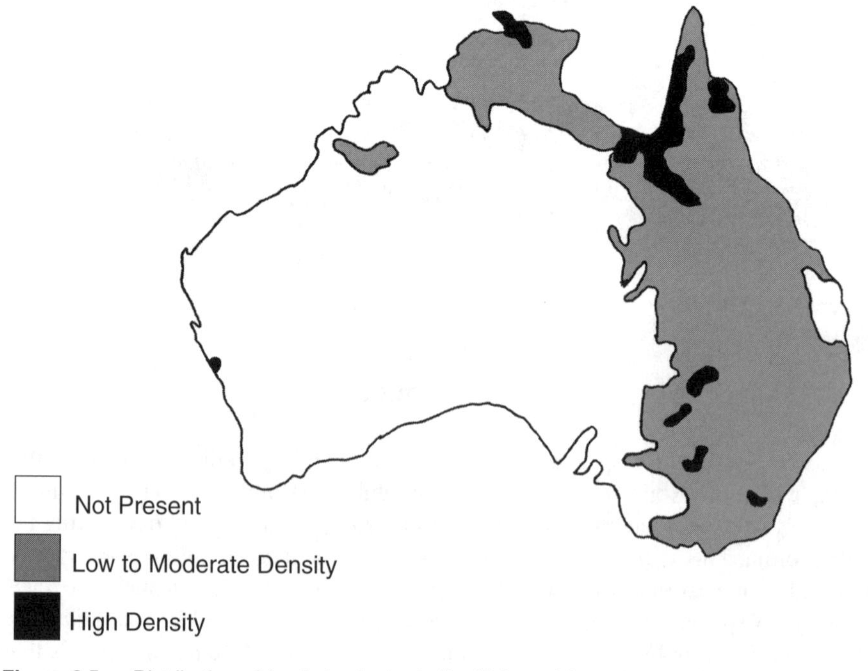

Figure 6.5 Distribution of feral pigs in Australia. (Adapted from Wilson et al. 1993.)

mensal species accidently introduced, and 1 species that escaped from a zoo (O'Brien 1993). Some of these exotic mammal populations have exploded in Australia. There are now millions of European rabbits, feral pigs, and foxes and thousands of feral goats, horses, donkeys, camels, and water buffalo on the continent (Figures 6.4 and 6.5). The introduction of so many exotic species has devastated Australia's native mammals. Over half of the world's mammals that became extinct during the last 200 years lived in Australia. Many other marsupials in Australia are currently endangered.

The first exotic predator brought to Australia by humans about 4000 years ago was the dingo. Two native predators, the Tasmanian devil and the thylacine, were eradicated from mainland Australia because they could not compete successfully with dingoes. Other mammalian predators arrived on the continent with European settlers, including red foxes, feral cats, and feral dogs. These exotic predators have had a devastating effect on Australia's native birds and mammals, which have not had time to evolve a suitable defense against them. It has proven impossible to eradicate these exotic predators but, in some areas, their numbers have been controlled. When this has happened, populations of some native species have rebounded.

NORTH AMERICA

In the U.S. and its territories, at least 75 free-ranging populations of exotic birds have become established since the arrival of European colonists (Figure 6.6; Temple 1990). Concomitantly, many native birds have expanded their range or moved to new locations due to environmental changes brought by humans. For instance, the Great Plains were once an effective barrier to the dispersal of woodland bird species. However, humans planted trees throughout the region, making it possible for many forest-adapted species to survive on the Great Plains (Knopf 1992). As an example, 90% of the birds which now breed in eastern Colorado were not there a century ago (Knopf 1986).

In California, feral horses and burros have existed at least since the California Gold Rush in the 1800s. In arid regions, these herbivores cause numerous environmental problems by competing with native herbivores, overgrazing the vegetation, and disturbing water sources (Weaver 1974; Carothers et al. 1977). Efforts to reduce feral horse and burro populations have been controversial because many people consider them to be native to the Southwest and symbols of the region's colorful history. Feral pigs have also spread throughout much of California, their numbers increasing to where they are the state's most popular big-game species (Howard and Marsh 1986). Unfortunately, feral pigs cause considerable environmental damage in areas where hunting is not intense enough to control their numbers. Feral goats and European rabbits have been released numerous times in California but have established populations only on islands where predators are absent. Efforts to eradicate exotic species from some of these islands in recent decades have been successful (Howard and Marsh 1986).

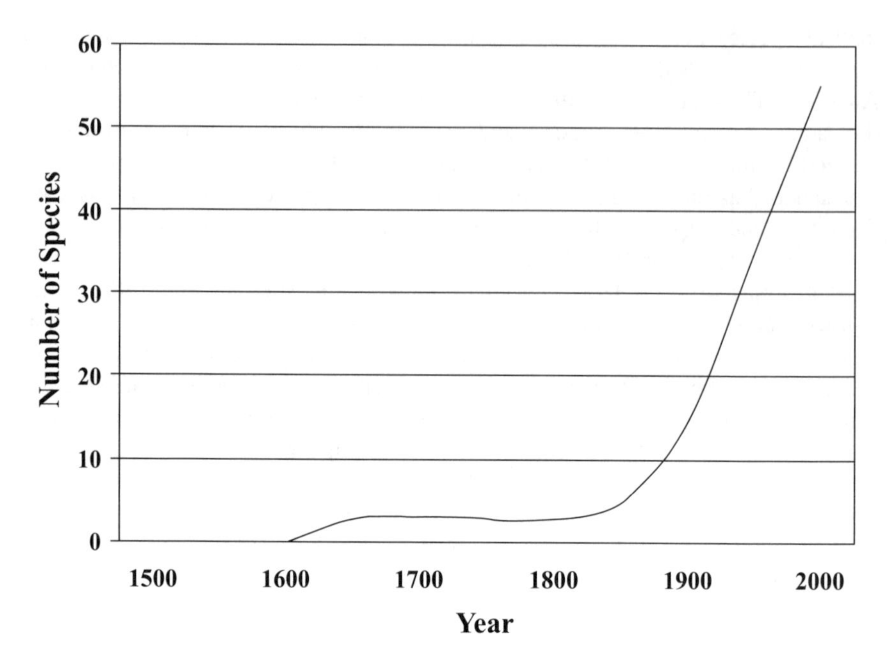

Figure 6.6 Cumulative number of exotic bird species which have established a free-ranging population in the U.S., based on the 55 species for which we know the date of their importation. (Adapted from Temple 1992.)

In south Florida, over 2 million people live on a slightly elevated strip of land previously covered with a continuous stand of pine trees. South Florida residents have replaced the native forest community with hundreds of colorful exotic plant species gathered from across the tropics. It is estimated that over 1000 exotic insect species and 900 exotic plant species have established free-ranging populations in Florida, and 25,000 plant species are cultivated in Florida, but have not yet escaped (Frank and McCoy 1995a, b). In turn, these new plant and insect communities created a favorable habitat and open niches for tropical birds, allowing several exotic species, including many parrots and parakeets, to become established in south Florida (Swain 1988).

RESOLVING ENVIRONMENTAL PROBLEMS CAUSED BY EXOTIC ANIMALS

Exotic species pose the greatest threat to biodiversity and natural ecosystem functions (Baskin 1996; Vitousek et al. 1996; Wilcove et al. 1998). In 1992, the United Nation's Convention on Biological Diversity called on participating nations to "as far as possible and as appropriate … prevent the introduction of, control or eradicate those alien species which threaten ecosystems, habitats or species" (Baskin 1996). However, many people rely on nonnative species for their livelihood, and so efforts to stop further invasions will need to include public education along with ways to evaluate economic and environmental trade-offs (Baskin 1996).

Preventing Exotic Animals from Reaching Foreign Shores

The simplest line of defense against exotic species is to prevent their spread to new areas. Most countries quarantine all animals, including livestock and pets, which are being brought into them, to make sure that they are free of disease. These quarantines have prevented the spread of innumerable plant and animal diseases into new areas. Many countries also ban the importation of live plants and animals without governmental approval. The problem for most countries is that people want to import a wide range of exotic animals, especially aquatic fish or pets. Deciding which exotic species can be safely brought into a country is difficult but should be based on the probability of escapees being able to establish a free-ranging population and the consequences if they do so (Table 6.1).

One of the most intensive efforts to curtail the spread of an exotic species is the program to prevent the dispersal of brown tree snakes from Guam to other Pacific islands. As part of this program, U.S.D.A. Wildlife Services employees control brown tree snakes near Guam airports and harbors. They also check all outbound planes and ships using dogs trained to detect snakes by smell to reduce the chances of a snake slipping out unnoticed. So far these efforts have been successful.

Preventing Exotic Animals from Establishing a Free-Ranging Population

Once exotic animals have reached a new shore, the best chance to eradicate them is before they can become established and spread (Usher 1989; Williams and Moore 1989). Often, exotic animals initially do poorly in their new environment and may die out. Alternatively, this period may be followed by one of explosive population growth (Simberloff 1969; Caughley 1977; Bomford 1991; Baskin 1996). Unfortu-

Table 6.1 A Risk-Assessment Model to Decide Whether an Application to Import an Exotic Species Should Be Approved (Bomford 1991)

Question	Answer	Decision
1. What is the probability of the species providing a substantial benefit for society?	Low	Reject
	Unknown	Collect necessary data
	High	Consider next question
2. What is the probability of an escaped animal causing substantial harm?	High	Reject
	Unknown	Collect necessary data
	Low	Consider next question
3. What is the probability of escaped animals establishing a free-ranging population?	High	Reject
	Unknown	Collect necessary data
	Low	Consider next question
4. What is the probability that a free-ranging population can be eradicated?	Low	Reject
	Unknown	Collect necessary data
	High	Consider next question
5. What is the probability of a free-ranging population having a major adverse impact?	High	Reject
	Unknown	Collect necessary data
	Low	Grant importation request

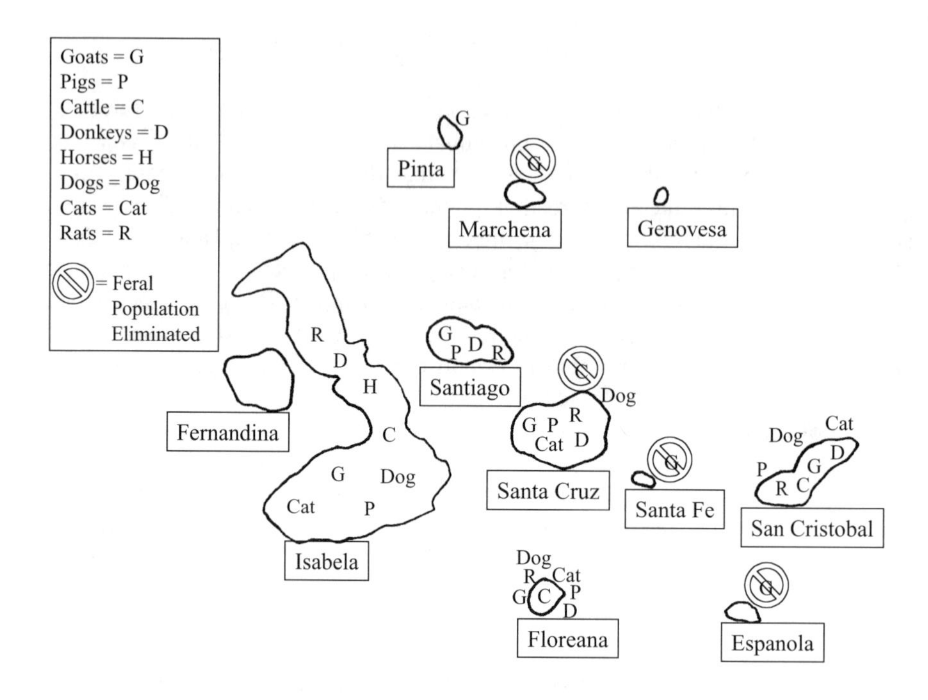

Figure 6.7 Distribution of exotic mammals on the Galapagos Islands. (Adapted from Barnett 1986.)

nately, it is hard to predict which outcome will occur once this initial period of adjustment and adaptation is over. Hence, we should err on the side of caution and not assume that a small exotic population is destined to die out on its own. It should also be noted that, when eradication programs have been successful, the native species often were able to recover from the deleterious effects of the exotic species (Newmann 1994; Cooper et al. 1995).

Efforts to eradicate exotic species generally are more successful against large and conspicuous animals (e.g., horses) than against small, secretive animals (e.g., mice). Feral populations of cattle, dogs, and goats have been exterminated on several Galapagos Islands, while exotic rats have proven harder to eliminate (Figure 6.7; Brockie et al. 1988). Off the coast of New Zealand, feral goats have been eradicated from 11 islands, feral cats from 5, sheep and cattle from 3, rabbits from 3, pigs and rats from 2, and stoats from 1 (Veitch 1985; Brockie et al. 1988; Clout and Lowe 2000).

The size of the island is also an important factor in the success of an eradication program. Efforts to eradicate exotic species on small Galapagos Islands have been more successful than on large ones. However, on large islands, progress can be achieved by dividing the island into manageable units using wildlife-proof fences. The advantage of this approach is that once a fenced unit has been cleared of an exotic animal, the fence can keep the area from being repopulated by immigrants from other parts of the island.

While eradication may seem to be the best option at first glance, it is important to recognize that exotic predators may actually prey upon populations of other introduced species. This could be the case with cats and rats introduced onto islands. Feral cats prey on island bird species, but also feed upon introduced rats. Rats directly affect birds through predation and also compete with them for shelter and food. If cats were eradicated, the rat population may explode, causing more severe impacts on island bird species. On the other hand, if rats were eradicated, the cats may start eating more native birds. Therefore, eradication of cats may not be the best way to protect endemic prey when another predator such as the rat is present (Courchamp et al. 1999).

Efforts to eradicate or control exotic species can be controversial. Some people object due to concerns about animal welfare, while others object because they like the exotic animals or feel that they belong there (Figure 6.8; Temple 1990). Attempts to control feral pig populations in the U.S. have been thwarted by people who like to hunt or see them (Howard and Marsh 1986; Peine and Farmer 1990). In Australia, feral pigs cause $80 million annually in agricultural damage, but they are also harvested commercially for their meat. About 150,000 feral pigs were exported annually during

Figure 6.8 Mute swans, an exotic species in North America, produce mixed emotions in people: some people love to watch them (A), whereas others worry that they threaten native waterfowl (B) due to their extreme aggressiveness. (Photos courtesy of Peter Picone.)

the 1980s, providing $15 million per year to the Australian economy (O'Brien 1993). Obviously, an Australian's perception of feral pigs will depend upon whether he or she is benefitting from them or being harmed by them. Hence, any effort to eradicate an exotic species needs to have public support and should be preceded by a program to inform the public about the dangers posed by the exotic species.

Controlling Populations of Exotic Animals

The advantage of eradicating an exotic population is that it is a permanent solution to the problem and only has to be successful once. On the other hand, efforts to control an exotic species have to be repeated constantly or the exotic population will rebound as soon as those efforts stop (Rainbolt and Coblentz 1997). Although eradication may be the preferred alternative, it may no longer be an option by the time an exotic population has started to expand. In these situations, it may be possible to reduce the deleterious effects of the exotic species on the local environment by controlling its population. Damage by feral goats in Hawaii's Haleakala National Park was reduced by confining these animals into small areas with 53 km of mammal-proof fencing. Once confined, they were controlled by shooting (Brockie et al. 1988). The program was successful in reducing goat numbers from about 15,000 to 1000, but the remaining goats learned to avoid hunters. "Judas goats" fitted with radio transmitters and helicopters are being used to locate the remaining goats for removal (Stone and Loope 1987). Judas goats were also used to eradicate feral goats from San Clemente Island, CA (Keegan et al. 1994). In parts of Hawaii, populations of black rats, Polynesian rats, and Norway rats were reduced by distributing poisoned baits, but rat populations simply returned to pretreatment levels as soon as the baiting programs ended (Stone and Loope 1987).

Efforts to protect the endangered Hawaiian goose from exotic predators have involved removing predators from sites where the geese were being raised in captivity or where the geese were released to the wild. These efforts reduced predation on geese, but did not stop it entirely (Stone and Loope 1987). The nesting success of another Hawaiian bird species, the dark-rumped petrel, was doubled after exotic predators were trapped from nesting areas (Simons 1983 as cited by Brockie et al. 1988).

Exotic species often reach densities much higher than in their native ranges, where their densities may be constrained by diseases, parasites, competitors, or predators. For this reason, efforts to control populations of exotic species often involve identifying the limiting factors in the species' native range and creating the same limiting factors in their new range. This is called biocontrol and involves introducing other exotic organisms (such as diseases or parasites) to control an exotic population (Freeland 1990). Both myxomatosis and rabbit calicivirus have been introduced into Australia to control rabbit numbers (Sidebar 6.2).

Sidebar 6.2 Using Myxomatosis to Control Exotic Rabbits in Australia

In the 1800s, colonists introduced European rabbits to Australia. By 1920, they had spread throughout the entire continent, causing an economic and environmental catastrophe. Rabbits reduced livestock productivity by competing with them for forage, caused soil erosion, and threatened the survival of many native plants and animals. For such a widespread problem involving the entire continent, site-specific methods of control would be ineffective. Hence, the rabbit virus, myxomatosis, was introduced into Australia in the 1950s. Initially, the disease had a profound impact on the rabbit population: up to 99% of the rabbits that contracted the virus in the first few years died. However, some individuals had an innate resistance to the disease and, because these individuals were the only survivors, the resistance quickly spread through the population. By 1990, even a highly virulent strain of the virus was killing only 50% of the rabbits, and rabbit populations had started to recover. Still, the rabbit populations in the 1980s and 1990s were only 25% of their size prior to the release of myxomatosis (Williams et al. 1995). In 1995, rabbit calicivirus was accidently released into Australia. It has reduced some rabbit populations by 90%, while having little impact on others.

Some attempts to use biocontrol have backfired and the introduced exotic organism intended to control another became a significant pest itself. For instance, the mongoose was introduced into Puerto Rico to control exotic rats. Unfortunately, it not only failed to control the rat population, but also became a major pest itself due to its predation on native birds and reptiles (Pimentel 1955). Introduced mongooses have also led to the extinction of three reptile species on the Caribbean island of St. Lucia. Likewise, mongooses were introduced to Hawaii in the 1880s to control rats in sugarcane fields. As in Puerto Rico, that effort failed to reduce rat densities, and the mongooses became established in the Hawaiian islands, where they pose a threat to native reptiles and ground-nesting birds. Cane toads introduced into Australia to control sugarcane beetles are another example of biocontrol gone awry (Freeland 1986; Tyler 1994).

Because of the danger of creating new environmental problems, parasites or diseases should only be introduced if it is known that the biocontrol agent is species-specific and poses no danger of infecting native flora or fauna. The deliberate introduction of an exotic predator, disease, or parasite to control an exotic species should only be taken with great caution. In one innovative approach to eradication, red foxes were sterilized and then introduced to some Alaskan islands where exotic arctic foxes were threatening native wildlife. The red foxes outcompeted the arctic foxes and excluded them from the islands. Because the sterilized red foxes could not reproduce, they died out within five years and left the islands free of predators (Bailey 1992).

CAN WE PREDICT WHEN AN EXOTIC SPECIES WILL CAUSE ENVIRONMENTAL DAMAGE?

Usually, when an exotic animal escapes into a new area, it does not survive. Williamson (1996) calls this the "tens rule." That is, roughly 10% of escaped animals will become established and 10% of those that become established will become a pest. Thus, most invasions can be ignored and nothing will result. However, by the time a population has become established or is causing environmental damage, it usually is too large to eradicate. What is needed is some way to predict 1. which animals are likely to invade, 2. when an invasive animal is likely to establish a free-ranging population, and 3. when a free-ranging population of an exotic species is likely to cause environmental damage. These topics are discussed next.

Which Animals Are Likely to Invade?

As humans have expanded across the globe, three groups of species have been able to take advantage of our mobility and have moved across the earth with us (Lockwood 1999). One group includes those plants and animals that are beneficial to humans and that we have deliberately taken with us when settling a new area. Fruits, vegetables, flowers, ornamental plants, and domestic animals have been transported across the world. Animals are also imported for recreation (e.g., pets), education (e.g., zoo animals), and medical research. Although our intention may have been to keep these animals in captivity, many were able to escape or were deliberately released (e.g., many people release aquarium fish into streams and lakes when they get tired of taking care of them).

A second group of exotic species includes those which were imported into a country with the intention of releasing them into the wild to establish free-ranging populations in their new surroundings. Plants and animals were often introduced into a new area by settlers who were homesick and longed to see the animals of their homeland. Others were deliberately released to provide food, fiber, or economic gains. For instance, sailors in the 1700s and 1800s often released livestock on oceanic islands in the hope that the animals would proliferate and provide a future food supply for themselves or other sailors.

The extent of these deliberate introductions of "beneficial animals" becomes more evident when one realizes that the European rabbit has been introduced to at least 598 islands (Flux and Fullagar 1983). Many game animals have been deliberately introduced into new areas either for aesthetic reasons (people enjoyed watching them) or to provide hunting opportunities. For instance, there were over 120,000 exotic big-game animals from over 94 different species or subspecies on Texas ranches during 1985 (Payne et al. 1987). Over 162 bird species have been released in Hawaii, 133 in New Zealand, and 56 in Tahiti (Eldredge 1992). Over 60% of the exotic birds in the U.S. were introduced on purpose (Temple 1992).

The third group of exotic species comprises those that people inadvertently move from place to place. These include diseases and parasites, which infected the people who traveled around the world and the plants and animals that went with them. They also include small, secretive animals (mice, rats, snakes, insects, etc.) that humans unknowingly transported with them. As one example of the extent of these inadvertent releases, rats have reached 82% of the world's islands (Atkinson 1985).

When Is an Invasive Animal Likely to Establish a Free-Ranging Population?

Being able to cross an immigration barrier is only one part of becoming a successful exotic species. The invasive animals must be able to survive and reproduce in their new environment and to establish a free-ranging population. Of 162 avian species introduced into Hawaii, only 45 were able to establish themselves, while 117 failed to establish a free-ranging population (Eldredge 1992).

Williamson and Brown (1986), Ehrlich (1989), and Bomford (1991) noted several characteristics of species that increase their probability of establishing a free-ranging population after they reach a new area. The most successful are those species which 1. originate from a similar area and climate as the new site, 2. have a large original range and are abundant within it, 3. successfully colonize elsewhere, 4. reach sexual maturity at an early age and have a high reproductive rate, 5. have a generalist diet, 6. are able to thrive in human-modified environments, 7. are nonmigratory, and 8. are social and live in flocks or herds. The probability of successful establishment also increases with the number of individuals that were released or escaped at the same time and place (Green 1997).

Other characteristics determine how hard it will be for an exotic population to be eradicated. Species with high intrinsic growth rates and the propensity to disperse over large distances will be much harder to eradicate than others. Small, secretive, or wary animals will be harder to eradicate than those that lack these traits. In addition, animals preferring remote or inaccessible habitats will be difficult to eradicate, as will those whose habitat preference is for thick cover (Bomford 1991).

When Is a Free-Ranging Exotic Population Likely to Cause Environmental Damage?

Not all exotic populations have a noticeable impact on their new environment. For instance, only about 10% of all exotic species in Britain have become pests (Williamson and Brown 1986). Lodge (1993) reported that, worldwide, most exotic populations have no detectable effects on the native biota. Whether an exotic species has a large impact depends upon the characteristics of both the exotic species and the community invaded. Predators, especially those that are arboreal mammals, have an unusually high probability of causing environmental damage. As noted previously, oceanic islands also have a high probability of being injured by an exotic species because their biota may have low diversity and may have evolved in the absence of mammals or predators (Lodge 1993).

WHICH SITES ARE VULNERABLE TO EXOTIC SPECIES?

Newsome and Noble (1986) found that habitat characteristics and environmental factors were often more important than species characteristics in determining whether an invasive species will become established. They reported that an invasion was more likely to succeed if it occurred in a habitat similar to the invasive species' native habitat or in a human-altered habitat. In general, an invasive animal is more likely to survive when it invades an isolated site, such as an oceanic island, than when it invades a site on a large continent. One reason for this is because remote, isolated sites are more likely to contain open niches that an invasive species can occupy (Cox 1999). It is harder for an exotic species to invade a large continent where there already is a large number of native species. For example, only 24 of the more than 600 vertebrate species that live in Chile are exotic species (Jaksic 1998). South Africa also has been resistant to exotic species because the continent already contains a large number of native animals. In contrast, the native biota of New Zealand evolved under isolated conditions and lacked mammalian predators, which made it vulnerable to exotic species (Ehrlich 1989). Furthermore, an exotic species is more likely to have a deleterious impact on the native flora and fauna of an oceanic island than on those of a continental site (Brockie et al. 1988). For example, 93% of all birds that have become extinct in the last few centuries have been endemic species on islands (King 1980).

Human endeavors can create unique habitats that local species may not be able to occupy or exploit because these species have not had time to adapt to the new conditions. However, some exotic species are able to occupy these open niches because they are preadapted to them. On Pacific islands, exotic lizards are abundant in agricultural habitats and in cities (which are newly created habitats), but few can successfully compete with the native lizards in the undisturbed forests (Case and Bolger 1991). Cities, with their cement cliffs, exotic landscaping plants, grass lawns, and humans to feed them, provide exotic lizards the means to flourish. Similarly, pigeons, starlings, house sparrows, house mice, and rats are able to occupy these converted habitats and thus to spread across the world. In Melbourne, Australia, exotic birds outnumber native birds by 2:1 (Green 1984). In Tucson, AZ, 95% of the birds are exotics (Emlen 1974; Knopf 1992).

DEVELOPING AN INTEGRATED PROGRAM TO STOP THE SPREAD OF EXOTICS

Brockie et al. (1988) recommended an eight-point program to protect endemic species on islands from exotic species:

1. The native flora and fauna need to be inventoried. Surprisingly, species lists have not been compiled for many islands. Furthermore, our efforts to catalog species often are concentrated on vertebrates, and less effort is spent surveying plants, insects, and other invertebrates.
2. Because we lack the resources to protect everything, islands with relatively intact native flora and fauna should be given the highest priority for preservation.
3. The conversion of native habitats to agriculture or other uses should be slowed because converted habitats are more vulnerable to invasions of exotic species.
4. Quarantines should be required before any plant or animal is imported onto islands.
5. Hunting and collecting of native species need to be regulated.

6. Predator-free islands need to be identified and used to harbor species that are endangered in their own home range. This should be done, however, only after ascertaining that our purposeful introduction of endangered animals will not create problems for the island's own native flora and fauna.
7. Exotic animals and plants need to be eradicated or controlled.
8. Research needs to be conducted to determine the impact exotic species have on the native biota. Furthermore, the success of eradication or control programs should be monitored.

Wildlife biologists can predict broad trends involving exotic species and their impacts, but they cannot predict with confidence the outcome of a single invasive threat or what environmental consequences might result if an exotic species becomes established. To gain more accuracy, Lodge (1993) argued that every potential invasion needs to be studied. However, this is impossible, given the thousands of introductions that occur each year. These studies also would require months or years to complete and many exotic species can make good use of this time to become established and to increase in number. Until we can sort out which of the thousands of invasions pose a threat and require our intervention and which can be safely ignored, exotic species will remain a serious environmental threat.

SUMMARY

A native species is one that currently occupies the same area where it existed before the arrival of humans. Conversely, an exotic species is one that exists outside its natural range as a result of human activity. Populations of exotic species often reach higher levels than in their native range because the diseases, parasites, competitors, or predators limiting their population in their native range may be absent in the new area that they colonize. In their worldwide travels, people have, either deliberately or accidently, brought many species into new areas. Exotic species often cause environmental changes by threatening the survival of native plants and animals in their new surroundings. This is especially true for isolated areas, such as oceanic islands, where many endemic species have become extinct after the arrival of an exotic predator or herbivore.

Countries usually have laws governing the importation of plants and animals that may become exotic species and may require a long quarantine to make sure imported animals are free of diseases or parasites. Despite these precautions, many exotic populations have become established throughout the world. When the potential threat posed by an exotic species is detected before the population has had time to spread, it may be possible to eradicate the exotic species. Normally, there is little demand to remove an exotic species until it begins to cause damage but, by that point, the exotic population is so large and dispersed that eradication is impossible. What is needed is a way to identify which species are likely to invade, which invaders are likely to establish free-ranging populations, and which free-ranging populations are likely to harm the native biota, so that we can act early.

LITERATURE CITED

Arano, B., G. Llorente, M. Garcia-Paris, and P. Herrero, Species translocation menaces Iberian waterfrogs. *Conserv. Biol.*, 9, 196–198, 1995.

Atkinson, I. A. E., The spread of commensal species of *Rattus* to oceanic islands and their effects on island avifauna. P. 35–83 in P. J. Moors, Ed., *Conservation of Island Birds: Case Studies for the Management of Threatened Island Species*. International Council for Bird Preservation, Cambridge, England, 1985.

Augustine, D. J. and S. J. McNaughton, Ungulate effects on the functional species composition of plant communities: herbivore selectivity and plant tolerance. *J. Wildl. Manage.*, 62, 1165–1183, 1998.

Bailey, G. P., Red foxes, *Vulpes vulpes*, as a biocontrol agent for introduced arctic fox, *Alopex lagopus*, on Alaskan islands. *Can. Field-Naturalist,* 106, 200–205, 1992.

Barnett, B. D., Eradication and control of feral and free-ranging dogs in the Galapagos Islands. In *Proc. Vert. Pest Conf.*, 12, 358–368, 1986.

Baskin, Y., Curbing undesirable invaders. *Biosci.*, 46, 732–736, 1996.

Bomford, M., Importing and keeping exotic vertebrates in Australia. *Department of Primary Industries and Energy, Bureau of Rural Resources Bulletin 12*, Australian Government Publishing Service, Canberra, Australia, 1991.

Brockie, R. E., L. L. Loope, M. B. Usher, and O. Hamann, Biological invasions of island nature reserves. *Biol. Conserv.*, 44, 9–36, 1988.

Brooks, R. T. and W. M. Healy, Response of small mammal communities to silvicultural treatments in eastern hardwood forests of West Virginia and Massachusetts. P. 313–318 in *Management of Amphibians, Reptiles, and Small Mammals in North America*. U.S. Forest Service, General Technical Report RM–166, 1988.

Carlton, J. T., Biological invasions and cryptogenic species. *Ecology*, 77, 1653–1655, 1996.

Carothers, S. W., M. E. Stitt, and R. R. Johnsen, Feral asses on public lands: an analysis of biotic impact, legal considerations and management alternatives. *Trans. North Am. Wildl. Nat. Res. Conf.*, 41, 396–406, 1977.

Case, T. J. and D. T. Bolger, The role of introduced species in shaping the distribution and abundance of island reptiles. *Evol. Ecol.*, 5, 272–290, 1991.

Casey, D. and D. Hein, Effects of heavy browsing on a bird community in deciduous forest. *J. Wildl. Manage.*, 47, 829–836, 1983.

Caughley, G., *Analysis of Vertebrate Populations*. John Wiley & Sons, New York, 1977.

Ceballos, G. and J. H. Brown, Global patterns of mammalian diversity, endemism and endangerment. *Conserv. Biol.*, 9, 559–568, 1995.

Clout, M. N. and S. J. Lowe, Invasive species and environmental changes in New Zealand. P. 369–383 in H. A Mooney and R. J. Hobbs, Eds., *Invasive Species in a Changing World*. Island Press, Covelo, CA, 2000.

Cooper, J., A. V. N. Marais, J. P. Bloomer, and M. N. Bester, A success story: breeding of burrowing petrels (Procellariidae) before and after eradication of feral cats (*Felis catus*) at subantarctic Marion Island. *Mar. Ornithol.,* 23, 33–37, 1995.

Courchamp, F., M. Langlais, and G. Sugihara, Control of rabbits to protect island birds from cat predation. *Biol. Conserv.*, 89, 219–225, 1999.

Cox, G. W., *Alien Species in North America and Hawaii*. Island Press, Washington, D.C., 1999.

deCalesta, D. S., Effect of white-tailed deer on songbirds within managed forests in Pennsylvania. *J. Wildl. Manage.*, 58, 711–717, 1994.

Demarais, S., D. A. Osborn, and J. J. Jackley, Exotic big game: a controversial resource. *Rangelands,* 12, 121–125, 1990.

Ehrlich, P. R., Attributes of invaders and the invading processes: vertebrates. P. 315–328 in J. A. Drake, H. A. Mooney, F. di Castri, R. H. Groves, F. J. Kruger, M. Rejmanek, and M. W. Williamson, Eds., *Biological Invasions: a Global Perspective*. John Wiley & Sons, Chichester, England, 1989.

Eldredge, L. G., Unwanted strangers: an overview of animals introduced to Pacific Islands. *Pac. Sci.,* 46, 384–386, 1992.

Emlen, J. T., An urban bird community in Tucson, Arizona: derivation, structure, and regulation. *Condor,* 76, 184–197, 1974.

Flux, J. E. C. and P. J. Fullagar, World distribution of the rabbit *Oryctolagus cuniculus*. *Acta Zool. Fennica,* 174, 75–77, 1983.

Frank, J. H. and E. D. McCoy, Invasive adventive insects and other organisms in Florida. *Fla. Entomol.,* 78, 1–15, 1995a.

Frank, J. H. and E. D. McCoy, Precinctive insect species in Florida. *Fla. Entomol.,* 78, 21–35, 1995b.

Freeland, W. J., Populations of cane toads, *Bufo marinus*, in relation to time since colonization. *Aust. Wildl. Res.*, 13, 321–329, 1986.

Freeland, W. J., Large herbivorous mammals: exotic species in northern Australia. *J. Biogeography,* 17, 445–452, 1990.

Frelich, L. E. and C. G. Lorimer, Current and predicted long-term effects of deer browsing in hemlock forests in Michigan, U.S.A. *Biol. Conserv.*, 34, 99–120, 1985.

Fritts, T. H. and G. H. Rodda, The role of introduced species in the degradation of island ecosystems: a case history of Guam. *Annul. Rev. Ecol. Syst.*, 29, 113–140, 1998.

Gebhardt, H., Ecological and economic consequences of introductions of exotic wildlife (birds and mammals) in Germany. *Wildl. Biol.*, 2, 205–211, 1996.

Goodrich, J. M. and S. W. Buskirk, Control of abundant native vertebrates for conservation of endangered species. *Conserv. Biol.*, 9, 1357–1364, 1995.

Green, R. E., The influence of numbers released on the outcome of attempts to introduce exotic bird species to New Zealand. *J. Anim. Ecol.*, 66, 25–35, 1997.

Green, R. J., Native and exotic birds in a suburban habitat. *Aust. Wildl. Res.*, 11, 181–190, 1984.

Hamann, O., Regeneration of vegetation on Santa Fe and Pinta Islands, Galapagos, after the eradication of goats. *Biol. Conserv.,* 15, 215–236, 1979.

Honegger, R. E., List of amphibians and reptiles either known or thought to have become extinct since 1600. *Biol. Conserv.*, 19, 141–158, 1981.

Howard, W. E. and R. E. Marsh, Implications and management of feral mammals in California. In *Proc. Vert. Pest Conf.*, 12, 226–229, 1986.

Ivey, G. L., J. E. Cornely, and B. D. Ehlers, Carp impacts on waterfowl at Malheur National Wildlife Refuge, Oregon. *Trans. North Am. Wildl. Nat. Res. Conf.*, 63, 66–74, 1998.

Jaksic, F. M., Vertebrate invaders and their ecological impacts in Chile. *Biodiversity Conserv.*, 7, 1427–1445, 1998.

Jones, C. G., Parrot on the way to extinction. *Oryx,* 15, 350–354, 1980.

Keegan, D. R., B. E. Coblentz, and C. S. Winchell, Feral goat eradication on San Clemente Island, California. *Wildl. Soc. Bull.*, 22, 56–61, 1994.

Kelly, S. T. and M. E. DeCapita, Cowbird control and its effect on Kirtland's warbler reproductive success. *Wilson Bull.,* 94, 363–365, 1982.

Kenward, R. E. and J. L. Holm, What future for British red squirrels? *Biol. J. Linnean Soc.*, 38, 83–89, 1989.

King, W. B., Ecological basis of extinctions in birds. In *Proc. Int. Ornithol. Congr.*, 17, 905–911, 1980.

Knopf, F. L., Changing landscapes and the cosmopolitism of the eastern Colorado avifauna. *Wildl. Soc. Bull.*, 14, 132–142, 1986.

Knopf, F. L., Faunal mixing, faunal integrity, and the biopolitical template for diversity conservation. *Trans. North Am. Wildl. Nat. Res. Conf.*, 57, 330–342, 1992.

Lockwood, J. L., Using taxonomy to predict success among introduced avifauna: relative importance of transport and establishment. *Conserv. Biol.*, 13, 560–567, 1999.

Lodge, D. M., Biological invasions: lessons from ecology. *Trends Ecol. Evol.,* 8, 133–137, 1993.

Mayfield, P. M., Kirtland's warblers benefit from large forest tracts. *Wilson Bull.*, 105, 351–353, 1993.

McNeely, J. A., The future of alien invasive species: changing societal views. P. 171–189 in H. A. Mooney, and R. J. Hobbs, Eds., *Invasive Species in a Changing World.* Island Press, Covelo, CA, 2000.

McShea, W., Herbivores and the ecology of forest understory birds. P. 298–309 in W. J. McShea, H. B. Underwood, and J. H. Rappole, Eds., *The Science of Overabundance: Deer Ecology and Population Management.* Smithsonian Institute Press, Washington, D.C., 1997.

Miller, S. G., S. P. Bratton, and J. Hadidian, Impacts of white-tailed deer on endangered and threatened vascular plants. *Nat. Areas J.,* 12, 67–74, 1992.

Moors, P. J. and I. A. E. Atkinson, Predation on seabirds by introduced animals, and factors affecting its severity. P. 667–690 in J. P. Croxall, P. G. H. Evans, and R. W. Schreiber, Eds., *Status and Conservation of the World's Seabirds.* International Council for Bird Preservation, Cambridge, England, 1984.

Newmann, D. E., Effects of a mouse, *Mus musculus*, eradication programme and habitat changes on a lizard population of Mana Island, New Zealand with special reference to McGregors skink, *Cyclodina macgregori. N.Z. J. Zool.*, 21, 443–456, 1994.

Newsome, A. E. and I. R. Noble, Ecological and physiological characters of invading species. P. 1–20 in R. H. Groves, and J. J. Burdon, Eds., *Ecology of Biological Invasions.* Cambridge University Press, Cambridge, England, 1986.

O'Brien, P. H., Introduced animals and exotic disease: assessing potential risk and appropriate response. *Aust. Vet. J.*, 66, 382–385, 1989.

O'Brien, P. H., Managing introduced pests. *Res. Sci. Interface,* 1, 4–11, 1993.

Olsen, S. L., Additional notes on subfossil bird remains from Ascension Island. *Ibis* 119, 37–43, 1977.

Ostfeld, R. S., C. G. Jones, and J. O. Wolff, Of mice and mast: ecological connections in eastern deciduous forests. *BioScience*, 46, 323–330, 1996.

Payne, J. M., R. D. Brown, and F. S. Guthery, Wild game in Texas. *Rangelands,* 9(5), 207–211, 1987.

Peine, J. D. and J. A. Farmer, Wild hog management program at Great Smoky Mountains National Park. In *Proc. Vert. Pest Conf.*, 14, 221–227, 1990.

Pimentel, D., Biology of the Indian mongoose in Puerto Rico. *J. Mammal.,* 36, 62–68, 1955.

Rainbolt, R. E. and B. E. Coblentz, A different perspective on eradication of vertebrate pests. *Wildl. Soc. Bull.*, 25, 189–191, 1997.

Rodda, G. H., T. H. Fritts, and D. Chiszar, The disappearance of Guam's wildlife: new insights for herpetology, evolutionary ecology, and conservation. *Bioscience*, 47, 565–574, 1997.

Sargeant, A. B. and P. M. Arnold, Predator management for ducks on waterfowl production areas in the Northern Plains. In *Vert. Pest Conf.*, 11, 161–167, 1984.

Savidge, J. A., Extinction of an island forest avifauna by an introduced snake. *Ecology,* 68, 660–668, 1987.

Simberloff, D. S., Experimental zoogeography of islands: a model for insular colonization. *Ecology,* 50, 296–314, 1969.

Simberloff, D., Hybridization between native and introduced wildlife species: importance for conservation. *Wildl. Biol.*, 2, 143–150, 1996.

Simons, T. R., *Biology and Conservation of the Endangered Hawaiian Dark-Rumped Petrel (Pterodroma phaeopygis sandwichensis)*. Cooperative National Park Studies Unit, University of Washington, Seattle, 1983.

Stone, C. P. and L. L. Loope, Reducing negative effects of introduced animals on native biotas in Hawaii: what is being done, what needs doing, and the role of national parks. *Environ. Conserv.*, 14, 245–258, 1987.

Swain, R. B., Palms and parrots. *Horticulture,* 66(6), 48–55, 1988.

Temple, S. A., The nasty necessity: eradicating exotics. *Conserv. Biol.*, 4, 113–115, 1990.

Temple, S. A., Exotic birds: a growing problem with no easy solution. *Auk*, 109, 395–397, 1992.

Tyler, M., *Australian Frogs: A Natural History.* Reed Books, New South Wales, Australia, 1994.

Usher, M. B., Ecological effects of controlling invasive terrestrial vertebrates. P. 463–489 in J. A. Drake, H. A. Mooney, F. di Castri, R. H. Groves, F. J. Kruger, M. Rejmanek, and M. W. Williamson, Eds., *Biological Invasions: A Global Perspective*. John Wiley & Sons, Chichester, England, 1989.

van Riper, C., III, S. G. van Riper, M. L. Goff, and M. Laird, The impact of malaria on birds in Hawaii Volcanoes National Park. *Technical Report 47*, University of Hawaii Cooperative National Park Resources Studies Unit, Honolulu, HI, 1982.

Veitch, C. R., Methods of eradicating feral cats from offshore islands in New Zealand. P. 125–142 in P. J. Moors, Ed., *Conservation of Island Birds: Case Studies for the Management of Threatened Island Species*. International Council for Bird Preservation, Cambridge, England, 1985.

Vitousek, P. M., C. M. D'Antonio, L. L. Loope, and R. Westbrooks, Biological invasions as global environmental change. *Am. Sci.*, 84, 468–478, 1996.

Walker, L. and L. Hudson, *Midway in Verse — or Worse*. Hester Colorgraphic Studios, San Diego, CA, 1945.

Waller, D. M. and W. S. Alverson, The white-tailed deer: a keystone herbivore. *Wildl. Soc. Bull.*, 25, 217–226, 1997.

Warren, R. J., Ecological justification for controlling deer populations in eastern national parks. *Trans. North Am. Wildl. Nat. Res. Conf.*, 56, 56–66, 1991.

Warren, R. J., The challenge of deer overabundance in the 21st century. *Wildl. Soc. Bull.*, 25, 213–214, 1997.

Weaver, R. A., Feral burros and wildlife. In *Proc. Vert. Pest Conf.*, 6, 204–209, 1974.

Wilcove, D., Silent Hawaii. *Living Bird Q.,* 8(3), 8–13, 1989.

Wilcove, D. S., D. Rothstein, J. Dubow, A. Phillips, and E. Losos, Quantifying threats to imperiled species in the U.S. *Bioscience,* 48, 607–615, 1998.

Williams, C. K. and R. J. Moore, Phenotypic adaptation and natural selection in the wild rabbit *Oryctolagus cuniculus* in Australia. *J. Anim. Ecol.*, 58, 495–507, 1989.

Williams, K., I. Parer, B. Coman, J. Burley, and M. Braysher, *Managing Vertebrate Pests: Rabbits*. Australian Government Publishing Service, Canberra, Australia.

Williamson, M. H. and K. C. Brown, The analysis and modeling of British invasion. *Philos. Trans. R. Soc. London, Series B, Biol. Sci.,* 314, 505–521, 1986.

Williamson, M., *Biological Invasions*. Chapman and Hall, London, 1996.

Wilson, G., N. Dexter, P. O'Brien, and M. Bomford, *Pest Animals in Australia: A Survey of Introduced Wild Animals*. Australian Bureau of Rural Resources, Kangaroo Press, Canberra, Australia, 1993.

CHAPTER 7

Lethal Control

"One way to change an animal's behavior is to kill it, but that is a rather crude way to accomplish the task."

Anonymous

"A single death is a tragedy, a million is a statistic."

Joseph Stalin
former leader of the U.S.S.R

"The only thing you need to do to manage elk is to manage the air-to-bullet ratio."

Anonymous

"Help preserve American forests, go deer hunting."

Automobile bumper sticker

Humans have used lethal means to reduce wildlife damage for thousands of years. Although it seems obvious that killing animals should reduce the amount of damage they cause, the relationship is rarely straightforward. This chapter examines several questions and issues involving the use of lethal methods to reduce wildlife damage.

INTRINSIC GROWTH RATES OF WILDLIFE POPULATIONS

Most wildlife populations have the potential to grow by an ever-increasing number. Consider a population where there is no mortality: each animal can breed when one year old, and each pair of animals can produce two young annually. If we start with 2 individuals in year 1, the population will grow to 4 in year 2, to 8

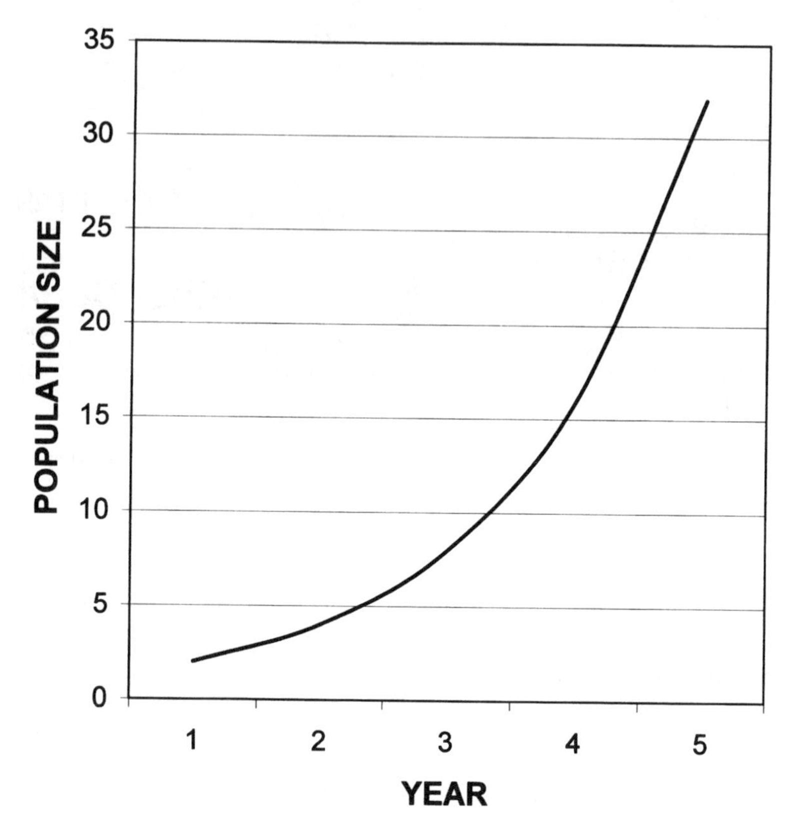

Figure 7.1 An exponential curve which shows how a wildlife population would increase over time if every female reaches sexual maturity in one year, each sexually mature female produces two young every year, and there is no mortality.

in year 3, to 16 in year 4, and to 32 in year 5 (Figure 7.1). If this rate is maintained, the population passes 1000 in year 10 and exceeds 1 million in year 20. The potential for a population to increase over time is not linear but exponential.

Exponential growth may be observed in populations recovering from a severe population crash or when an exotic species invades a new habitat where food is abundant and there are no predators or diseases to check its growth (McCallum and Singleton 1989). This rate of population growth under ideal conditions is called the intrinsic growth rate. It varies among species depending upon the age at which individuals reach sexual maturity and how many offspring each female can produce annually. In our hypothetical population described in Figure 7.1, offspring were able to breed after one year. If two years were required to reach sexual maturity, the growth rate of the population would be slower (Figure 7.2). Obviously, the longer it takes for an individual to reach sexual maturity, the slower the population will increase. Growth rates also will increase with the number of young each female can produce in a year. If instead of producing two offspring per year, a pair of adults can produce four, then only 13 years will be required for the population to increase

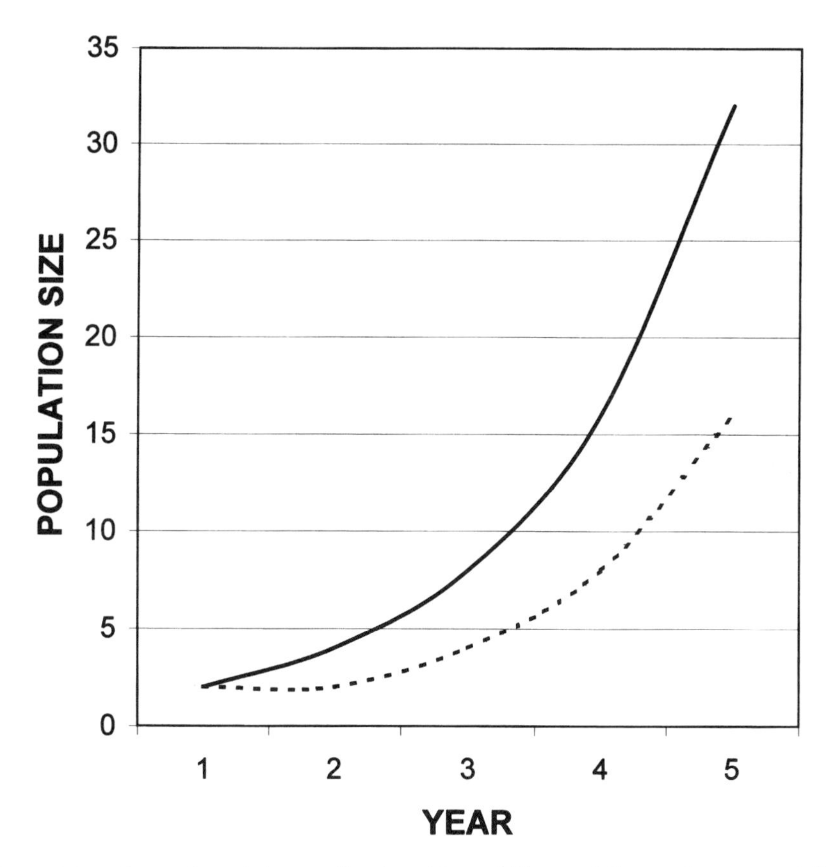

Figure 7.2 Comparison of the growth rates of two wildlife populations — in one population, the females reach sexual maturity in one year (solid line), and in the other (dotted line), the females need two years to reach sexual maturity (assumes that each female produces two young per year and no mortality).

to a million (Figure 7.3), providing females reach sexual maturity in one year. The population would increase to over 3 million in 9 years if each female produced 10 offspring annually and all survived.

WHAT EFFECT DOES LETHAL CONTROL HAVE ON A WILDLIFE POPULATION'S BIRTH AND MORTALITY RATES?

When wildlife densities in a suitable habitat are low and food is plentiful, birth rates generally are high because females are in good health and can produce large litters of young or clutches of eggs. For the same reason, parents also have little difficulty feeding several young. As populations increase and competition for food intensifies, the health of females may deteriorate and they may produce smaller litters. It is also more difficult for the parents to obtain enough food to keep all of their young alive. Hence, birth rates decrease as populations increase. For example,

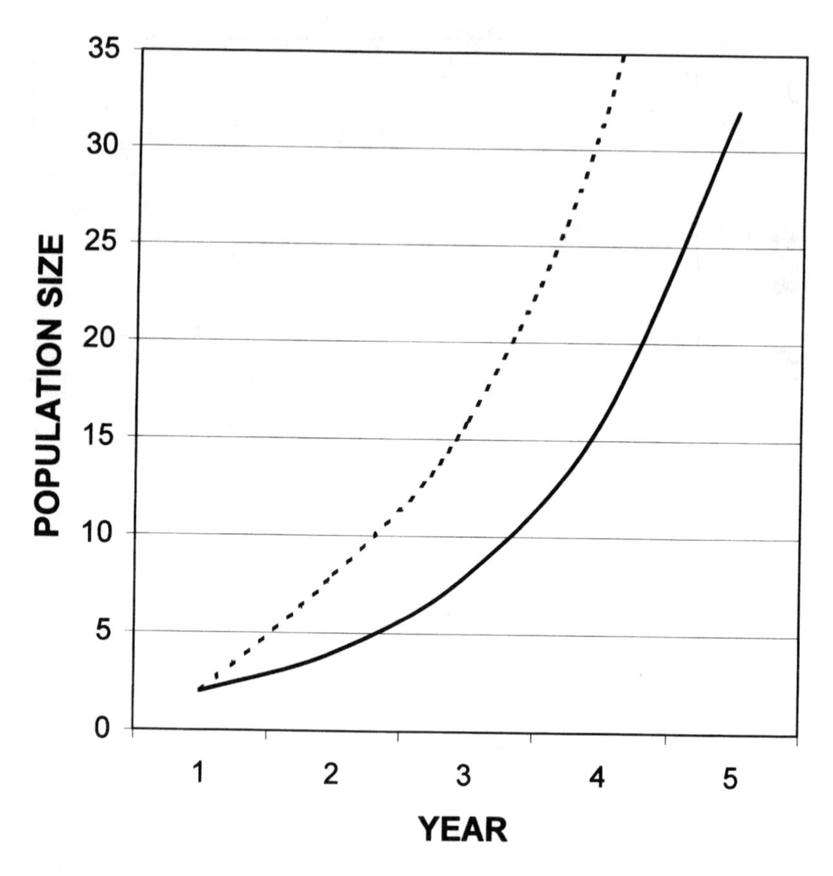

Figure 7.3 Comparison of the growth rates of two wildlife populations — in one population, each female produces two young per year (solid line), while females in the other population (dotted line) produce four young per year (assumes that females reach sexually maturity in one year and no mortality).

coyote litter size can increase from three pups when food is scarce to eight when food is abundant. If lethal control reduces a wildlife population, competition for food will decrease and birth rates will increase. Most wildlife populations can recover quickly from any population reduction due to lethal control. This is especially true for species, such as rodents, that normally have high birth rates.

When wildlife populations and animal densities increase, mortality rates rise because food starts to become scarce and animals begin to starve. At the same time, diseases spread more easily among higher densities of animals. Predation also increases because predator numbers increase in response to the abundance of food and because individual predators develop a search image for the abundant prey and concentrate their feeding activities on that species. Ultimately, the mortality rate from starvation, disease, and predation increases until it equals the birth rate and the population stabilizes (Figure 7.4). The point at which the population stabilizes is called the biological carrying capacity. It is a characteristic of the habitat and can

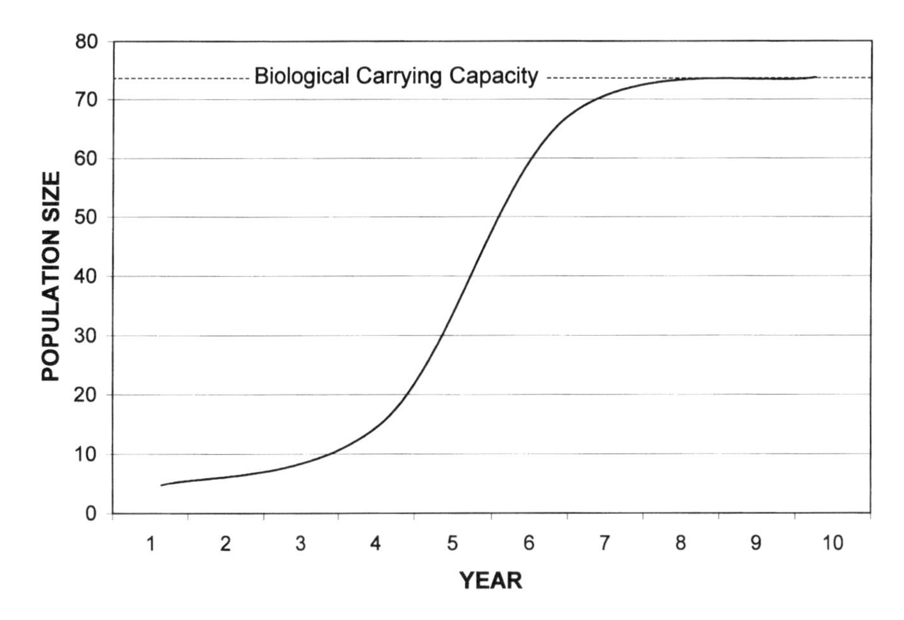

Figure 7.4 A logistic growth curve which shows a wildlife population that grows exponentially when it is low, but as the population grows, its rate of increase slows because of increasing mortality from predation, disease, and/or starvation.

be viewed as the number of animals which can be sustained by the annual production of food or the number of animals that can find safe places to hide from predators. The resource (e.g., food or shelter) in the shortest supply acts to limit the size of the wildlife population. This type of equation, where the population initially grows exponentially and then slows as it approaches the carrying capacity, is called a logistical equation (Caughley 1980; Eberhardt 1988).

A wildlife population is usually large by the time it causes enough damage that lethal techniques are considered. Therefore, the population is probably already experiencing a high rate of natural mortality. If, by using lethal methods, we are only killing animals that would have died anyway before the next breeding season, then lethal control may have no impact on the population. In this case, the mortality caused by lethal control is considered to be "compensatory" and simply replaces a natural form of mortality (Errington 1946, 1956; Errington et al. 1963). Hence, the population during the next breeding season would be the same whether or not lethal means are employed (Sidebar 7.1). For example, Murton et al. (1974) examined whether shooting wood-pigeons in England was an effective method to reduce their damage to agricultural crops. For a decade, up to 60% of the wood-pigeon population were removed each winter. However, by late spring, wood-pigeon populations and the amount of damage they caused had returned to pre-control levels. The loss of birds from shooting was compensated for by a decrease in natural mortality rates and immigration by birds from nearby areas. Murton et al. (1974) found that the food supply, not shooting, determined the amount of winter mortality from year to year.

Sidebar 7.1 Effect of Lethal Control on Coyote Populations

In areas such as Yellowstone National Park where coyotes are not subjected to hunting or lethal control, their population dynamics are very different than in areas where they are subjected to heavy control. In protected areas, females normally do not breed before two years of age, and litter sizes are three to four pups. Coyotes often live in packs containing a dominant breeding pair and several subordinate individuals that do not breed but help feed and raise the pups produced by the dominant pair. These helpers are often young produced by the dominant pair in previous years. The dominant pair and their pack keep other coyotes out of their territory. At high densities, all suitable habitat is filled up with territories. Coyotes unable to secure a territory of their own become transient individuals and confine their movements to the spaces between occupied territories or areas of poor habitat. Like helpers, transient coyotes do not breed. Hence, at high densities, only a small proportion of the coyotes are able to breed. In protected areas, such as in national parks, over half of the coyotes die annually from starvation, disease, parasites, and injuries.

In areas where coyotes are subjected to high rates of lethal control and coyote densities are low, coyotes leave their parents' territory during their first year of life, establish a territory of their own, and breed. They do not become helpers or transient coyotes. In these areas, litters may average seven to eight pups. When most female coyotes produce large litters every year, the coyote population is capable of tripling or quadrupling its numbers each year. Furthermore, in a heavily controlled population, few coyotes starve to death or die from diseases. Instead, natural forms of mortality have been replaced by human-caused mortality. For this reason, more than 60% of the coyote population must be killed by humans annually before we would expect to see any reduction in coyote densities (Wagner 1988).

Sometimes, animals removed by lethal methods may not have simply replaced animals that would have died anyway. Instead, these deaths may be "additive" to those from natural causes. In these cases, wildlife populations during the next breeding season would be smaller than what would have occurred in the absence of control.

Usually, lethal control is compensatory to a point and additive thereafter, meaning that, below some threshold, lethal control has little effect on populations, but as its intensity increases past that threshold, it will reduce populations (Burnham and Anderson 1984; Nichols et al. 1984; Clark 1987). Although it may be possible to kill many animals using lethal control methods, it usually is very difficult to suppress wildlife populations to a low enough level where the population cannot recover within a short period. For instance, Crosby and Graham (1986) calculated that if the number of black-tailed prairie dogs in a colony of 100 were reduced by 55% during the winter, reproduction by the remaining animals would return the colony to its original size by the end of spring. Hone and Pedersen (1980) reduced a feral pig population by 58%, but the population returned to its original level within a year. Because free-ranging populations are dynamic, lethal control often must be repeated annually to be effective.

Most wildlife populations go through a critical period sometime during the year when much of the mortality occurs. These critical periods, or bottlenecks, often occur in winter when food resources are lowest. Often, any mortality which occurs before the bottleneck is compensatory because there are only enough resources for a certain number of animals to survive the winter. For this reason, hunting normally occurs in the fall, because wildlife agencies want hunters to have a minimal impact on game populations. However, if we want to use lethal control to suppress a population, the control is much more likely to have an effect if it is delayed until after the population has passed through the bottleneck. In most cases, lethal control is more effective if conducted in late winter or early spring. For example, Anthony et al. (1991) increased the nesting success of black brant by shooting and trapping Arctic foxes during or immediately before the brant started to nest. Similarly, Knowlton (1972) suggested that coyote control should occur during the spring, when their populations and immigration rates are lowest.

WHAT EFFECT DOES LETHAL CONTROL HAVE ON A WILDLIFE POPULATION'S IMMIGRATION RATE?

If lethal control reduces the density of a wildlife population in an area, individuals from outside this area typically move in and occupy any void. Immigration of individuals into an area helps nullify any population reduction produced by the lethal technique. This is especially true for populations that are close to carrying capacity and contain transient individuals that are unable to secure a territory of their own. For example, brown tree snakes quickly reoccupied small areas on Guam where they were being removed. Although snake densities were 37 per ha, 151 snakes were removed from a 1.4 ha plot during a two-week trapping period. Biologists had predicted that this period of time would be sufficient to eradicate the population. Instead, the number of snakes captured daily did not even decline during the study, due to high rate of snake dispersal into the small area (Witmer et al. 1996). In contrast, eradication was achieved during the same time period in plots surrounded by snake-proof barriers.

Immigration rates will be fastest for those species that disperse long distances. Rabbits have high rates of dispersal and individual rabbits may move more than 20 km. Because of this, controlling rabbits in small areas is difficult when there are high rabbit densities in adjacent areas (Williams et al. 1995). When the climate is favorable, dispersing rabbits can quickly recolonize even large areas following lethal control (Parer and Parker 1986).

Efforts to reduce populations over large areas are usually more effective than those in small areas because the sources of immigrants are more remote and more immigrants are required to fill the larger void. Garrettson et al. (1996) doubled the nesting success of ducks when predators were trapped from large areas (each 16 square miles or 4150 ha), whereas predator trapping in small areas (142 ha) had little effect (Sargeant et al. 1995).

IS THERE A CORRELATION BETWEEN WILDLIFE POPULATION LEVELS AND WILDLIFE DAMAGE?

We often assume that there is a direct one-to-one relationship between any change in a wildlife population and the amount of damage it produces (e.g., a 50% reduction in a deer population will decrease the damage they cause by 50%). This assumption often is incorrect; a 50% reduction in a deer population may have little impact on deer damage or may stop it completely. The actual relationship between a reduction in a wildlife population and a reduction in wildlife damage depends, in part, on why the animals are causing the damage (i.e., what is motivating them). To illustrate this, consider the problem of deer–automobile collisions. If most collisions result from deer just roaming throughout their home range as part of their daily activities, then reducing the deer population by half will probably reduce deer–automobile collisions by the same amount. However, if most of the collisions are caused by hungry deer which have eaten all the food within their home ranges and are searching for new food sources, then most of the collisions will occur when deer populations exceed the biological carrying capacity. In this case, reducing the deer population by half may stop almost all deer–automobile collisions because deer will not have to wander in search of food.

For another example, consider a population of 100 deer that are causing problems for local farmers by consuming crops. Let us assume the deer have 20 different plant species in their home ranges that they can consume and that they will selectively forage on the plant species that is most palatable. Under this scenario, deer concentrate their foraging on the most palatable plant until that plant becomes scarce and then will turn their attention to the second most palatable plant. The deer will continue this pattern: grazing each plant species in turn, based on the plant's palatability relative to the other plants that are still available. Let us also assume that foraging by the 100 deer is intense enough that they have overgrazed nine plant species. Hence, they are concentrating their foraging on the tenth most palatable plant species and grazing the more palatable plants as they grow back. Now, assume that the farmers complain about deer damage and the state has sent hunters that have reduced the deer population from 100 deer to 50 deer. What impact will this have on reducing deer damage to the farmers' crops? That will depend upon how the palatability of the farmers' crops compares to that of the other 20 plant species within the deer's home range. If the farmers are growing alfalfa, which typically is very palatable, any deer in the area will forage preferentially on it. For this plant, reducing the deer herd from 100 to 50 animals will have little impact on damage if 50 deer are sufficient to consume all of the alfalfa. To reduce herbivory on alfalfa, almost all deer in the area would have to be removed. However, if farmers are growing wheat, which deer find only moderately palatable, and deer are just starting to forage on it when their population rises to 100 individuals, reducing the deer population from 100 to 50 may completely stop foraging on wheat because there will be enough food for 50 deer from the more palatable plant species. Hence, reducing a deer population may not significantly reduce damage to highly palatable crops but may completely stop damage to less palatable ones (Figure 7.5).

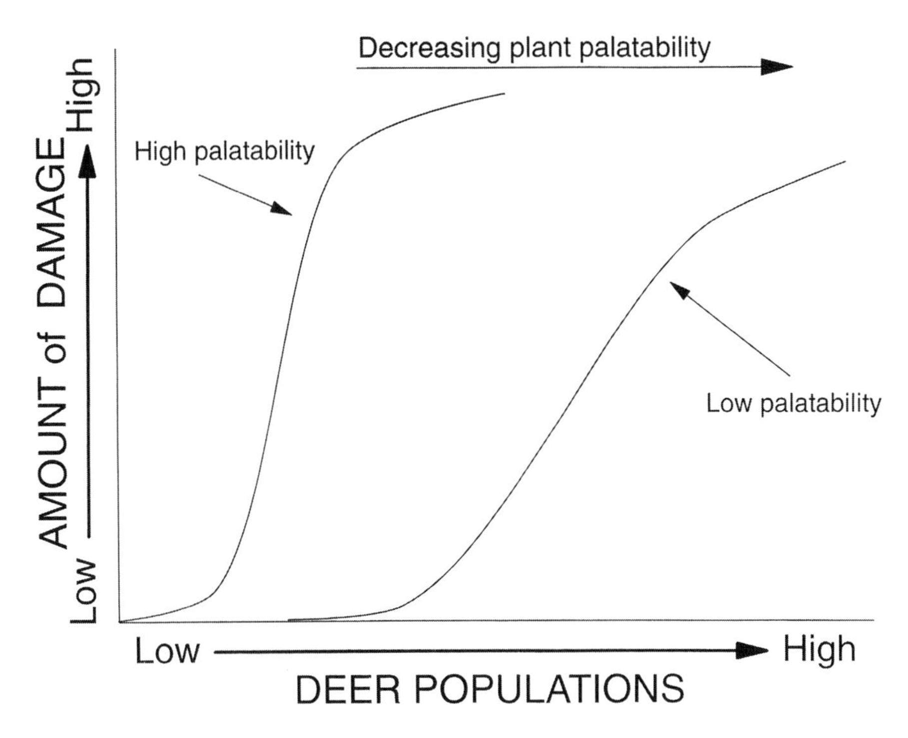

Figure 7.5 A small reduction in a deer population may significantly reduce damage to less palatable plants but a large reduction may be required to reduce damage to a palatable plant species.

HOW DO VALUES PROVIDED BY WILDLIFE CHANGE AS THEIR POPULATIONS INCREASE?

Although there may not be a direct one-to-one relationship between an increasing wildlife population and the severity of a specific wildlife problem, human–wildlife conflicts become more intense and more frequent as wildlife populations increase. In general, the negative values of wildlife increase exponentially as wildlife populations expand (Figure 7.6). This occurs because as wildlife populations increase, there are more animals to cause damage, and humans become less tolerant as their losses increase. For instance, suburban residents are willing to tolerate some risk of contracting Lyme disease or of being involved in a deer–automobile collision because of the positive values deer provide. However, as the risk increases for them and their families, they become more concerned by increasing deer density.

As wildlife populations increase, the positive values of wildlife also increase but at a decreasing rate. With more deer, for instance, there are more opportunities for people to see deer and more hunters will have the opportunity to shoot one. Hence, positive values rise. But the amount of pleasure people receive from seeing their 1000th deer will probably be less than what they experienced from seeing their first.

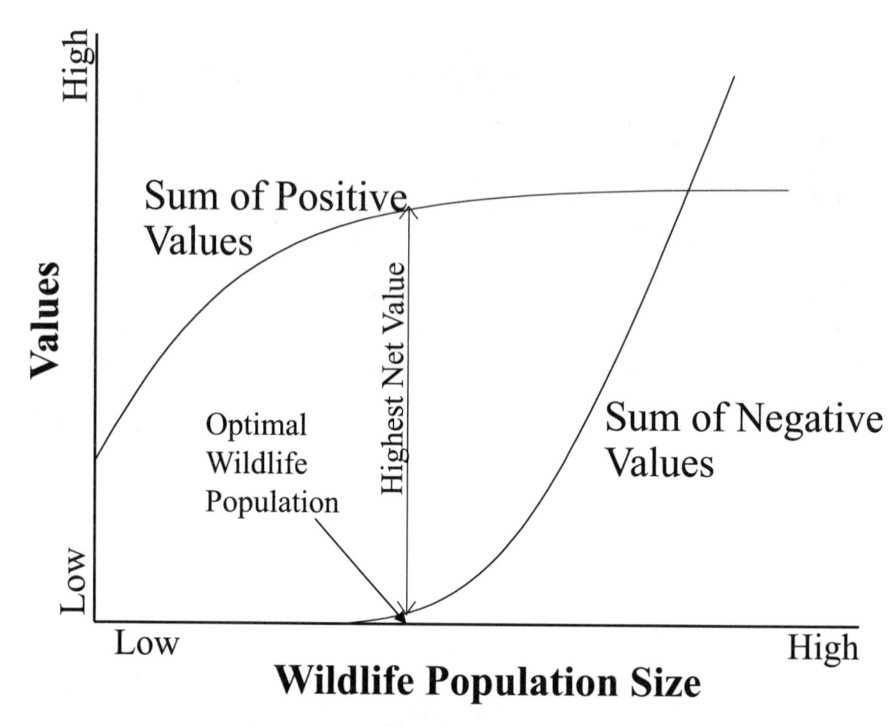

Figure 7.6 How the positive and negative values provided by a wildlife population change as the wildlife population increases.

There also are some positive values of wildlife, such as its existence value, which are independent of size of the wildlife population and do not change as populations increase (Figure 7.6).

Hence, at low populations, positive values provided by a wildlife species greatly exceed the negative values, but as a wildlife population increases, negative values increase faster than positive values. Ultimately, a wildlife population can increase to the point where the sum of its negative values exceeds its positive values. At this point, it could be considered a pest and both nonlethal and lethal methods might be employed to reduce the population (Sidebar 7.2). The optimal wildlife population is the one with the greatest net value (positive benefits minus negative benefits; Figure 7.6).

Sidebar 7.2 Impact of Goose Removal on Minnesota's Urban Goose Population

In Minneapolis–St. Paul, MN, the urban goose population had increased from 480 urban geese in 1968 to 14,000 in 1984. The population was growing exponentially (Figure 7.7), and it was estimated that it would reach 500,000 before the carrying capacity was reached (Cooper and Keefe 1997). In response to numerous complaints about nuisance problems caused by these birds, geese were rounded up during the summer molt and relocated, beginning in 1982. From 1982 to 2000, 56,000 geese were

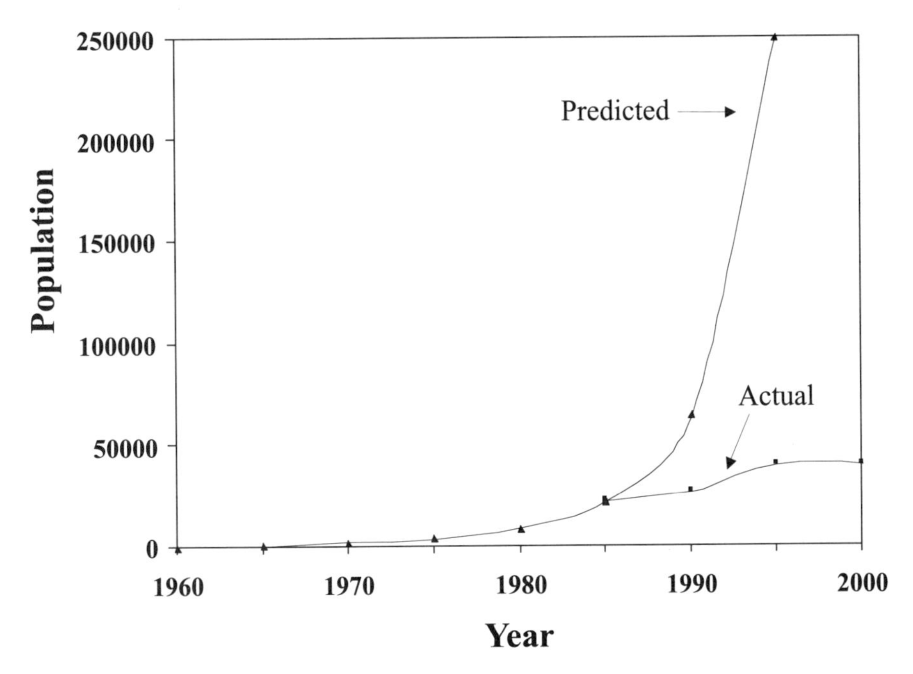

Figure 7.7 Predicted and actual changes in the urban Canada goose population in Minneapolis–St. Paul, MN, as a result of a program involving both hunting and translocating the geese.

translocated. Special goose-hunting seasons also were established to reduce the population of urban geese. To avoid shooting migrants, these special hunting seasons were conducted in September before the migratory geese arrived in Minnesota and in December after the migrants had left. From 1987 to 1999, 88,000 urban geese were shot by hunters (Cooper and Keefe 1997; R. A. Dolbeer, personal communication). As a result of these efforts, the growth rate of the urban goose population in Minneapolis–St. Paul slowed considerably (Figure 7.7).

SHOULD LETHAL TECHNIQUES BE DIRECTED AT SPECIFIC INDIVIDUALS, SPECIFIC SUBPOPULATIONS, OR THE ENTIRE POPULATION?

A wildlife population is a group of animals of the same species that is isolated from other groups by physical barriers. There can be considerable movement of individual animals within a population through dispersal and also an exchange of genetic material through breeding. For instance, gray squirrels in England can breed with each other, but they cannot breed with those in North America. Therefore, the squirrels in England constitute a separate population.

Each population is composed of many subpopulations (Figure 7.8). When wildlife damage is occurring at a farm, the wildlife subpopulation can be defined as the

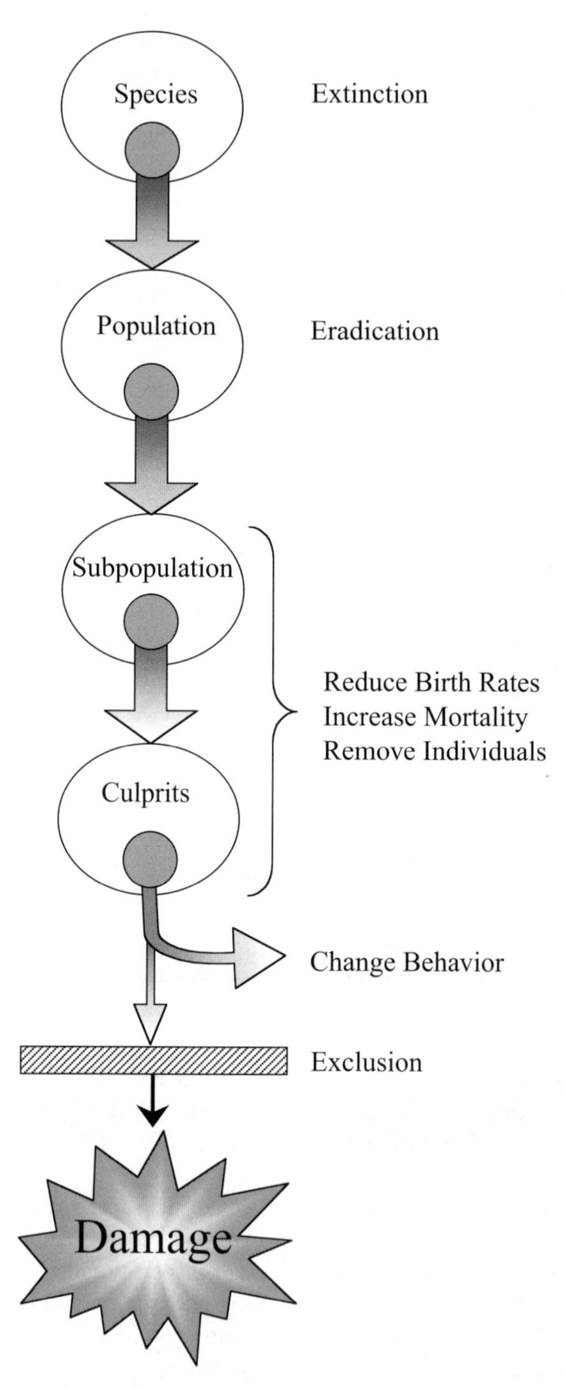

Figure 7.8 The relationship between culprits, subpopulations, and populations and how lethal control might be directed at any of these levels.

group of animals whose home ranges include the farm. They are the group of animals that have the potential to cause damage because they can travel to the farm. Individuals in other subpopulations are located too far from the site to cause damage. Not all members of the local subpopulation may actually be causing damage. Instead, only a small number of individuals may be responsible for the damage, and I will refer to these individuals as the "culprits" (Figure 7.8).

When lethal means are used in wildlife damage management, it is best to be as selective as possible. The goal should be to stop the damage by killing as few animals as possible because wildlife have high positive values for society (Chapter 1). Ideally, we should remove only the culprits from the local subpopulation and leave all other individuals alone. When animals are shot while actually in the process of causing damage, we can be assured that only culprits are being removed. For instance, some states only allow farmers to kill deer that are seen foraging in their fields. Other deer, which may be in the surrounding woodlots, are left alone.

One lethal technique that removes only the culprits is the Livestock Protection Collar® that is placed around the neck of sheep or lambs (Connolly and Burns 1990; Burns et al. 1996). This collar has pouches filled with a toxicant, sodium monofluoroacetate (also called 1080). Coyotes normally kill sheep by biting hard enough on the throat to close the sheep's windpipe and to suffocate it. When a coyote bites the throat of a sheep wearing a Livestock Protection Collar, its teeth puncture one or more of the pouches and the coyote gets some of the toxicant in its mouth and dies. Hence, these collars only remove culprits. After all, if a coyote has its teeth around the throat of a sheep, it is pretty good evidence that it was trying to kill that sheep. Unfortunately, Livestock Protection Collars have not lived up to their potential to solve the problem of predation on livestock. They are too expensive ($15 to 20 each) to put one on every sheep in a flock, and it is impossible to predict which sheep a coyote is going to attack next (Knowlton et al. 1999). Thus, this technique is more effective if most of the flock can be temporarily moved, leaving only a few animals with collars. Unfortunately, this requires additional work and a safe area where the rest of the sheep can be moved. Another limitation is that the U.S. Environmental Protection Agency requires states to have registration, training, and record-keeping programs before these collars can be used. Most states find these programs too expensive and so Livestock Protection Collars were registered for use in only seven states in 1999.

In some situations, we cannot identify, locate, or target only the culprits, but we may be able to identify members of the subpopulation most likely to be culprits and target that group. For instance, Till and Knowlton (1983) demonstrated that most sheep and lamb killing was done by adult coyotes that had pups to feed. Apparently, the additional stress caused by hungry pups drove their parents to shift their hunting efforts to large food items, such as lambs and sheep, that they might have avoided otherwise. We can target coyotes with pups by killing adults at the den sites or by selectively removing territorial coyotes (only territorial adults produce pups). Territorial coyotes can be targeted because they will chase small dogs from their territories, and shooters can use this knowledge to bring territorial coyotes within gun range (when the dog returns to the shooter, the coyote often comes with it). A high proportion of coyotes removed by aerial gunning during the winter also are territorial (Wagner and Conover 1999).

Sometimes it is impossible to selectively remove only culprits because we cannot identify them or cannot target them, or because almost all members of the local subpopulation are culprits. In these cases, lethal control can be directed at the subpopulation of animals whose home ranges encompass the specific site where damage is occurring by limiting lethal control to that specific site. One example of this practice is issuing depredation permits to farmers and ranchers to shoot deer and elk on their property. Most efforts to control predators by the U.S. Department of Agriculture (U.S.D.A.) Wildlife Services are targeted at subpopulations. For instance, coyotes are not killed unless they are in the immediate vicinity of a sheep herd that is suffering from predation. Likewise, when coyotes are killed through aerial gunning, the only area where they are targeted is the land immediately around a ranch or the specific mountain meadow where the sheep will be located in the summer.

Today, almost all lethal control is targeted at the culprits or subpopulations, and there are few attempts to control wildlife at a population level. The main exceptions are efforts to control exotic species which have escaped into new areas where they are causing environmental harm. Examples include recent attempts to control arctic foxes on some of the Aleutian Islands where they were introduced early in the 20th century. These foxes threaten the nesting success of the islands' native birds, including Aleutian Canada geese. Likewise, U.S.D.A. Wildlife Services employees are trying to remove an exotic population of roof rats from Buck Island in the Virgin Islands, where the rats are preying upon newly hatched sea turtles (Allen Gosser, personal communication).

Sometimes, lethal control is used not just to reduce a population of an exotic species but to eradicate it. The advantage of eradication is that it only has to be successful once; after the last animal is removed, the problem is resolved permanently. In contrast, attempts to suppress a wildlife population require consistent efforts because as soon as lethal control ends, the wildlife population will return to its former levels. Most efforts to eradicate a wildlife population fail. While it may be relatively easy to kill most individuals in a population, it becomes difficult to kill the last ones, which are the most wary and live in remote areas.

It is also possible to target lethal control efforts at the species level and drive an entire species to extinction. There have been a few attempts to do this in the past. For instance at one time, the U.S. government wanted to drive bison to extinction as a means of depriving Native Americans of their food resources and forcing them back to reservations (Sidebar 2.3). Today, I know of no attempts to drive a wildlife species to extinction; they are too valuable. However, there are some human diseases that we attempt to drive to extinction. In fact, we accomplished this with the smallpox virus; it no longer occurs in the wild.

ARE LETHAL METHODS LEGAL?

Society generally allows people a free hand in using most nonlethal techniques to resolve wildlife problems. For instance, governmental approval is not required if someone wants to place a fence around his alfalfa field to protect it from deer, to put a scarecrow in the garden, or to manipulate the habitat on his or her property to reduce animal numbers. Likewise, no permits are required if a person wants to kill insects. However, state and federal laws protect most wildlife species because of wildlife's high value, and governmental approval is usually required before free-ranging animals are killed. A state permit is required for most wildlife species. The only exceptions are a few species classified as "vermin" in particular states and not protected by state law. Starlings, pigeons, house sparrows, rats, house mice, and coyotes are the species most often excluded from state protection. Additionally, a federal permit is required if migratory birds or endangered species are going to be killed. In most municipalities, local permission from the police is required to discharge a firearm.

Many lethal techniques are dangerous if used by inexperienced people. For this reason, most states require people employing specific lethal techniques to have a state and/or federal permit to use them. For instance, a pesticide applicator's license is required to use toxicants. People must pass a hunter safety course and have a hunting license before they can use a firearm to shoot most animals. Suffice it to say that everyone should check with local, state, and federal agencies before using any lethal technique to make sure that what they want to do is safe and legal, and all necessary permits have been obtained.

ARE LETHAL METHODS EFFECTIVE AT REDUCING WILDLIFE DAMAGE?

One of the most important questions regarding the use of lethal methods is whether they alleviate wildlife damage. If not, there is no reason to use them. Intuitively we know that if all of a damaging wildlife species is removed then damage by that species will stop. However, the issue is the shape of the curve between no removal and total eradication (Figure 7.9).

Many studies have reported that lethal control was effective in alleviating wildlife problems. For instance, nesting success of waterfowl has been increased by reducing predator populations through the use of traps (Garrettson et al. 1996) or toxicants (Duebbert and Lokemoen 1980). Predator control also has increased nesting success of several endangered bird species (Witmer et al. 1996). Several studies compared livestock losses at ranches with and without predator control and found that loss rates on ranches with predator control were lower than those observed on ranches without predator control (Nass 1980; Wagner 1988). In one interesting experiment, all efforts to kill predators were stopped at a Montana ranch to see what effect this might have on livestock losses to predators (Sidebar 7.3).

Sidebar 7.3 Predation at a Montana Sheep Ranch When Lethal Control of Predators Was Terminated

Some people believe that high rates of coyote predation on sheep result from poor management. Their assumption is that overgrazing leads to sick and weak sheep, and this leads to predation (Allen 1962). To test this hypothesis and determine what would happen if there was no predator control, scientists convinced the owner of the Eight Mile Ranch not to use lethal control to protect his sheep from predators, with the understanding that the scientists would compensate him for the value of any sheep killed by predators. This ranch was selected because it was well managed, the pastures lightly grazed, shed lambing was practiced, and coyotes had been a problem in prior years. After the first year of the study, the scientists had to change the study design and reinitiate predator control because sheep and lamb losses were so high that the scientists were in danger of running out of money to compensate the owner. The study was discontinued after 30 months because the researchers had already paid the owner over $50,000 in compensation. During the entire study, predators (mostly coyotes) killed 1223 sheep. Before predator control was reimposed, approximately 25% of the lambs and 6% of the ewes were killed annually by predators (O'Gara et al. 1983).

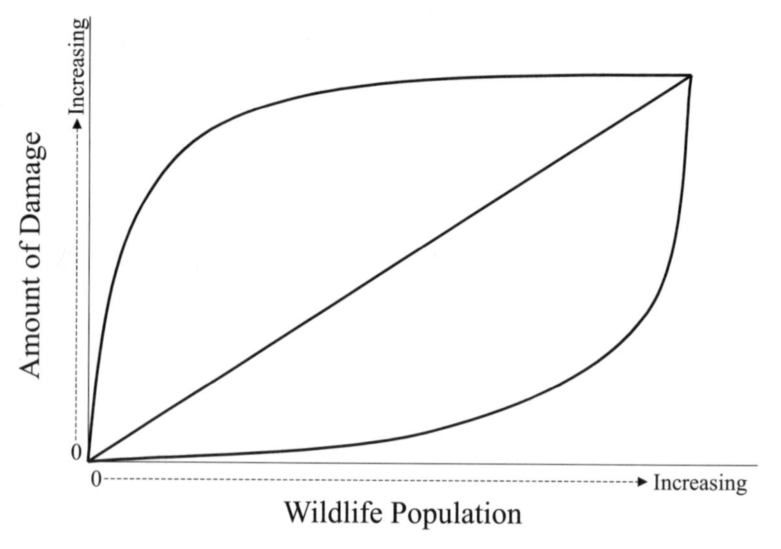

Figure 7.9 Three potential relationships between an increasing wildlife population and the amount of damage it causes.

During the 1950s and 1960s, a major effort was made in the western U.S. to reduce predation problems by reducing coyote populations over large areas through the distribution of large pieces (20 to 50 kg) of sheep or horse carcasses which had been poisoned with thallium or compound 1080. To reduce the nontarget hazard to birds, these baits were distributed in the winter after most birds had migrated to warmer climates. Only one poisoned carcass was placed in each township (36 mi^2 or 93 km^2). It was assumed that most coyotes would find a poisoned carcass while roaming within their large home ranges, as would some smaller predators (such as foxes and skunks). However, these small predators have smaller home ranges than coyotes and are less mobile; hence, most of them would not have a poisoned carcass within their home range and would be spared (Wagner 1988). These poisoned carcasses were distributed in most of the townships of Utah, Idaho, and Wyoming. During years when these poisoned carcasses were used, coyote densities in these states were lower than they had been previously (Linhart and Robinson 1972;

Wagner 1988). In other western states (Colorado, Nevada, New Mexico, and Texas), poisoned carcasses were distributed more sporadically and the results were more ambiguous. Although coyote populations were reduced in some states, Wagner (1988) concluded that the program had little effect on reducing predation on sheep and lambs. The widespread distribution of poisoned carcasses ended in 1972 when President Richard M. Nixon issued an executive order banning the use of toxicants by federal employees and on federal land. Since then, the use of lethal control to reduce predation on livestock has become site specific. Current efforts target culprits or subpopulations around individual ranches rather than trying to reduce coyote populations over large areas.

SHOULD LETHAL METHODS BE USED AHEAD OF TIME TO PREVENT WILDLIFE DAMAGE OR ONLY AFTER DAMAGE HAS BEGUN?

Lethal control is normally used in a "corrective" mode, meaning that it is employed only after damage has already begun. Furthermore, lethal control is normally used only after potential nonlethal means to solve the problem have been tried unsuccessfully. Hence, lethal methods are often employed only for intractable problems of a long-standing nature. The liability of this approach is that some people endure considerable damage before lethal methods are employed.

Sometimes, lethal control is employed in a "preventive" mode, meaning that lethal methods will be used to prevent a potential problem from starting. We often encourage farmers to start using nonlethal techniques before wildlife damage begins because we have learned that once animals start feeding in a field or causing problems, it is much more difficult to change their behavior than it is to prevent them from feeding in the field initially. However, lethal methods are rarely employed in a preventive mode, due to society's reluctance to kill wildlife unless absolutely necessary.

The main use of lethal methods in a preventive mode occurs when there is a threat to human health or safety or when economic losses occur so quickly that they reach unacceptable levels before lethal control can be employed in a corrective sense. In cases where wildlife damage can be predicted with accuracy, and there are valid reasons for the use of lethal means, we allow these to be used to prevent a problem from beginning. By using lethal control against wolves in a preventive model, researchers in Alberta were able to reduce cattle losses by more than 50% (Bjorge and Gunson 1985).

In the Rocky Mountains, coyote predation on livestock often occurs at the same places every year, usually during the spring or summer when the sheep are in remote pastures. Ranchers employ several methods (traps, snares, and shooting) to kill the coyotes responsible for killing their livestock. However, these methods are often labor intensive, and it may take weeks or months to kill the culprits. An easier, faster, and cheaper method of killing coyotes is to shoot them from an airplane or helicopter during the winter, when snow makes coyotes easier to spot and track. When aerial gunning occurs in the winter, it is used as a preventive mode of control because several months may elapse before the sheep arrive at these pastures. This raises the question of whether killing coyotes in the winter really reduces livestock losses six months later when the sheep and lambs arrive. Wagner and Conover (1999) tested this and found that aerial gunning was effective in reducing predation on sheep and lambs (Sidebar 7.4).

Sidebar 7.4 Can Aerial Gunning of Coyotes during Winter Reduce Livestock Losses during Summer?

In the Intermountain West, sheep and lambs spend the summer grazing in mountain meadows above 2500 m in elevation. Grazing conditions are often excellent at such high elevations because the grass grows all summer, whereas at lower elevations it becomes dormant during the dry summer. Unfortunately, many sheep and lambs are killed by predators while at these high elevations and their remoteness can make it difficult to track and kill the culprits during the summer. One alternative is to use helicopters to shoot coyotes in these high mountain meadows during the winter. Some people question whether killing coyotes several months before the sheep arrive makes any sense. They note that the snow is so deep at these elevations that most birds and mammals (including deer and elk) migrate to lower elevations during the winter. If coyotes also spend the winters at lower elevations, then aerial gunning of these mountain meadows during the winter would be pointless because coyotes would be absent or the wrong ones would be there. To answer these questions, scientists with the U.S.D.A. Wildlife Services'–National Wildlife Research Center and the Jack H. Berryman Institute conducted a series of experiments in Utah. First, Glen Gantz and Fred Knowlton showed that coyotes occupy their territories year-round, even those at high elevations. Hence, the coyotes present during the winter when aerial gunning would occur are the same ones that would be there during the summer. Next, Kimberly Wagner compared coyote predation rates on sheep and lambs on summer pastures after half of them had been subjected to preventive aerial gunning during the prior winter and half had not. An average of 12 lambs were killed by coyotes in pastures where aerial gunning had taken place vs. 35 where it had not. Hence, aerial gunning was effective in reducing coyote predation on sheep and lambs even when used before the predation started. Another benefit of aerial gunning was that it reduced the amount of time required in the summer to deal with coyote predation (Gantz 1990; Wagner and Conover 1999).

Preventive hunting of coyotes is also occasionally used to improve fawn survival of free-ranging ungulates. Killing coyotes prior to the fawning season increased fawn survival in pronghorn antelope (Smith et al. 1986) and white-tailed deer (Guthery and Beasom 1977; Stout 1982; Teer et al. 1991).

ARE LETHAL TECHNIQUES COST EFFECTIVE?

The benefits provided by any method to reduce wildlife damage should always exceed the cost of the method. In other words, the benefit-to-cost ratio should be greater than 1.0. Usually, lethal techniques are cost efficient and less expensive than other alternatives for reducing wildlife damage (Table 7.1). For instance, in

Table 7.1 Benefit-to-Cost Ratios for Different Uses of Lethal Control to Reduce Wildlife Damage

Problem and approach	Benefit:cost ratio	Reference
Blackbird control to protect sunflowers	9:1	Besser and Guarino (1976)
Predator control to reduce predation on sheep and goats	5:1	Pearson and Caroline (1981)
Predator control every other year to reduce predation on pronghorn	6:1	Smith et al. (1986); Hone (1994)
Predator control to reduce predation on game species	3:1	Beasom (1974a)
Aerial hunting of coyotes to reduce predation on livestock	2:1	Wagner and Conover (1999)
Rodent control during a mouse plague to protect sunflowers	16:1	Saunders and Robards (1983)
Rodent control during a mouse plague to protect corn	6:1	Saunders and Robards (1983)
Rodent control during a mouse plague to protect sorghum	6:1	Saunders and Robards (1983)

Minneapolis and St. Paul, MN, the cost of removing one Canada goose was $0 using sport hunters, $10 using relocation, $24 killing and processing it for food, $45 using egg destruction, $100 using sterilization, and even more expensive using habitat modification (Cooper and Keefe 1997). Campa et al. (1997) estimated that it annually cost less than $3 per ha to protect an agricultural field from deer using hunters, $54 per ha using a multi-wire electric deer fence, and $77 per ha using a woven-wire fence.

The money spent to employ a lethal technique is only one of the costs of lethal control. Other costs occur because valuable animals are killed (Chapter 2). For example, an experimental predator control program in Texas cost $7000, but produced $21,000 in game (deer and turkeys), so on a strictly economic basis, the program was cost effective (Beasom 1974a). But while the program produced an extra 87 male turkeys, 170 deer, and unknown numbers of nongame birds, it also cost the lives of 188 coyotes, 120 bobcats, and over 140 other predators (skunks, badgers, and opossums). Was the increase in deer and turkeys worth the number of predators killed? That is not an easy question to answer and depends upon one's values and perspective. One problem is that wildlife are owned by society but the damage they inflict is usually to private property. For individual property owners who are suffering from wildlife damage, it is in their self-interest to stop the damage by any means possible, including lethal methods. After all, property owners receive the economic benefit if the damage stops but are only remotely impacted by the loss of the wildlife because they do not have to compensate society for killing them.

The perspective of other people can be different. What they see is their wildlife, which they own as members of society, being destroyed for someone else's private gain. These people are only remotely impacted by the wildlife damage (it is not their livestock which are being preyed upon or their crops that are being eaten). Hence, their sentiments may lie with the wildlife and not with the property owners.

That is, it often is in their self-interest for the wildlife to be left unharmed, even if this means that the damage will continue. This problem occurs because the costs and benefits of wildlife do not fall evenly on everyone. For example, ranchers place a higher value on their livestock than on predators and may want to use lethal methods for livestock protection; other people may place a higher value on the predators than on the livestock and not want predators killed. In this case, the various stakeholders are pursuing their own self-interest but differ in where their self-interest lies. These differences in self-interests and perspectives are a major reason why lethal control of wildlife is controversial (see Chapter 15 for more details).

In contrast, when nonlethal techniques are used to stop wildlife damage, there is no market externality and little controversy. In this case, the rancher has to pay the cost of the nonlethal technique and also reaps the benefit if it is successful (all of the costs and benefits fall upon the same individual, so he or she can make rational decisions). When a rancher uses nonlethal techniques, wildlife are rarely harmed and society allows ranchers to make their own decisions about whether the cost of using a nonlethal technique is economically justified. Likewise, if a lethal technique is used to protect another wildlife species, there may be no market externality because the wildlife species being killed and the wildlife species that benefits are both owned by society. Hence, society can make a decision about whether a lethal control program is worthwhile by comparing the relative value of the different wildlife species. For instance, by killing an average of 13 arctic foxes annually, one predator control program in Alaska saved approximately 3000 brant eggs each year from depredation (Anthony et al. 1991). Because brant were uncommon and declining in the area while arctic foxes were abundant, the authors believed that this trade-off of foxes for brants was justified. At Gray's Lake National Wildlife Refuge in Idaho, killing about 80 predators annually from 1976 to 1984 resulted in saving about 4 to 7 whooping crane eggs and young from predation each year (Drewien et al. 1985). Most people supported this predator program because whooping cranes were threatened with extinction. However, conflicts still arise when segments of society place different values on various species.

DO LETHAL TECHNIQUES POSE A RISK TO NONTARGET SPECIES?

A problem with any lethal technique is the danger that the wrong species will be killed. This is called a "nontarget hazard." The species that we are trying to kill is referred to as the "target species" while all other animals are "nontarget species." When animals are shot, the nontarget hazard is low because the shooter must see the animal before it can be killed. Thus, the shooter has an opportunity to avoid killing the wrong species by simply not pulling the trigger.

Traps, however, are less selective because no one is present when an animal is captured. Hence, traps pose a danger that a nontarget species will be killed. Several steps can be taken to make traps more selective. Some selectivity can be obtained by only using traps of the proper size. Additionally, some leghold traps have a pan-tension device which can be adjusted so that the trap will not close unless an animal of a certain weight steps on it. Snares can be equipped with a break-away mechanism so that if a deer or other large animal gets caught, the animal can break free. Some traps are designed to catch only specific species. For instance, the egg trap consists of an egg-shaped plastic container which encloses a trap. It is designed to catch only raccoons because an animal has to reach inside a small hole to get caught and raccoons are one of the few species having the dexterity to do this.

Traps also can be set in locations where only the target species can access them. Traps for squirrels should be set on tree branches where only arboreal mammals can reach them, mole traps should be set in mole burrows, and beaver traps set in water. As another example, the USDA Wildlife Services has a strict policy of not setting leghold traps within 10 m of a carcass to avoid catching scavenging birds and mammals (U.S. Department of Agriculture 1994).

Traps can also be made more species-specific by using bait that is attractive to only the target species. Many species use chemicals (called pheromones) to mark their territories. These pheromones are of great interest to members of a particular species, but seldom attract others. Hence, they can be used to target selectively that species. For instance, beaver traps are often baited with some of the secretion from a beaver's castor gland.

Through careful selection of the proper trap, location, and bait, traps can be made quite selective, but this requires considerable expertise. In the hands of an inexperienced individual, traps pose a great threat to nontarget species. This is one of the major benefits of having professional trappers, such as those employed by the USDA Wildlife Services, be responsible for using lethal means to control predators rather than expecting landowners to do it themselves.

The nontarget hazard also can be high when toxicants are used to kill animals because no one is present when an animal consumes a poisoned bait. Nontarget species may receive a lethal dose of a toxicant by consuming the poisoned bait intended for the target species or by eating the carcass of a poisoned animal (called secondary poisoning). For instance, Balcomb (1983) showed that red-shouldered hawks occasionally received a toxic dose of carbofuran by feeding on animals that had been killed by it. A secondary poisoning hazard also exists for barn owls that eat rodents killed by anticoagulant rodenticides (Mendenhall and Pank 1980). On the other hand, Eastland and Beasom (1986) found that mammalian scavengers were unlikely to be killed if they fed on 1080-killed coyotes. Some toxicants may also be absorbed by plants and pose a risk to herbivores. To reduce these and other environmental hazards, the U.S. Environmental Protection Agency only approves toxicants which decompose rapidly and do not accumulate in the food chain.

Like traps, there are many ways to make toxicants species-specific, such as placing toxic baits in locations where they are accessible only to the target species (Hygnstrom et al. 1998). For instance, farmers place poisonous baits for voles under boards or in furrows where only small mammals can gain access to them (Hygnstrom et al. 2000). Pocket gophers and ground squirrels are targeted by placing the poison in their burrows, rather than on the ground. Whisson (1999) designed and tested bait stations which could be used to deliver toxicants to California ground squirrels but would pose no risk to endangered kangaroo rats. The openings to her bait stations were elevated, excluding kangaroo rats, which are poor climbers, but not ground squirrels, which are agile climbers.

Toxicants can also be made more specific by using a poison that the target species is more vulnerable to than nontarget species. Canids, such as coyotes, are very susceptible to compound 1080. Therefore, baits can be placed out that contain enough 1080 to kill a coyote, but not enough to kill another predator. Also, baits can be used which will not attract the attention of nontarget species. Baits intended for small mammals can be dyed green so that they will not attract the attention of birds (Brunner and Coman 1983; Bryant et al. 1984). Repellents can also be applied to poisonous baits to reduce their appeal to nontarget species.

In assessing the risk a toxicant poses to nontarget species, some of the first steps involve determining how much of the toxicant the nontarget species might obtain, how long the toxicant persists in the animal before it is excreted or detoxified, and what the average lethal dose is for that species (referred to as the LD_{50} or the dose which kills 50% of the individuals receiving it). As one example of this approach, O'Brien et al. (1986) showed that feral pigs poisoned with 1080 often vomited and that concentrations of the poison in the vomitus exceeded the LD_{50} for some bird species based on how much of the vomitus an individual bird might consume.

Receiving a sublethal dose of a pesticide may not kill an animal immediately, but the dose may be sufficient to cause illness or impair the animal's ability to survive. The hazards of a sublethal dose can be difficult to assess. For instance, if a small dose causes a slight impairment of motor function, its effects might not be discernible in captive animals, but could prevent a free-ranging animal from evading a predator.

Whenever lethal techniques are employed, it is important to assess the nontarget hazard. It is easy to tabulate the total number of nontarget animals killed by traps and shooting. For instance, Newsome et al. (1983) reported the number of nontarget animals killed in dingo traps. The U.S.D.A. Wildlife Services keeps track of how many target and nontarget animals it kills yearly during its control operations (Table 7.2). It is also possible to count the number of individuals killed by fumigating dens or burrows or by spraying bird roosts (Meinzingen et al. 1989). Dolbeer et al. (1991) evaluated the nontarget hazard associated with using gas cartridges to kill woodchucks in their burrows by digging up the burrows to ascertain what other

Table 7.2 Comparison between the Numbers and Species of Mammals Killed Intentionally (Target Species) and Unintentionally (Nontarget Species) by the U.S.D.A. Wildlife Services Program during 1991 (U.S. Department of Agriculture 1994)

Number of target mammals killed		Number of nontarget mammals killed	
Species	**Number**	**Species**	**Number**
Coyotes	96,000	Raccoons	1280
Beaver	16,000	Gray Fox	897
Striped skunk	9000	Peccary	703
Red fox	8000	Porcupine	627
Raccoon	7000	Red fox	446
All other mammals	17,000	All other mammals	3117
Total	153,000	Total	7070

animals were inside. They reported that the nontarget hazard was minimal — one cottontail rabbit and three mice killed from 98 burrows.

It is harder to determine the number of nontarget animals killed from oral toxicants because dead animals might be dispersed over an extended area, the size of which would depend upon the home range of the nontarget species and how quickly the animal is killed by the toxicant. For instance, strychnine kills quickly and most dead animals will be found within a few meters of the poisoned bait. In contrast, Avitrol kills within 10 to 30 minutes (Conover 1994), and DRC-1339 kills within a day or two. Animals killed by slow-acting toxicants might be located over a wide region and only a small proportion of them will be found during a search for carcasses. Hence, it is difficult to determine the actual number of animals killed by the toxicant. Estimating the true number of nontarget kills is made even more difficult because carcasses of small mammals and birds are easy to overlook during a search and can be quickly consumed by scavengers (Tobin and Dolbeer 1990; Linz et al. 1991). In some areas, scavengers are so numerous and efficient that carcasses can disappear within a day (Balcomb 1986). However, the disappearance rate can be measured and data on nontarget kills can be corrected for it.

ARE LETHAL TECHNIQUES HUMANE AND SOCIALLY ACCEPTABLE?

Many people object to the use of lethal means for reducing wildlife damage. Some people empathize with the animals that will be killed by the lethal technique and are concerned about the animal's pain and suffering. However, there is no clear division between lethal and nonlethal techniques because even "nonlethal" techniques may result in the death of animals. For instance, translocation is considered a nonlethal technique but many relocated animals do not survive (as we discover in Chapter 9). Excluding bats from an attic by closing the hole they used for access is considered a nonlethal technique (at least if the bats are outside when the hole is closed), but we do not know how many of the displaced bats will die before they are able to find a new roosting site.

If animals must be killed, everyone wants them to be killed as humanely as possible to prevent undue suffering. However, there are debates about whether one technique is more humane than another (e.g., are traps that cause animals to drown less humane than snares that use a noose to suffocate animals?) or whether a lethal technique is too inhumane to use (Ludders et al. 1999). For example, there have been several efforts to ban the use of leghold traps because some people view them as cruel.

Some people feel that wildlife species are too precious to permit their killing and that animals have a right to exist despite the damage they may cause. Other people sympathize with those individuals suffering damage and believe that lethal control should be an option whenever there is a human–wildlife conflict. Some people think the suffering caused by overpopulation (e.g., starvation) is worse than death from lethal control because the latter is usually quicker.

Sidebar 7.5 Using Lethal Control of Gulls to Reduce the Frequency of Bird–Aircraft Collisions at John F. Kennedy Airport in New York City

Adjacent to JFK Airport is a laughing gull colony which grew from less than 10 nests during 1972 to over 7500 nesting pairs during 1990. These gulls posed a significant hazard to aircraft as they flew over the runways or loafed on them. Control of the gulls at the colony was impossible because it is located within a wildlife refuge. As a last resort, the airport contracted with U.S.D.A. Wildlife Services to shoot gulls flying over the runways. From 1991 to 1997, over 47,000 gulls were shot. This action reduced the number of aircraft collisions with gulls at JFK airport from over 150 per year before the program to less than 20 per year afterwards (Figure 7.10). Yet, it had no impact on the regional population of laughing gulls; in fact, the population increased during this same period (Dolbeer 1998).

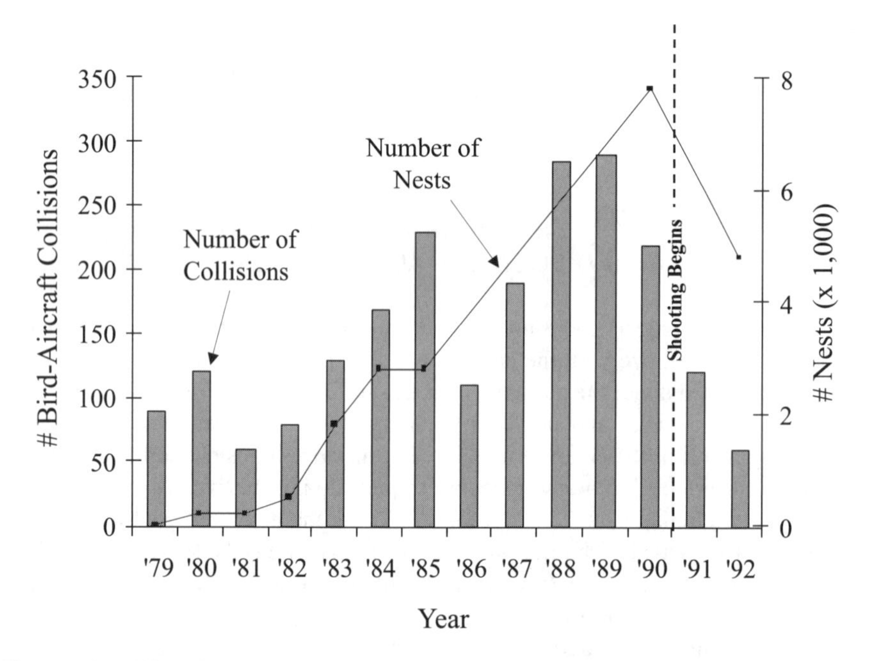

Figure 7.10 Effect of a gull shooting program beginning in 1991 at John F. Kennedy International Airport in New York City on both the number of bird–aircraft collisions (solid bars) and the number of gulls nesting in a colony at the end of the runway (black line).

Most people, however, believe that lethal control is acceptable under some circumstances, but not others. These people do not like the use of lethal methods but support its use for serious problems. For instance, many people would support killing wildlife if there is a risk to human life or health (such as protecting airplane passengers from airplane crashes caused by bird–aircraft collisions), but would not support lethal control to alleviate a nuisance problem (Sidebar 7.5). All leaders of animal rights and animal welfare groups interviewed by Hooper (1992) felt that it was acceptable to kill animals if a human life swayed in the balance (e.g., it was acceptable for someone lost in the wilderness to eat animals to survive). Many people support killing predators to protect endangered species from becoming extinct but not to stop predation on an abundant species. People also change their opinions about lethal control based on what species are going to be killed. Many people are less upset if lethal control is going to be directed at a mouse or snake than at an elephant or mountain lion.

Hence, society is divided over the question of using lethal techniques in wildlife damage management. It is a personal question and a matter of an individual's perspective and values. However, the following principles are generally agreed upon: 1. nonlethal techniques which are efficacious and efficient should be used in lieu of lethal techniques, and 2. if lethal techniques are used, we should select those that are as humane, selective, and socially acceptable as possible.

COMMON METHODS USED IN LETHAL CONTROL

Cage Traps

Cage or box traps usually consist of a rectangular box made of wire, sheet metal, netting, wood, or plastic. Most have a trap door at one or both ends which closes when the animal steps on a trigger inside the trap. These traps are designed to hold the animal, which can then be released unharmed, relocated, or euthanized (Figure 7.11). Cage traps of various sizes are used to catch everything from mice to moose (VerCauteren et al. 1999).

The Australian crow trap is a modified box trap. It is usually large (over 2 m × 2 m × 2 m) and made of woven wire. It has narrow slits cut in the top so that birds can drop down into the trap but cannot fly out (Figure 7.12). It functions as a decoy trap and is effective against birds that normally forage in flocks, such as crows, blackbirds, and starlings (Dolbeer 1994; Johnson and Glahn 1994). To catch a particular species, a few individuals of that species are placed inside the trap along with food and water. These individuals then attract conspecifics that try to enter the trap. The trapped birds can then be translocated or euthanized.

Leghold Traps

Leghold traps are one of the most commonly used methods in North America for catching mammalian predators (e.g., wolves, coyotes, foxes, and raccoons) and aquatic mammals (e.g., beaver, nutria, and muskrats). These traps are designed to hold the animal until the trapper returns (Figure 7.13). This allows the trapper to release nontarget animals that might have been captured. Some people consider such traps to be inhumane because the animal is held (sometimes for an extended period) rather than being killed immediately. They note that animals are often injured during this holding period (Van Ballenberghe 1984; Olsen et al. 1988). Currently, there is much research aimed at reducing this injury rate. For instance, a laminated foothold trap (Houben et al. 1993) and a padded-jaw trap have been developed (Olsen et al.

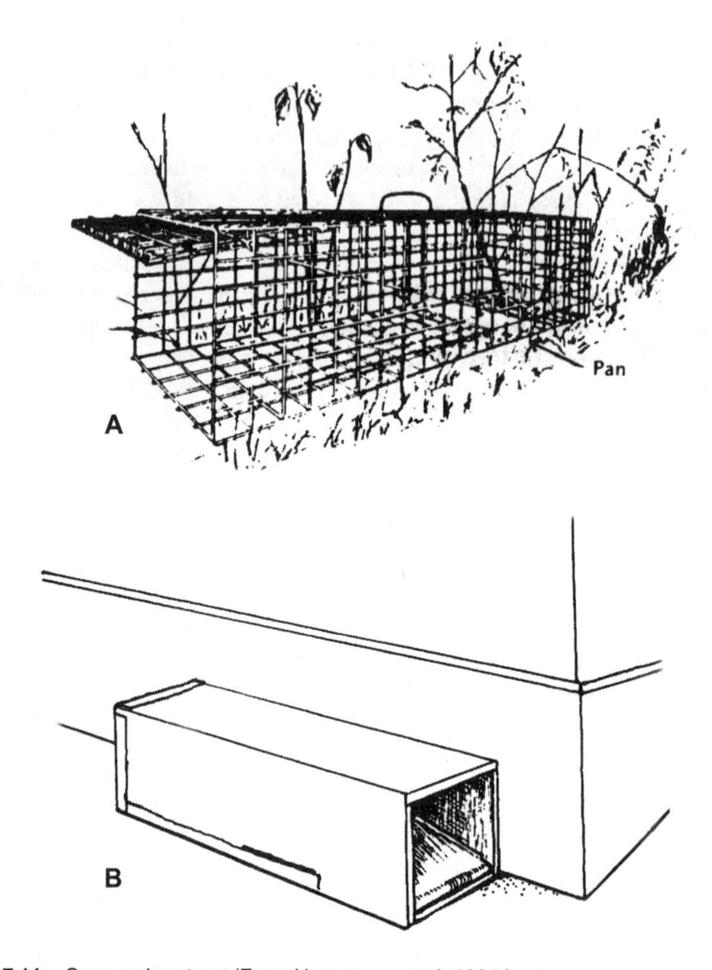

Figure 7.11 Cage or box trap. (From Hygnstrom et al. 1994.)

1986, 1988; Phillips et al. 1996) that help cushion the animal's foot. Another approach is to attach a tranquilizer tablet on the trap. Trapped animals normally bite at the trap and, in doing so, consume some of the sedative (Sahr and Knowlton 2000). Leghold traps used to catch aquatic mammals can be set so that a trapped animal will dive for deeper water and drown. This eliminates the holding period, but does not mollify critics of leghold traps, because many people consider drowning to be an inhumane way to kill animals (Ludders et al. 1999).

Killing Traps and Snares

Some traps are designed to kill, rather than hold, the animal. Snap traps are the most common traps in the U.S. and are used by many households to kill mice and rats in buildings (Figure 7.14). They have a baited trigger that, when moved by the animal, releases a spring-powered bar that snaps over and breaks the animal's spine. Spring-powered harpoon traps are used to control moles in yards and gardens (Figure 7.14).

Figure 7.12 Photos of a modified Australia crow trap taken from the side (A) and from the top (B) showing the narrow slit through which birds can jump into the trap but cannot fly out.

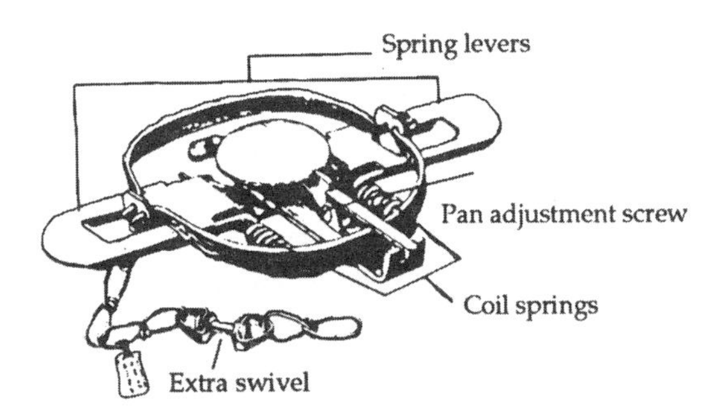

Figure 7.13 Leg-hold trap. (From Hygnstrom et al. 1994.)

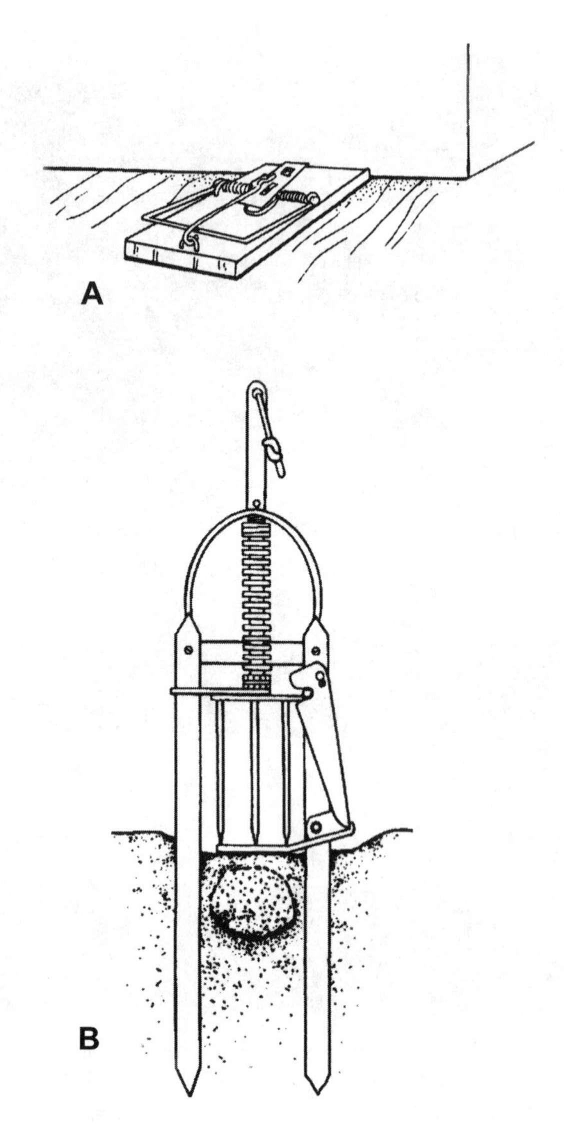

Figure 7.14 Snap trap commonly used to kill mice inside buildings (A) and a harpoon-type trap (B) used to kill moles. (From Hygnstrom et al. 1994.)

These are placed over a section of a mole's tunnel that the trapper has smashed. When the mole returns to repair the tunnel, its activity sets off the trap and the animal is speared.

Conibear traps are rectangular loops and are placed along a trail or burrow (Figure 7.15). When an animal attempts to walk through one of these loops, two spring-powered bars snap shut, crushing the animal. These traps are commonly used in shallow water to catch beaver, muskrats, and nutria.

Snares are a wire noose with a locking device that can only move in one direction. As a result, the noose can be tightened by the trapped animal but not loosened. Snares are set at sites where the target animal must pass through a restricted path or crawl

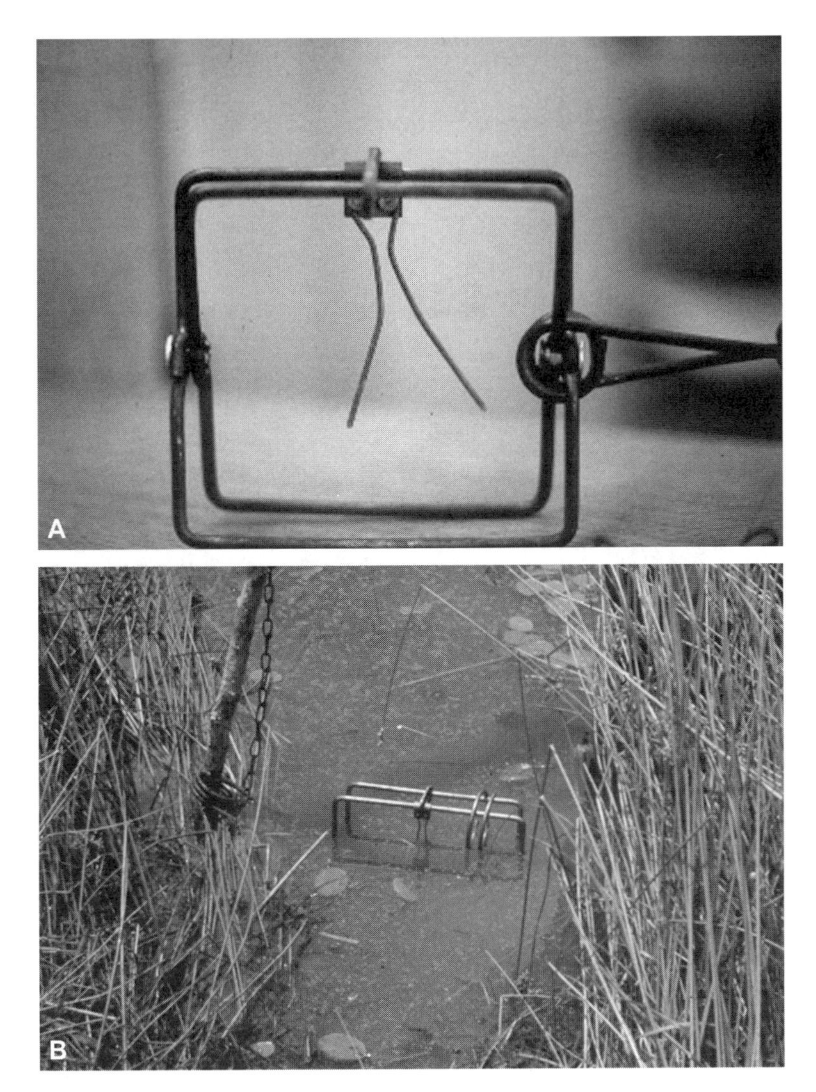

Figure 7.15 Conibear traps. (Photos courtesy of Gary San Julian and C. R. Madsen. With permission.)

under a fence. When an animal does so, the snare closes on its neck and the animal's attempts to free itself only serve to tighten the noose. Death typically is caused by suffocation. Snares can also be set to hold an animal without killing it. There also are foot snares, which are spring-powered devices that propel a noose around the animal's leg when it is triggered. Foot snares hold the animal much like leghold traps.

Denning

One way to kill animals that use dens (e.g., coyotes and foxes), warrens (e.g., European rabbits), or burrows (e.g., ground squirrels, prairie dogs, and woodchucks)

Figure 7.16 A red fox den (note the birds' wings in front of the burrow, some of the fox's prior meals).

is to trap the animals in their dens and burrows and then dig them out and euthanize them (Figure 7.16; Hygnstrom and VerCauteren 2000). More commonly, occupants of the burrow are asphyxiated when the den is fumigated and den openings are closed. The dens and burrows can then be destroyed so that they do not attract other animals (Williams et al. 1995). Many opponents of predator control find this method of control distasteful (Wagner 1988). In 1994, gas cartridges and aluminum phosphide were registered with the U.S. government as den fumigants to control the following species: pocket gophers, commensal rodents, voles, prairie dogs, ground squirrels, woodchucks, yellowbelly marmots, chipmunks, coyotes, skunks, and moles (Jacobs 1994).

Roost Sprays

Starling and blackbird roosts in North America and quelea roosts in Africa may contain over a million birds each. Sometimes, birds associated with these roosts cause so much damage that lethal means are used to reduce their number. This can be accomplished through the use of a chemical surfactant which breaks down the oil in feathers. When sprayed on roosting birds along with water during cold weather, the feathers become saturated with water and the birds die from hypothermia (Stickley et al. 1986).

Shooting

Shooting is a common way to kill large- and medium-sized animals that are causing problems. This method is labor intensive but has the advantage of being species specific. Shooting has been used successfully to reduce the hazard of aircraft colliding with cattle egrets (Fellows and Paton 1988) or gulls (Dolbeer et al. 1993; Sidebar 7.5).

Sometimes, animals are shot from aircraft. This has been used to reduce coyote predation on livestock (Connolly and O'Gara 1987). Various techniques can be used to lure an animal close enough so that it can be shot (Coolahan 1990). These might include using a territorial call or a distress call (this process is called "calling and shooting"). Another approach is to use "Judas animals." For instance, to track down and kill feral goats, wildlife damage managers can place a radio collar on a goat and let it go. Goats are social animals and the Judas goats will soon join up with other goats. The shooter can then locate the goatherd by homing in on the radio signal (Taylor and Katahira 1988).

Toxicants

A number of chemicals have been used to kill depredating animals (Table 7.3). Most toxicants have to be consumed in order to be fatal. Hence, they are placed in a bait so target species will ingest them. The advantage of toxicants is that they do not require much labor and can be used to kill large numbers of animals even in remote areas. Their disadvantage is that poisons frequently are not as species specific as we might prefer.

A few toxicants are administered by methods other than mixing them with a food bait. Fenthion is a toxicant that is readily absorbed through the skin. Hollow wick perches, such as the Rid-A-Bird® perch, are saturated with this toxicant. Thus, when a bird lands on the perch, it gets a lethal dose of fenthion on its foot. These perches are used around buildings to control starlings, house sparrows, and pigeons (Jackson 1978).

When coyotes attack sheep, USDA Wildlife Services employees sometimes use a spring-powered device called the M-44 to deliver a lethal dose of sodium cyanide to kill the coyotes. A hollow stake is driven into the ground and a spring-powered ejector placed in the stake and armed with a cyanide capsule. A bait designed to only attract canids is then spread on the capsule. When a coyote grabs the bait with its mouth and pulls, the ejector fires the cyanide capsule into the coyote's mouth (U.S. Department of Agriculture 1994). M-44s are selective for coyotes and pose less of a risk to nontarget species than the use of either toxicants or leghold traps (Beasom 1974b).

Diseases and Parasites

Diseases and parasites, such as sarcoptic mange, can spread rapidly through wildlife populations when animal densities are high and can limit wildlife populations (Pence and Windberg 1994). When exotic wildlife species escape into a new area, their populations can increase to levels much higher than in their native range due to the absence of the diseases or other natural constraints (competitors or predators) which limit their numbers elsewhere (see Chapter 6 for more details). For this reason, diseases (Williams et al. 1995) or parasites (Singleton et al. 1995) are sometimes released where an exotic species is causing destruction in an effort to control the

Table 7.3 Toxicants Registered as of 2001 with the U.S. Environmental Protection Agency (EPA) for Use in Resolving Human–Wildlife Conflicts (commensal rodents include Norway rat, roof rat, house mouse) (U.S. EPA Pesticide Product Information System 2001)

Chemical	Trade Names	Examples of Target Species	Product Status
4-aminopyridine	Avitrol	Blackbirds, grackles, cowbirds, starlings, crows, pigeons, and gulls	9 currently registered products
3-chloro-p-toluidine	DRC-1339, Starlicide	Blackbirds, grackles, cowbirds, starlings, pigeons, gulls, crows, ravens, and magpies	7 currently registered products
Brodifacoum	Talon, Havoc, D-Con, Final	Commensal rodents, rats, and mice	23 currently registered products
Bromadiolone	Contrac, Maki	Commensal rodents, rats, and mice	27 currently registered products
Bromethalin	Assault, Vengeance, Fastrac, Hot Shot	Commensal rodents, rats, and mice	26 currently registered products
Chlorophacinone	Rozol	Commensal rodents, rats, mice, pocket gophers, moles, ground squirrels, jackrabbits, chipmunks, muskrats, and voles	18 currently registered products
Difethialone		Commensal rodents, rats, and mice	8 currently registered products
Diphacinone	Liqui-Tox Ramik, Ditrac, Diphacin, Promar, Kill-Ko, ProZap	Commensal rodents, rats, mice, voles, pocket gophers, ground squirrels, jackrabbits, muskrats, and mongooses	45 currently registered products
Pivalyl Valone	Pindone	Technical grade material — No registered pests	1 currently registered product
Sodium Cyanide	M-44 Cyanide Capsules	Coyotes, foxes, and feral dogs	8 currently registered products
Sodium fluoroacetate	Compound 1080 Livestock Protection Collar	Coyotes	7 currently registered products
Strychnine Alkaloid	Strychnine	Pocket gophers and moles	36 currently registered products
Warfarin	Warfarin	Commensal rodents	38 currently registered products
Sodium Warfarin			1 currently registered product
Zinc Phosphide	Zinc phosphide	Commensal rodents, rats, mice, pocket gophers, ground squirrels, prairie dogs, voles, muskrats, nutria, field rats, and mice	33 currently registered products

exotic population. In Australia, the virus responsible for myxomatosis was released to control the European rabbit. This virus caused a substantial reduction in rabbit populations, but resistance to the disease subsequently developed in these populations.

A free-ranging population of domestic cats became established on Marian Island in the Pacific Ocean and were having an adverse impact on nesting seabirds. Feline leucopenia disease was introduced to the island and caused a substantial decrease in the cat population. The few remaining cats were shot or trapped and the cat population was eradicated from the island.

SUMMARY

For many wildlife damage problems, we lack the ability to alleviate the problem in an effective and economical way using nonlethal techniques. Lethal control offers a potential means to solve some of these problems. It is usually employed after a wildlife problem has already begun. Sometimes, when problems are predictable, lethal control is used as a preventive strategy, especially when there are threats to human safety or when unacceptable economic losses can occur quickly. When lethal control is employed, it is best to focus on those individuals actually causing the problem (the culprits) or, if this is not possible, to target the group of animals whose home range includes the site where the problem is occurring (the subpopulation). Only rarely are attempts made to reduce entire wildlife populations, and these usually target an exotic species that is causing environmental destruction. Lethal techniques often produce ephemeral results because a population reduction often results in an increase in the birth rate, a decrease in natural forms of mortality, and an increase in immigration of animals into the area. Consequently, lethal control can result in the killing of many animals without having any impact on the population. Once a lethal control program ends, wildlife populations can quickly recover. Therefore, most lethal control programs have to be continued annually and can become quite costly. Lethal control often is controversial because it is in the self-interest of people suffering damage to solve their problem even if they have to kill wildlife to accomplish this. In contrast, the self-interest of other members of society may lie in prohibiting the killing of wildlife, because they are not harmed by the damage but may be harmed if the wildlife is killed because the wildlife belongs to them. The use of any lethal control technique raises questions about whether it is humane and the threat it might pose to nontarget species. When used by trained personnel, however, the risk of such techniques to nontarget species is usually minimal.

LITERATURE CITED

Allen, D. L., *Our Wildlife Legacy*. Funk and Wagnalls Co., New York, 1962.

Anthony, R. M., P. L. Flint, and J. S. Sedinger, Arctic fox removal improves nest success of black brant. *Wildl. Soc. Bull.*, 19, 76–184, 1991.

Balcomb, R., Secondary poisoning of red-shouldered hawks with carbofuran. *J. Wildl. Manage.*, 47, 1129–1132, 1983.

Balcomb, R., Songbird carcasses disappear rapidly from agricultural fields. *Auk*, 103, 817–820, 1986.

Beasom, S. L., Intensive short-term predator removal as a game management tool. *Trans. North Am. Wildl. Nat. Res. Conf.*, 39, 230–240, 1974a.

Beasom, S. L., Selectivity of predator control techniques in south Texas. *J. Wildl. Manage.*, 38, 837–844, 1974b.

Besser, J. F. and J. L. Guarino, Protection of ripening sunflowers from blackbird damage by baiting with Avitrol FC Corn Chops-99S. In *Proc. Bird Control Sem.*, 7, 200–203, 1976.

Bjorge, R. R. and J. R. Gunson, Evaluation of wolf control to reduce cattle predation in Alberta. *J. Range Manage.*, 38, 483–487, 1985.

Brunner, H. and B. J. Coman, The ingestion of artificially coloured grain by birds, and its relevance to vertebrate pest control. *Aust. Wildl. Res.*, 10, 303–310, 1983.

Bryant, H., J. Hone, and P. Nicholls, The acceptance of dyed grain by feral pigs and birds. I. Birds. *Aust. Wildl. Res.*, 11, 509–516, 1984.

Burham, K. P. and D. R. Anderson, Tests of compensatory versus additive hypotheses of mortality in mallards. *Ecology*, 65, 105–112, 1984.

Burns, R. J., D. E. Zemlicka, and P. J. Savarie, Effectiveness of large livestock protection collars against depredating coyotes. *Wildl. Soc. Bull.*, 24, 123–127, 1996.

Campa, H., III, S. R. Winterstein, R. B. Peyton, and G. R. Dudderar, An evaluation of a multidisciplinary problem: ecological and sociological factors influencing white-tailed deer damage to agricultural crops in Michigan. *Trans. North Am. Wildl. Nat. Res. Conf.*, 62, 431–440, 1997.

Caughley, G., *Analysis of Vertebrate Populations*. John Wiley and Sons, London, 1980.

Clark, W. R., Effects of harvest on annual survival of muskrats. *J. Wildl. Manage.*, 51, 265–272, 1987.

Connolly, G. E. and R. J. Burns, Efficacy of compound 1080 Livestock Protection Collars for killing coyotes that attack sheep. In *Proc. Vert. Pest Conf.*, 14, 269–276, 1990.

Connolly, G. E. and B. W. O'Gara, Aerial hunting takes sheep-killing coyotes in western Montana. In *Proc. Great Plains Wildl. Damage Control Workshop*, 8, 184–188, 1987.

Conover, M. R., Behavioral responses of red-winged blackbirds (*Agelaius phoeniceus*) to viewing a conspecific distressed by 4-aminopyridine. *Pest. Sci.,* 41, 13–19, 1994.

Coolahan, C., The use of dogs and calls to take coyotes around dens and resting areas. In *Proc. Vert. Pest Conf.*, 14, 260–262, 1990.

Cooper, J. A. and T. Keefe, Urban Canada goose management: policies and procedures. *Trans. North Am. Wildl. Nat. Res. Conf.*, 62, 412–430, 1997.

Crosby, L. A. and R. Graham, Population dynamics and expansion rates of black-tailed prairie dogs. In *Proc. Vert. Pest Conf.*, 12, 112–115, 1986.

Dolbeer, R. A., Blackbirds. P. E25–E32 in S. E. Hygnstrom, R. M. Timm, and G. E. Larson, Eds., *Prevention and Control of Wildlife Damage*. University of Nebraska Cooperative Extension, Lincoln, 1994.

Dolbeer, R. A., Population dynamics: the foundation of wildlife management for the 21st century. In *Proc. Vert. Pest Conf.*, 18, 2–11, 1998.

Dolbeer, R. A., J. L. Belant, and J. L. Sillings, Shooting gulls reduces strikes with aircraft at John F. Kennedy International Airport. *Wildl. Soc. Bull.*, 21, 442–450, 1993.

Dolbeer, R. A., G. E. Bernhart, T. W. Seamans, and P. P. Woronecki, Efficacy of two gas cartridge formulations in killing woodchucks in burrows. *Wildl. Soc. Bull.*, 19, 200–204, 1991.

Drewien, R. C., S. H. Bouffard, D. D. Call, and R. A. Wonacott, The whooping crane cross-fostering experiment: the role of animal damage control. In *Proc. East. Wildl. Damage Control Conf.*, 2, 7–13, 1985.

Duebbert, H. R. and J. T. Lokemoen, High duck nesting success in a predator-reduced environment. *J. Wildl. Manage.*, 44, 428–437, 1980.

Eastland, W. G. and S. L. Beasom, Potential secondary hazards of compound 1080 to three mammalian scavengers. *Wildl. Soc. Bull.*, 14, 232–233, 1986.

Eberhardt, L. L., Testing hypotheses about populations. *J. Wildl. Manage.*, 52, 50–56, 1988.

Errington, P. L., Predation and vertebrate populations. *Q. Rev. Biol.*, 21, 144–177, 1946.

Errington, P. L., Factors limiting higher vertebrate populations. *Science,* 124, 304–307, 1956.

Errington, P. L., R. J. Siglin, and R. C. Clark, The decline of a muskrat population. *J. Wildl. Manage.*, 27, 1–8, 1963.

Fellows, D. P. and P. W. C. Paton, Behavioral response of cattle egrets to population control measures in Hawaii. In *Proc. Vert. Pest Conf.*, 13, 315–318, 1988.

Gantz, G. F., Seasonal movement patterns of coyotes in the Bear River Mountains of Utah and Idaho. M.S. thesis, Utah State University, Logan, 1990.

Garrettson, P. R., F. C. Rohwer, J. M. Zimmer, B. J. Mense, and N. Dion, Effects of mammalian predator removal on waterfowl and non-game birds in North Dakota. *Trans. North Am. Wildl. Nat. Res. Conf.*, 61, 94–101, 1996.

Guthery, F. S. and S. L. Beasom, Responses of game and non-game wildlife to predator control in south Texas. *J. Range Manage.*, 30, 404–409, 1977.

Hone, J., *Analysis of Vertebrate Pest Control.* Cambridge University Press, Cambridge, England, 1994.

Hone, J. and H. Pedersen, Changes in a feral pig population after poisoning. In *Proc. Vert. Pest Conf.*, 9, 176–182, 1980.

Hooper, J. K., Animal welfarists and rightists: insights into an expanding constituency for wildlife interpreters. *Legacy*, (Nov./Dec. 1992), 20–25, 1992.

Houben, J. M., M. Holland, S. W. Jack, and C. R. Boyle, An evaluation of laminated offset traps for reducing injuries to coyotes. In *Proc. Great Plains Wildl. Damage Control Workshop*, 11, 148–155, 1993.

Hygnstrom, S. E., P. M. McDonald, and D. R. Virchow, Efficacy of three formulations of zinc phosphide for managing black-tailed prairie dogs. *Int. Biodeterioration Biodegradation*, 41, 1–6, 1998.

Hygnstrom, S. E., R. M. Timm, and G. E. Larson, *Prevention and Control of Wildlife Damage*. University of Nebraska Cooperative Extension, Lincoln, 1994.

Hygnstrom, S. E. and K. C. VerCauteren, Efficacy of five fumigants for managing black-tailed prairie dogs. *Int. Biodeterioration Biodegradation*, 45, 159–168, 2000.

Hygnstrom, S. E., K. C. VerCauteren, R. A. Hines, and C. W. Mansfield, Efficacy of in-furrow zinc phosphide pellets for controlling rodent damage in no-till corn. *Int. Biodeterioration Biodegradation*, 45, 215–222, 2000.

Jackson, W. B., Rid-A-Bird perches to control bird damage. In *Proc. Vert. Pest Conf.*, 8, 47–50, 1978.

Jacobs, W. W., Pesticides federally registered for control of terrestrial vertebrate pests. P. G1–G22 in S. E. Hygnstrom, R. M. Timm, and G. E. Larson, Eds., *Prevention and Control of Wildlife Damage*. University of Nebraska Cooperative Extension, Lincoln, 1994.

Johnson, R. J. and J. F. Glahn, European starlings P. E109–E120 in S. E. Hygnstrom, R. M. Timm, and G. E. Larson, Eds., *Prevention and Control of Wildlife Damage.* University of Nebraska Cooperative Extension, Lincoln, 1994.

Knowlton, F. F., Preliminary interpretations of coyote population mechanisms with some management implications. *J. Wildl. Manage.*, 36, 369–382, 1972.

Knowlton, F. F., E. M. Gese, and M. M. Jaeger, Coyote depredation control: an interface between biology and management. *J. Range Manage.*, 52, 398–412, 1999.

Linhart, S. D. and W. B. Robinson, Some relative carnivore densities in areas under sustained coyote control. *J. Mammal.,* 53, 880–884, 1972.

Linz, G. M., J. E. Davis, Jr., R. M. Engeman, D. L. Otis, and M. L. Avery, Estimating survival of bird carcasses in cattail marshes. *Wildl. Soc. Bull.*, 19, 195–199, 1991.

Ludders, J. W., R. H. Schmidt, F. J. Dein, and P. N. Klein, Drowning is not euthanasia. *Wildl. Soc. Bull.*, 27, 666–670, 1999.

McCallum, H. I. and G. R. Singleton, Models to assess the potential of *Capillaria hepatica* to control population outbreaks of house mice. *Parasitology*, 98, 425–437, 1989.

Meinzingen, W. W., E. S. A. Bashir, J. D. Parker, J. Heckel, and C. C. H. Elliott, Lethal control of quelea. P. 293–316 in R. L. Bruggers and C. C. H. Elliott, Eds., *Quelea quelea: Africa's Bird Pest.* Oxford University Press, Oxford, England, 1989.

Mendenhall, V. M. and L. F. Pank, Secondary poisoning of owls by anticoagulant rodenticides. *Wildl. Soc. Bull.*, 8, 311–315, 1980.

Murton, R. K., N. J. Westwood, and A. J. Isaacson, A study of wood-pigeon shooting: the exploitation of a natural animal population. *J. App. Ecol.*, 11, 61–81, 1974.

Nass, R. D., Efficacy of predator damage control programs. In *Proc. Vert. Pest Conf.*, 9, 205–208, 1980.

Newsome, A. E., L. K. Corbett, P. C. Catling, and R. J. Burt, The feeding ecology of the dingo. I. Stomach contents from trapping in south-eastern Australia, and the non-target wildlife also caught in dingo traps. *Aust. Wildl. Res.*, 10, 477–486, 1983.

Nichols, J. D., M. J. Conroy, D. R. Anderson, and K. P. Burnham, Compensatory mortality in waterfowl populations: a review of the evidence and implications for research and management. *Trans. North Am. Wildl. Nat. Res. Conf.*, 49, 535–554, 1984.

O'Brien, P. H., R. E. Kleba, J. A. Beck, and P. J. Baker, Vomiting by feral pigs after 1080 intoxication: non-target hazard and influence of anti-emetics. *Wildl. Soc. Bull.*, 14, 425–432, 1986.

O'Gara, B. W., K. C. Brawley, J. R. Menoz, and D. R. Henne, Predation on domestic sheep on a western Montana ranch. *Wildl. Soc. Bull.*, 11, 253–264, 1983.

Olsen, G. H., S. B. Linhart, R. A. Holmes, and C. B. Male, Injuries to coyotes caught in padded and unpadded steel foothold traps. *Wildl. Soc. Bull.*, 14, 219–223, 1986.

Olsen, G. H., R. G. Linscombe, V. L. Wright, and R. A. Holmes, Reducing injuries to terrestrial furbearers by using padded foothold traps. *Wildl. Soc. Bull.*, 16, 303–307, 1988.

Parer, I. and B. S. Parker, Recolonisation by rabbits (*Oryctolagus cuniculus*) after warren destruction in western New South Wales. *Aust. Rangelands J.,* 8, 150–152, 1986.

Pearson, E. W. and M. Caroline, Predator control in relation to livestock losses in central Texas. *J. Range Manage.*, 34, 435–441, 1981.

Pence, D. B. and L. A. Windberg, Impact of a sarcoptic mange epizootic on a coyote population. *J. Wildl. Manage.*, 58, 624–633, 1994.

Phillips, R. L., K. Suver, and E. S. Williams, Leg injuries of coyotes captured in three types of foothold traps. *Wildl. Soc. Bull.*, 24, 260–263, 1996.

Sahr, D. P. and F. F. Knowlton, Evaluation of tranquilizer trap devices (TTDs) for foothold traps used to capture gray wolves. *Wildl. Soc. Bull.*, 28, 597–605, 2000.

Sargeant, A. B., M. A. Sovada, and T. L. Shaffer, Seasonal predator removal relative to hatch rate of duck nests in waterfowl production areas. *Wildl. Soc. Bull.*, 23, 507–513, 1995.

Saunders, G. R. and G. E. Robards, Economic considerations of mouse-plague control in irrigated sunflower crops. *Crop Prot.,* 2, 153–158, 1983.

Singleton, G. R., L. K. Chambers, and D. M. Spratt, An experimental field study to examine whether *Capillaria hepatica* (Nematoda) can limit house mouse populations in eastern Australia. *Wildl. Res.*, 22, 31–53, 1995.
Smith, R. H., D. N. Neff, and N. G. Woolsey, Pronghorn response to coyote control — a benefit:cost analysis. *Wildl. Soc. Bull.*, 14, 226–231, 1986.
Stickley, A. R., Jr., D. J. Twedt, J. F. Heisterberg, D. F. Mott, and J. F. Glahn, Surfactant spray system for controlling blackbirds and starlings in urban roosts. *Wildl. Soc. Bull.*, 14, 412–418, 1986.
Stout, G. G., Effects of coyote reduction on white-tailed deer productivity on Fort Sill, Oklahoma. *Wildl. Soc. Bull.*, 10, 329–332, 1982.
Taylor, D. and L. Katahira, Radio telemetry as an aid in eradicating remnant feral goats. *Wildl. Soc. Bull.*, 16, 297–299, 1988.
Teer, J. G., D. L. Drawe, T. L. Blackenship, W. F. Andelt, R. S. Cook, J. G. Kie, F. F. Knowlton, and M. White. Deer and coyotes: the Welder experiments. *Trans. North Am. Wildl. Nat. Res. Conf.*, 56, 550–560, 1991.
Till, J. A. and F. F. Knowlton, Efficacy of denning in alleviating coyote depredations upon domestic sheep. *J. Wildl. Manage.*, 47, 1018–1025, 1983.
Tobin, M. E. and R. A. Dolbeer, Disappearance and recoverability of songbird carcasses in fruit orchards. *J. Field Ornithol.*, 61, 237–242. 1990.
U.S. Department of Agriculture, *Animal Damage Control Program Final Environmental Impact Statement*. U.S. Department of Agriculture, Animal Plant Health Inspection Service, Washington, D.C., 1994.
Van Ballenberghe, V., Injuries to wolves sustained during live-capture. *J. Wildl. Manage.*, 48, 1425–1429, 1984.
VerCauteren, K. C., J. Beringer, and S. E. Hygnstrom, Use of netted cage traps in population management and research of urban white-tailed deer. P. 168–178 in G. Proulx, Ed., *Mammal Trapping*. Alpha Wildlife, Sherwood Park, Alberta, Canada, 1999.
Wagner, F. H., *Predator Control and the Sheep Industry*. Regina Books, Claremont, CA, 1988.
Wagner, K. K. and M. R. Conover, Effect of preventive coyote hunting on sheep losses to coyote predation. *J. Wildl. Manage.*, 63, 606–612, 1999.
Whisson, D. A., Modified bait stations for California ground squirrel control in endangered kangaroo rat habitat. *Wildl. Soc. Bull.*, 27, 172–177, 1999.
Williams, K., I. Parer, B. Coman, J. Burley, and M. Braysher, *Managing Vertebrate Pests: Rabbits*. Bureau of Resource Sciences/CSIRO Division of Wildlife and Ecology, Australian Government Publishing Service, Canberra, Australia, 1995.
Witmer, G. W., J. L. Bucknall, T. H. Fritts, and D. G. Moreno, Predator management to protect endangered avian species. *Trans. North Am. Wildl. Nat. Res. Conf.*, 61, 102–108, 1996.

CHAPTER 8

Fertility Control

"As for rabbits, God must have loved them because He made so many of them."

Anonymous

"Sex hormones are invented by doctors to explain things which otherwise would require harder thinking."

Adapted from Jerome LeHuin

"You want to sterilize coyotes? Perhaps you don't understand the problem: the coyotes are killing my sheep, they are not mating with them."

Response of a rancher upon hearing about research on fertility control of coyotes

Fertility control is used to reduce the fertility of a wildlife population to keep that population in check. The goal is to resolve a human–wildlife conflict. Fertility control can be accomplished by interrupting reproduction at various stages. A contraceptive (meaning "against conception") method is one that prevents reproduction by preventing egg fertilization. An abortive method is one that kills the embryo before birth. Fertility control methods in wildlife can be categorized into three general groups: 1. mechanical and surgical techniques, 2. endocrine disruption, and 3. immunocontraception. Each of these methods has advantages and disadvantages that affect its practicality in managing wildlife populations. Ideally contraceptive agents should: 1. be reversible (for some species), 2. be suitable for field delivery, 3. be effective with a single dose, 4. pose no hazard to nontarget species, 5. have no harmful side effects, and 6. have no effect on the social behavior of the animals (Becker and Katz 1997).

NORMAL REPRODUCTIVE FUNCTION

Reproduction is dependent upon a complex interplay of hormones which integrate body functions and regulate the necessary changes required for the production of eggs and sperm. One hormone regulating reproduction, gonadotropin-releasing hormone (GnRH), is produced in a section of the brain called the hypothalamus. GnRH travels in the bloodstream to the anterior pituitary gland at the base of the brain and stimulates release of gonadotropins, which include luteinizing hormone (LH) and follicle-stimulating hormone (FSH). In females, these hormones influence the release of progesterone and estradiol from the ovaries. FSH also stimulates the growth and maturation of eggs, while LH induces ovulation. In males, LH and FSH stimulate the release of testosterone and estradiol from the testes (DeLiberto et al. 1998). The hormones secreted from the ovaries and testes travel in the bloodstream back to the hypothalamus and pituitary gland where they regulate the production of GnRH and gonadotropins through a feedback mechanism (Becker and Katz 1997; Figure 8.1).

MECHANICAL AND SURGICAL TECHNIQUES TO REDUCE FERTILITY

Mechanical devices are one approach to suppressing reproduction in wildlife. In males, plugs or clips can block sperm from exiting the testes. In females, intrauterine devices can be used. These are implanted in the uterus and prevent fertilization of the egg (Asa et al. 1996). Surgery has been used for many years to sterilize domestic pets, zoo animals, and some free-ranging wildlife populations (Neville and Remfry 1984; Bailey 1992; Kennelly and Converse 1997).

The primary advantages of mechanical or surgical sterilization are that the animals are permanently incapable of reproducing after a single treatment and there is no danger to nontarget species. However, surgical sterilization is rarely used in wildlife management because each animal must be live-trapped and transported to a veterinarian for surgery. Thus, it is impractical for most wildlife damage problems.

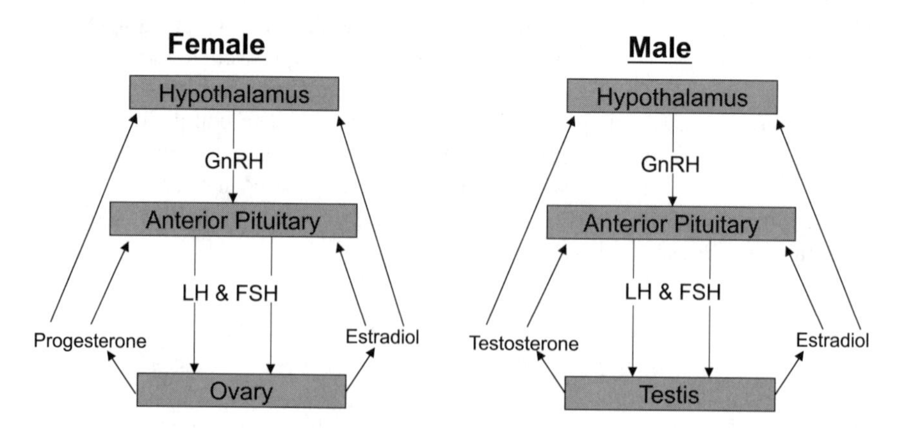

Figure 8.1 Diagram of how different reproductive hormones are regulated in mammals.

At present, surgical sterilization of wildlife is used mainly as a research tool to determine if induced sterility will be effective in controlling a population (Kennelly and Converse 1997).

REDUCING FERTILITY BY DISRUPTING ENDOCRINE REGULATION

Normal hormonal function can be disrupted through the use of natural or synthetic steroid hormones (such as progesterone, estradiol, or testosterone), antiprogestins (which act in much the same way as steroids), GnRH, or GnRH analogs. These analogs are molecules which show an affinity for GnRH receptors in the pituitary glands and compete with GnRH for these receptors.

In the past, fertility regulation in animals has been primarily accomplished through the use of steroids, such as progestins (which prepare the uterus for pregnancy), to disrupt female reproduction. Progestins, however, have been reported to cause abnormal cell growth of the uterine lining along with mammary development and lactation (Asa and Porton 1991; Nettles 1997). Some steroid hormones, such as androgens (a group of male sex hormones including testosterone), can interfere with both male and female reproduction. Unfortunately, androgens can cause changes in sexual and social behavior and may induce external masculinization when administered to females (Tyndale-Biscoe 1994).

Synthetic steroids can also be used to disrupt normal reproductive function. Melengestrol acetate is a synthetic progesterone that is widely used in implants to control reproduction in female zoo animals with no apparent influence on social behavior (Asa 1997); it also can be given orally. Other synthetic progestins used for female contraception include diethylstilbestrol, prostaglandins, mibolerone, medroxyprogesterone acetate (Depo-Provera®), levonorgestrel (Norplant®), and megestrol acetate (Megace® or Oviban®). Quinestrol®, testosterone cypionate, testosterone propionate (Turner and Kirkpatrick 1996), and bisdiamine (Asa 1997) are synthetic steroids that have been shown to be effective in suppressing reproduction in male animals. Potential side effects, high cost, and the administration requirements of steroids usually make them impractical for use in wildlife fertility control (DeLiberto et al. 1998).

Controlling the amount of GnRH to the pituitary gland affects the production of gonadotropins, allowing fertility control of both male and female animals. Large, continuous doses of GnRH or GnRH analogs can inhibit gonadotropin secretion by overloading the pituitary gland. The overload of GnRH fills all of the receptors on the pituitary gland, leading to an initial increase in pituitary stimulation and gonadotropin secretion. However, after all of the GnRH receptors become occupied, the pituitary gland cannot be stimulated any further and gonadotropin secretion along with reproductive function is suspended (Conn and Crowley 1991).

Targeting GnRH function also has several benefits. GnRH and its analogs should not cause environmental problems or secondary poisoning because they are rapidly metabolized into amino acids when consumed (Becker and Katz 1997; De Liberto et al. 1998). Treatment is also reversible and suitable for remote delivery. In the future, the development of long-lasting microcapsules may allow a single administration to be sufficient for fertility suppression over long periods of time (Becker and Katz 1997). The major problem with using GnRH-related methods is that they suppress sexual and social behavior and, consequently, can change the social structure of a population (Becker and Katz 1997). Treatment may also prolong the breeding season or simply delay it, resulting in off-season births (e.g., during winter). Unfortunately, little work has been conducted to test the effectiveness of GnRH-related methods on wildlife (Becker and Katz 1997).

IMMUNOCONTRACEPTION

Immunocontraception uses the individual's own immune system to disrupt reproduction. A vaccine is administered to fool the animal's immune system into producing antibodies against its own reproductive hormones or gametes. Contraceptive vaccines are directed against reproductive hormones, such as GnRH, LH, and FSH, to prevent them from functioning correctly (Hunter and Byers 1996). For instance, to inhibit GnRH, copies of this hormone are bound to a carrier protein and injected into the animal's blood stream. Once there, the body's immune system identifies the carrier protein and the GnRH as foreign invaders, and antibodies are developed to fight them. These antibodies, however, also attack the body's own GnRH. As antibody levels (called titer levels or titers) increase, GnRH levels in the blood decrease below the level necessary to stimulate the pituitary gland to produce LH and FSH, causing ovulation to cease (Becker and Katz 1997; De Liberto et al. 1998). Immunocontraception has also been used to target pituitary gonadotropins. Ovulation in female macaque monkeys has been suppressed by using immunocontraception to attack LH, but attempts to control fertility by targeting FSH in males have produced mixed results.

Hormones produced by the placenta or embryo offer yet another potential target for immunocontraception (Hunter and Byers 1996). In some species, chorionic gonadotropin hormone is produced by the embryo to induce the continued production of progesterone required to maintain pregnancy. When immunocontraception is used to cause an animal to produce antibodies against these hormones, fertilization or gestation can be interrupted (Miller et al. 1998).

There are several problems with using immunocontraception to target hormones. First, normal hormonal function is disrupted and this often affects the sexual and social behavior of the animal (Tyndale-Biscoe 1994). Another problem is that these hormones serve many functions in the body; therefore, the risk of unwanted side effects is high. Additionally, it is difficult to deliver immunocontraceptives orally because they are usually destroyed by the digestive system (Hunter and Byers 1996; DeLiberto et al. 1998).

More promising is the use of immunocontraceptive vaccines to disrupt sperm formation, egg development, fertilization, or implantation. Vaccines that target the egg's zona pellucida (ZP), the layer surrounding a mature mammalian egg, have received the most attention (DeLiberto et al. 1998). Sperm must penetrate the ZP before fertilization can occur (Hunter and Byers 1996). Multiple injections of ZP proteins into the bloodstream cause the immune system to identify the proteins as foreign bodies and to develop antibodies targeted against them. This in turn will interfere with sperm binding. Injections of ZP proteins have been used to decrease fertility in a number of different species (Hunter and Byers 1996). A common source of ZP is pig ovaries, and this type of ZP is referred to as porcine zona pellucida (PZP).

It is also possible to target sperm with an immunocontraceptive vaccine to induce sterility. In normal males, there is a blood–testes barrier that prevents sperm antigens from being detected by the immune system. If this barrier is broken, antibodies are produced against certain proteins in the testes, resulting in sperm destruction. If sperm proteins are injected into the female bloodstream, antibodies to those proteins are produced, which then bind to and kill sperm. This approach shows promise for use in wildlife contraception because there is no change in the behavior of the treated animal (Hunter and Byers 1996; Miller et al. 1998).

Gamete-based immunocontraception has several advantages for controlling wildlife fertility, including its reversibility, lack of interference with the birthing process or milk production, and lack of effect on sexual or social behaviors. It has, however, caused abnormal reproductive cycles in some species. Another drawback is that multiple injections of an individual are required to be effective. Furthermore, long-term effects of immunocontraception using PZP are poorly understood. In the case of feral horses, research indicates that fertility may not be fully restored if PZP is administered for long periods of time (Kirkpatrick et al. 1997b).

ADMINISTERING FERTILITY DRUGS TO ANIMALS

One problem limiting fertility control in wildlife is the difficulty of administering a drug to a free-ranging animal (Kreeger 1997). Remote delivery systems (RDS) are mechanical devices designed to administer a contraceptive to a free-ranging animal. This is usually accomplished by firing a hollow dart tipped with a syringe needle and filled with a liquid drug. The dart can be shot at an animal using a longbow, crossbow, blowgun, pistol, or rifle. The effective range of darts is from 1 to 100 m, and they can deliver up to 25 ml of a drug (Kreeger 1997).

Another RDS, biobullets, are hollow projectiles that can be used to deliver drugs to free-ranging animals by firing them from guns. There has been increasing interest in the use of biobullets to administer contraceptive agents. After a biobullet is shot into an animal, its casing degrades over time, releasing the drug. Biobullets have a range of 25 m and can deliver 125 to 300 mg of a contraceptive (Kreeger 1997).

Another type of RDS is the remote capture collar. It contains a radio transmitter, which allows the animal to be tracked, and a device that administers a drug when signaled to do so. The transmitter can be used to collect data, such as location, ambient temperature, battery life, and dart status. This collar is useful in fertility control when boosters are required and there is a need to collect information about the animal. Once the drug has been administered, the collar falls off and emits a radio signal so it can be recovered. The primary drawback with the use of remote capture collars is their high cost (Kreeger 1997).

A contraceptive can be delivered to an animal through the use of treated bait (oral delivery). An oral drug must be protected from destruction in the digestive tract. This can be accomplished by placing the drug inside micro-spheres or liposomes (Miller 1997). Difficulties with oral delivery techniques include poor bait acceptance, the need for precise timing relative to breeding cycle, drug instability, the inability to control dosage, and the risk of non-target animals receiving a dose (Bomford 1990).

Another type of RDS is genetically modifying an infectious organism so that it contains portions of the target species' own reproductive proteins. When an animal is infected by the genetically modified organism, it will develop antibodies not only to the organism but also to its own reproductive hormone or protein. To be successful, the vector must be a virus or bacterium that is natural to the host, be capable of holding the DNA responsible for the reproductive immunogen, and be able to produce the reproductive hormone or protein (Tyndale-Biscoe 1994). The rabbit disease *Myxoma* has been suggested for contraception in rabbits and hares because it is species specific. Viral-vectored immunocontraception has been proposed for rabbits, red foxes, and brushtail possums (Bradley 1997; Jolly et al. 1997).

CONTROLLING THE RELEASE OF AN ANTIFERTILITY DRUG TO THE BODY

One problem with remotely delivered drugs is ensuring that the drug is released at the correct rate and for the proper period of time. Controlled-release systems are designed to accomplish these tasks. There are four types of controlled-release systems that are appropriate for wildlife: mechanical pumps, osmotic pumps, chemically controlled systems, and liposomes (Kesler 1997). Mechanical pumps use vapor pressure to force the drug through a filter and into the blood at a constant rate. They require surgical implantation and are expensive but can be refilled and precisely control the release of a drug into an animal. Osmotic pumps are devices that contain a drug reservoir surrounded by a polymer and a water permeable membrane. The polymer uses hydrostatic pressure to put enough pressure on the drug reservoir to force the drug through the membrane. These pumps are less expensive, but require capture and restraint of the animal. Chemically controlled systems consist of a polymer matrix with a contraceptive drug either dispersed throughout it or encapsulated within it. When injected into the animal, the polymer matrix gradually degrades, releasing the drug at a constant rate. Liposomes are cellular structures that contain one or multiple lipid bilayers surrounding a liquid core. They discharge their contents through cellular mechanisms.

WHEN SHOULD WE USE CONTRACEPTION IN WILDLIFE?

Fertility control is not a panacea for all human–wildlife conflicts, but it can be used to reduce some wildlife populations. However, before we employ these techniques, we need to understand their long-term effects on the target population (Kennelly and Lyons 1983).

Influence of Mating Systems on Contraception

Male sterilization may be a useful strategy to control polygynous species because the number of breeding males is small relative to the number of breeding females (Nelson 1980). If the dominant male can be sterilized without affecting his sexual and social behavior, all females in his harem may experience a reduction in fertility. However, male sterility would be ineffective in species with promiscuous mating behavior.

Fertility control methods could also be successful when target animals are monogamous and pair bonds are long lasting (Kennelly and Lyons 1983). This would be especially true for species with low immigration rates or that breed only once per year (Eagle et al. 1993; Haight and Mech 1997).

Species that hold and defend territories could also be vulnerable to reproductive control. In the case of the gray wolf, Mech et al. (1996) found that vasectomized wolves continue to hold territories for years. By sterilizing territorial males, the fertility of the population can be reduced for an extended period of time.

Influence of Population Dynamics on Contraception

Many wildlife populations are regulated by density-dependent factors. If the target population is at or near carrying capacity, reducing the birth rate may simply prevent the birth of offspring that would otherwise die. In such cases, reducing the birth rate may have little effect on the size of the population (Bomford 1990). To enhance the effectiveness of a fertility control program, it may be necessary to cull populations before implementing a fertility management program (Kennelly and Lyons 1983; Bomford 1990; Oogjes 1997).

The life history of a species must be considered before contraceptive management plans can be implemented. The reproductive rate, age at which animals reproduce, and survival rate all influence the outcome of any management strategy. Population models can be used to understand the population dynamics of wildlife species as well as assist managers in developing, evaluating, and defending a management decision involving fertility control (Sidebar 8.1).

Sidebar 8.1 Using Population Models to Assess the Impact of Fertility Control on a Wildlife Population.

Population models can demonstrate the impact of a management plan on the wildlife population over both short and long periods of time. For example, the giant fruit bat and the black rat are two species of mammals on the Republic of Maldives archipelago in the Indian Ocean that differ markedly in their reproductive rates. Comparatively, the black rat has higher reproductive and lower survival rates than the giant fruit bat. Computer models predicted that the fruit bat population could be controlled four to six times more efficiently with lethal control than with fertility control. On the other hand, rat populations could be controlled two to three times more efficiently with reproductive control. The validity of this model was supported when lethal control measures reduced populations of fruit bats by 46 to 70% after one year, while rat populations recovered fully in the same year (Dolbeer 1998).

Occasionally, it may be possible to reduce wildlife damage by reducing the fertility of specific animals without having any impact on the size of the population. Beaver damage is usually caused by dispersing young rather than by adults (Kennelly and Lyons 1983). Fertility control could be used to reduce the number of dispersing young each year (Kennelly and Lyons 1983).

Fertility control of coyotes also may reduce their predation on livestock without having any impact on coyote populations. Till and Knowlton (1983) demonstrated that most sheep and lambs are killed by adult coyotes feeding pups. Hence, a program to sterilize the few coyotes with territories near a sheep ranch may end sheep losses. Yet, doing so would not reduce the local subpopulation of coyotes, only the number of adults with pups to feed.

Public perception is another factor that needs to be considered before a program of fertility management is employed (Kennelly and Lyons 1983). Many people believe that contraception is a more humane form of population control than shooting, trapping, or poisoning. This depends, in part, on the delivery mechanism. For example, fertility control may not be considered humane if it is achieved by using a genetically modified virus which causes illness (Oogjes 1997). Capture and handling of animals cause panic and stress, along with possible injuries. This raises the question as to at what point capture and treatment are so traumatic that euthanasia is more humane (McCullough 1996).

USES OF CONTRACEPTION IN WILDLIFE

Canids

Interest in the use of contraceptives in free-ranging canids began as a method to control the spread of rabies by reducing canid densities. Administering diethylstilbestrol (DES) to foxes (Linhart and Enders 1964) or coyotes (Balser 1964) reduced their fertility. Unfortunately, these animals had to ingest DES at precise times for it to work, and neither coyotes nor foxes consumed baits regularly. Balser (1964) and Linhart and Enders (1964) also discovered that coyotes administered DES remained sexually active for a longer period of time and produced pups later in the season than usual. More recently, Allen (1982) found that consumption of DES-treated bait reduced the litter size of foxes from 4.6 to 3.0. Captive red foxes given an oral progestin, Provera, during the breeding season were less likely to produce a litter than untreated foxes (Storm and Sanderson 1969). Soon after this study, melengestrol acetate (another oral progestin closely related to Provera) was tested on domestic dogs. Estrus was completely inhibited. This finding led to the approval of the first commercial progestin for dogs (Kirkpatrick and Turner 1985).

Other researchers attempted to control canine fertility by immunizing dogs against LH (Al-Kafawi et al. 1974). This experiment failed because human gonadotropin was used for immunization and the antibodies produced did not react with canine gonadotropin. Vickery et al. (1996) used GnRH antagonists to suppress reproductive cycling in female dogs and inhibit sperm production in males.

Felids

A synthetic progestin, megestrol acetate, reduces fertility in feral cats with only minimal side effects (Remfry 1978). This drug is commercially available in England under the name Ovarid®. Melengestrol acetate implants were tested on free-ranging lions. No treated lions became pregnant during the study. Upon removal of the implants, the effect was reversed and the animals ovulated and became pregnant within a short amount of time (Orford 1996). Another synthetic progestin, medroxyprogesterone acetate (Provera), has been used along with megestrol acetate to inhibit reproduction in captive lions, jaguars, tigers, and leopards.

Elephants

As a result of international legislation banning the sale of ivory, elephant populations are increasing at an alarming rate in parts of Africa. Large populations of these animals have caused agricultural and environmental damage due to their destructive feeding habits. In the past, elephant overpopulation problems have been alleviated by culling and translocation programs. However, some people oppose the use of lethal methods to manage elephant populations, and sites willing to accept relocated elephants are dwindling.

Contraception experiments on elephants began in Kruger National Park in 1996. Ten female elephants were given estrogen implants to reduce fertility. The treated elephants began to experience social problems immediately. Some seemed to be permanently in heat and had trouble fighting off bulls attempting to mate with them, and the program was soon abandoned (Butler 1998). At the same time, 21 elephants were vaccinated with ZP; boosters were later administered with darts shot from a helicopter. Elephants receiving an initial dose of ZP followed by two booster shots had lower pregnancy rates than untreated elephants and did not experience negative side effects (Fayrer-Hosken et al. 1997; Butler 1998). This project is ongoing, and it will be several years before researchers know exactly how the elephants are reacting to the ZP (Butler 1998).

Deer, Elk, and Mountain Goats

Ungulates often live in places, such as national parks and urban areas, where lethal means cannot be used for population control. As early as 1964, researchers were attempting to use DES fertility control to manage elk herds in Yellowstone National Park. Injections of DES terminated pregnancy in 30% of treated elk (Greer et al. 1968). DES delivered orally or as an implant to female white-tailed deer produced a reduction in pregnancy rates (Harder and Peterle 1974; Bell and Peterle 1975). Other experiments in deer have not been as successful. Matschke (1977) orally administered microencapsulated DES to white-tailed deer. Deer did not take the bait well, required high doses of the drug, and experienced postabortion pregnancy, making this technique impractical.

Roughton (1979) successfully prevented reproduction in female deer by using orally administered melengestrol acetate. While there were no unwanted side effects, the drug required daily administration and, therefore, was impractical for use in free-ranging populations. Melengestrol acetate implants produced two years of infertility in deer (Plotka and Seal 1996). In contrast, levonorgestrel implants were tested in two separate studies but did not reduce fertility sufficiently (White et al. 1994; Plotka and Seal 1996).

Another synthetic progestin, norgestomet, has successfully inhibited reproduction in white-tailed and mule deer (Jacobsen et al. 1995; Kesler 1997). This hormone was incorporated into silicone rods and remotely delivered with a biobullet. In both species of deer, norgestomet was nearly 100% successful in preventing pregnancy but was only effective for one year. This technique might be useful where deer are in small, isolated populations (Warren et al. 1997).

Current efforts to control deer reproduction rely on the use of immunocontraception, specifically PZP. Turner et al. (1992) used blowdarts to remotely deliver an initial vaccine of PZP and two boosters to captive white-tailed deer. None of the treated does produced fawns, while 86% of the does in the control group did.

Encouraged by PZP experiments with captive white-tailed deer, researchers began testing the efficacy of this technique in free-ranging white-tailed deer (Sidebar 8.2). McShea et al. (1997) gave one group of does an injection of PZP followed by a booster shot. Another group was given a single injection of PZP, while the third group was given a placebo. After one year, none of the does that received the booster and 78% of does that were given a single injection of PZP produced fawns. Of the does that received the placebo, 82% produced offspring. Although deer in the second group experienced limited contraceptive efficacy from just one injection the first year, they experienced full contraceptive response from a single booster administered the following year. Both groups receiving PZP experienced a prolonged breeding season when they failed to get pregnant during the normal season (McShea et al. 1997). Another study was conducted in New York (Sidebar 8.2).

Sidebar 8.2 Testing the Effectiveness of PZP on Free-Ranging Deer

The effectiveness of PZP to suppress the fertility of free-ranging white-tailed deer was tested on Fire Island National Seashore in New York. The vaccine was delivered to deer with darts fired from a blowgun. In the first year, does were given either one or two inoculations of PZP. Only 39% of all treated does produced fawns, compared to the 90% fawning rate of the same group before they were treated with PZP. While there was a marked decrease in reproductive success, researchers believe that some of the darts did not function properly. The following year, half of the treated does were given a booster inoculation. Another group of previously untreated does received two inoculations with an improved dart. Only 15% of deer given boosters produced fawns, and 28% of the does receiving two inoculations produced fawns (Kirkpatrick et al. 1997b).

Another method of fertility control in deer uses prostaglandins. DeNicola et al. (1997) examined the efficacy of prostaglandins in terminating pregnancy in white-tailed deer. More recently, Waddell (2000) examined the effectiveness of prostaglandin treatment when given at different times during gestation of captive white-tailed deer. Treatment with prostaglandins was found to be more effective at inducing abortion during late gestation compared to early gestation.

Due to success in reducing deer reproduction with PZP, studies were conducted to determine how elk respond to treatment with PZP. Heilmann et al. (1998) treated cow elk with PZP to determine the behavioral side effects of this treatment. They found that treated cow elk continued to show mating behaviors well into the post-breeding season, indicating that treated cows that failed to become pregnant during the normal breeding season continued to cycle (Heilmann et al. 1998).

Mountain goats have also been the subject of contraceptive studies. Hoffman and Moorhead (1996) placed silicone implants containing melengestrol acetate into female mountain goats. They also injected sclerosing agent into the epididymis of five male mountain goats. For the next five years, the average kid production of treated females was 10%, compared to 68% for control females. Males were examined after two years, and their epididymis remained blocked. However, kid production had not been reduced in areas with treated males. Because this treatment requires the capture and handling of each animal, it is impractical to use over large areas (Hoffman and Moorhead 1996).

Equids

Control of feral horse populations has been difficult and controversial since feral horses received federal protection in 1971. This legislation bars the use of lethal means to manage horse populations, and relocating horses is ineffective. In response to the constraints imposed by this legislation, wildlife damage managers have been investigating the use of contraceptives as an alternative method of controlling horse populations. One method of controlling fertility in feral horses is to surgically sterilize dominant males. This technique is effective because horses are polygynous and dominant males try to keep other males away from females in their harems. However, this approach may not be sufficient to reduce horse populations, due to band instability and breeding by peripheral males (Eagle et al. 1993). Computer models simulating horse population dynamics and breeding cycles indicate that population growth would only be stopped if a large proportion of males could be sterilized (Garrott and Siniff 1992).

Another contraceptive technique that has been attempted with horses is the use of testosterone cypionate and microencapsulated testosterone propionate. After either of these drugs was administered, stallions experienced a significant decrease in sperm count while still retaining their normal social and sexual behaviors. The drugs were effective for about six months, after which the stallions regained their fertility (Turner and Kirkpatrick 1996). Disadvantages of this method are high cost and logistical problems associated with having to inject a contraceptive into free-ranging animals.

Kirkpatrick et al. (1997a) remotely administered a microencapsulated synthetic progestin to mares, but instead of inhibiting reproduction, progestin enhanced it (100% of the treated mares became pregnant). Plotka et al. (1996) placed implants containing steroid hormones into mares. Despite problems, such as the loss of implants, this method suppressed ovulation in feral mares for at least two breeding seasons.

Another approach was the use of intrauterine devices (IUDs) in mares. Daels and Hughes (1995) compared pregnancy rates among six mares with IUDs and six untreated mares. All of the untreated mares and none of the treated ones were pregnant the next year. Treated mares maintained normal reproduction cycles and became pregnant after the IUD was removed (Daels and Hughes 1995). While insertion of the IUD is simple and quick, it still requires capture of individual animals and, therefore, is impractical for widespread use in free-ranging populations.

Several studies have examined the use of immunocontraceptives in feral horses (Sidebar 8.3). During 1992, Turner et al. (1997) began another field study in which 131 free-ranging mares were caught and assigned to one of five treatment groups. The first group was administered two inoculations of PZP about three weeks apart, the second group received one inoculation, the third group received one inoculation of microencapsulated PZP, the fourth group received a placebo injection, and the fifth group was left untreated. The sham-treated mares had a reproductive success of 55% and the untreated mares 54%. Reproductive success of the treated mares was 5% (two injections), 29% (one injection), and 20% (one injection with microencapsulated PZP). Other species for which PZP vaccines have been tested include captive donkeys, Przewalski's horses, and zebras (Kirkpatrick et al. 1997b). The vaccine was 100% effective in preventing pregnancy in Przewalski's horses and more than 80% effective in zebras.

Sidebar 8.3 Using PZP to Reduce the Fertility of Horses

Kirkpatrick et al. (1990) tested the effectiveness of a PZP vaccine in free-ranging horses on the Assateague Island National Seashore. Using dart guns, they inoculated 26 mares with PZP and later administered a booster shot. Eighteen of the treated mares subsequently received a third booster. None of the 18 mares inoculated three times produced young, and only one of the mares receiving two inoculations produced a foal. Half of the untreated mares produced foals during the same year. PZP did not affect social behavior or herd organization and its effects were reversible when treatment stopped. The treatment did not interfere with existing pregnancies; 14 of the 26 treated mares were pregnant at the time of inoculation and, subsequently, all gave birth to healthy foals.

Rodents

Promiscuous behavior, opportunistic breeding, short life spans, and high turnover rates of individuals within a population make controlling rodent population a difficult proposition (Bomford 1990). Despite this, rodent fertility control programs began in 1969 when Marsh and Howard (1969) reduced pregnancy rates in free-ranging rats fed mestranol baits. However, the baits were poorly accepted by rats, and the method was soon abandoned. Marsh and Howard (1969) discovered that voles regularly accepted treatment bait and experienced a reduction in pregnancy rate. Female voles that consumed mestranol passed it to their nursing pups, making them irreversibly sterile. Another compound, quinestrol, was found to inhibit reproduction in rodents.

As with mestranol, the treated baits were poorly accepted, making the approach impractical for use in free-ranging populations (Brooks and Bowerman 1971).

Alpha-chlorohydrin, also known as Epibloc®, is the only drug approved by the U.S. Food and Drug Administration (F.D.A.) for fertility control in rodents. It is less expensive than other steroids, such as DES and mestranol, and renders male rodents permanently sterile without affecting sexual or social behavior. However, its usefulness is limited due to its toxicity and poor bait acceptance (Kirkpatrick and Turner 1985).

Brooks et al. (1980) and Kennelly and Lyons (1983) investigated the potential of surgical sterilization to control beaver populations. They sterilized either the adult male or female in colonies composed of one adult pair and several of their litters. This markedly reduced reproduction, but resulted in the disruption of pair-bonds in several colonies, particularly those where the male was castrated. Castrated males exhibited abnormal behavior and left the colonies. Surgical sterilization, however, is a very labor intensive approach and may be impractical in field situations (Kennelly and Lyons 1983).

Hormonal-based contraception was tested in black-tailed prairie dogs, using DES-treated oats (Garrett and Franklin 1983). Reproductive suppression was nearly 100%, and the effects were totally reversible once the treatment was discontinued.

Birds

Many drugs have been used to reduce fertility in birds. Reproduction in feral pigeons was suppressed using Provera, Arasan® (a common seed disinfectant and fungicide), and Ornitrol® (Elder 1964). Reproductive inhibition using Ornitrol in pigeons ranged from 87% to 100% in rural areas, but reached only about 10% in cities (Wofford and Elder 1967). Ornitrol suppressed reproduction by preventing sperm production but also affected sexual behavior (Sturtevant and Wentworth 1970).

Converse and Kennelly (1994) evaluated the efficacy of vasectomizing male Canada geese. Only 16% of nesting attempts made by females paired with a sterilized male were successful. Furthermore, vasectomized males were able to maintain their territories for several years following their vasectomy. Bray et al. (1975) vasectomized male red-winged blackbirds to determine effectiveness of male fertility control. They observed a significant reduction in fertility and believed that chemosterilization was a feasible option in blackbirds. Triethylene melamine reduced fertility in male starlings and red-winged blackbirds in some tests, but not in others (Vandenbergh and Davis 1962).

OILING, ADDLING, OR PUNCTURING EGGS

In addition to chemical fertility control and sterilization, other methods such as egg oiling, addling, or puncturing can be used to impair reproduction in birds such as Canada geese and mute swans (Figure 8.2). Impairing reproduction through these approaches requires a long-term commitment and federal and state permits (Smith et al. 1999).

Figure 8.2 The mute swan is an exotic species in North America, and egg oiling has been used in the past to reduce the number of young produced every year.

Addling eggs involves forcefully shaking the eggs until sloshing is heard, which destroys the embryo. Puncturing eggs is accomplished by piercing the eggshell with a needle. After puncturing or addling, the eggs are placed back in the nest, otherwise the birds will renest and lay another clutch of eggs (Smith et al. 1999).

Oiling eggs prevents respiratory gases from diffusing through the shell and causes the developing embryo to die of asphyxiation (Blokpoel and Hamilton 1989). Treating eggs with mineral oil, corn oil, or other types of food oils is effective in reducing the hatchability of eggs (Blokpoel and Hamilton 1989; Pochop et al. 1998). Oil is applied to the eggs by taking them out of the nest, then dunking, spraying, or brushing the entire egg with oil. The timing of egg oiling is critical. Canada goose eggs should be oiled between the fifth day of incubation and at least five days before hatching (Smith et al. 1999). Experiments on gulls indicate that only eggs oiled in the first eight days of the incubation period tend to hatch (Blokpoel and Hamilton 1989; Pochop et al. 1998). Blokpoel and Hamilton (1989) suggested four visits to a gull colony to oil eggs for maximum results. One advantage to oiling eggs is that birds continue incubating eggs past the normal hatching time, which prevents renesting (Christens and Blokpoel 1991).

EFFORTS IN AUSTRALIA TO RESOLVE HUMAN–WILDLIFE CONFLICTS USING IMMUNOCONTRACEPTIVES

Many exotic vertebrates, including European rabbits, red foxes, horses, cats, dingoes, goats, pigs, buffaloes, donkeys, house mice, and starlings have invaded Australia. Rabbits cause agricultural damage, as they compete with native animals for food, and destroy habitat. As a result, fertility control research in Australia is largely focused on rabbits. Because rabbits create problems throughout the continent and number in the hundreds of millions, normal methods of contraceptive delivery (e.g., baits or darts) are impractical. Instead, Australian researchers are using a virus capable of spreading on its own throughout the rabbit population but which will not infect other species. The approach is to genetically modify a virus to contain a protein found on the surface of rabbits' eggs or sperm. Rabbits infected with the modified virus produce antibodies against the virus and the inserted protein; consequently they will be rendered sterile (Tyndale-Biscoe 1997).

The *Myxoma* virus has been proven to be specific to rabbits and hares. It was originally introduced into Australia about 50 years ago in an effort to eradicate them (Tyndale-Biscoe 1994). At first, the mortality rate caused by the *Myxoma* virus was around 99%. However, in just a few years, the mortality rate declined because the virus evolved into a less virulent form, and rabbits developed some resistance to it (Morell 1993; Tyndale-Biscoe 1994). Nevertheless, use of a genetically modified *Myxoma* virus could offer an option for sterilizing rabbits to decrease their population in Australia.

In an effort to control rabbit populations, the red fox was introduced into Australia. However, the fox began feeding not only on rabbits but also on many native animals and has caused the extinction of many local marsupials (Morell 1993). Because a fox-specific virus has not been identified, virus-vectored immunocontraception cannot be applied. Instead, researchers use baits that deliver an immunocontraceptive agent to the foxes. It is important that foxes and rabbits be sterilized simultaneously. Otherwise, foxes will turn to native marsupials as a food source and decimate their populations when rabbit densities decline (Morell 1993).

Some native marsupials, such as kangaroos, also cause problems in Australia. Due to the provisioning of water for livestock and other factors, kangaroo populations and their range have increased (Bomford and O'Brien 1997). They compete with livestock for food and can cause environmental degradation. Attempts to reduce kangaroo fertility include immunizing them against certain reproductive hormones and using a lactation suppressant called bromocriptine (Bomford 1990). In gray kangaroos, a single injection of bromocriptine stopped lactation, causing the death of young within the pouch. The effect of bromocriptine was reversed when the treatment was stopped. Bromocriptine, however, must be given to kangaroos in the winter when the pouch young are not yet fully developed. To date, there is no oral formulation suitable for field delivery (Hinds and Tyndale-Biscoe 1994).

Koalas are overabundant in parts of Australia, where they cause environmental damage by defoliating and eventually killing entire stands of trees (Menkhorst et al. 1998). In an attempt to reduce their birth rate, Menkhorst et al. (1998) implanted koalas with slow-release silicon tubes containing estradiol or a progestin. None of the koalas that had received estradiol or levonorgestrel produced young, compared to 94% of untreated females.

Levonorgestrel implants have also been administered to female tammar wallabies. Implants did not prevent birth in females that were already pregnant but inhibited subsequent estrus so that no other births occurred in treated females. The treatment did not affect lactation for nursing young already present, and the effects of the treatment were shown to be reversible. In addition, a single levonorgestrel implant lasted at least three years (Nave et al. 2000).

WHAT ARE THE DRAWBACKS TO WILDLIFE CONTRACEPTION?

As with any strategy to resolve human–wildlife conflicts, there are drawbacks to the use of fertility control. For example, contraceptives may cause adverse side effects to the target animals, such as a reduction in body weight (Harder and Peterle 1974), changes in secondary sexual characteristics (Matschke 1977; Gardner et al. 1985), and behavioral changes (Gardner et al. 1985; Turner et al. 1992). Also, some contraceptives are toxic (Nettles 1997) or may produce birth defects if consumed by pregnant animals (Greer et al. 1968; Harder and Peterle 1974; Bell and Peterle 1975; Matschke 1977).

Contraceptives also may adversely affect current offspring of target animals. For instance, a contraceptive might reduce lactation, causing nursing young to starve (Bomford 1990; Hinds and Tyndale-Biscoe 1994). Sterility, gonadal abnormalities, changes in sex ratios, and delayed sexual development have been reported in offspring of animals treated with contraceptives (Marsh and Howard 1969; Storm and Sanderson 1969; Harder and Peterle 1974; Bomford 1990).

Contraceptives may lengthen the annual breeding season due to females going through another cycle when they do not get pregnant the first time (Balser 1964; Linhart and Enders 1964; McShea et al. 1997; Butler 1998; Heilmann et al. 1998). If the breeding period is lengthened, animals could deplete their energy reserves participating in prolonged breeding activities. This could disrupt herd demography, which in turn affects normal social behavior of the population (Heilmann et al. 1998). Additionally, if a female conceives during the extended breeding season due to repeated cycling, her offspring will not have sufficient time to develop before the next winter and may die. Another concern is that contraceptives could affect nontarget animals if they consume treated baits or prey upon sterilized animals (DeLiberto et al. 1998).

PUBLIC PERCEPTIONS OF WILDLIFE FERTILITY CONTROL

Public perceptions of wildlife fertility control are varied. Some people note that pets are commonly sterilized and see no reason why the same methods would be inappropriate for wildlife. Others believe that fertility control is humane but object to the use of certain delivery mechanisms (e.g., darts and diseases) which might produce pain or injury to the animal (Oogjes 1997). Some people oppose it because they do not believe that humans should be managing wildlife. Others support fertility control if it is going to be used as an alternative to lethal control, but they are opposed to using fertility control for species that would otherwise be unharmed. Some hunters and fur trappers are wary of fertility control because one justification for hunting and fur trapping is that it reduces wildlife damage. They worry that if fertility control is used to limit wildlife populations, there would be fewer opportunities to hunt or trap.

LAWS GOVERNING THE USE OF FERTILITY CONTROL TO MANAGE WILDLIFE

Wildlife is managed by state and federal governments. Therefore, any use of fertility control for wildlife requires a permit from the state government. A federal permit is also required if migratory birds or endangered species are impacted. All drugs must be registered with both federal and state authorities. For instance, all drugs used in the U.S. must be approved by the F.D.A., including contraceptives and sterilants. At present, no contraceptive vaccine has been approved by the F.D.A. for widespread use in wildlife (Guynn 1997).

SUMMARY

Fertility control of wildlife species is a tool that can be used to manage some wildlife problems. It usually functions by reducing population size, but sometimes can be effective by merely changing an animal's behavior. Methods to reduce the fertility of wildlife include mechanical and surgical techniques, endocrine disruption, and immunocontraception. Ideally, contraceptive agents should be reversible (depending on the species), suitable for field delivery, and effective with a single dose. Furthermore, treated animals should still attempt to breed and maintain their territories. Contraceptives should pose no risk to humans or nontarget species.

Contraception in wildlife management is a promising technology, but it is still in the experimental stage. Several problems have to be overcome before these techniques come into widespread use. Current delivery methods are not feasible for treating free-ranging wildlife, and there is a need to develop drugs that could be placed in baits and orally delivered to animals. The long-term effects of using contraceptive agents need further consideration. Questions about cost effectiveness and public acceptance of fertility control methods also have to be addressed.

Given all of the problems that are associated with fertility control in wildlife, why then are we still trying to develop effective contraceptive techniques? Contraception may be a useful technique when other forms of population control are ineffective or unavailable. For some species, such as coyotes and beaver, reducing fertility can reduce wildlife damage without actually reducing the population size. Furthermore, contraception offers the promise of a flexible management tool that can be discontinued if need be. It also may be perceived as more humane than other methods of resolving human–wildlife conflicts.

LITERATURE CITED

Al-Kafawi, A., M. L. Hopwood, M. H. Pineda, and L. C. Faulkner, Immunization of dogs against human chorionic gonadotropin. *Am. J. Vet. Res.*, 35, 261–264, 1974.

Allen, S. H., Bait consumption and diethylstilbestrol influence on North Dakota red fox reproductive performance. *Wildl. Soc. Bull.*, 10, 370–374, 1982.

Asa, C. S., The development of contraceptive methods for captive wildlife. P. 235–240 in T. J. Kreeger, Ed., *Contraception in Wildlife Management.* U.S. Government Printing Office, Washington, D.C., 1997.

Asa, C. S. and I. Porton, Concerns and prospects for contraception in carnivores. *Annual Proc. Am. Assoc. Zoo Vet.*, 298–303, 1991.

Asa, C. S., I. Porton, A. M. Baker, and E. D. Plotka, Contraception as a management tool for controlling surplus animals. P. 451–467 in D. G. Kleiman, M. E. Allen, K. V. Thompson, S. Lumpkin, and H. Harris, Eds., *Wild Mammals in Captivity: Principles and Techniques*. University of Chicago Press, Chicago, IL, 1996.

Bailey, E. P., Red foxes, *Vulpes vulpes*, as biological control agents for introduced arctic foxes, *Alopex lagopus*, on Alaskan Islands. *Can. Field-Naturalist*, 106, 200–205, 1992.

Balser, D. S., Management of predator populations with antifertility agents. *J. Wildl. Manage.*, 28, 352–358, 1964.

Becker, S. E. and L. S. Katz, Gonadotropin-releasing hormone (GnRH) analogs or active immunization against GnRH to control fertility in wildlife. P. 11–20 in T. J. Kreeger, Ed., *Contraception in Wildlife Management.* U.S. Government Printing Office, Washington, D.C., 1997.

Bell, R. L. and T. J. Peterle, Hormone implants control reproduction in white-tailed deer. *Wildl. Soc. Bull.*, 3, 152–156, 1975.

Blokpoel, H. and R. M. G. Hamilton, Effects of applying white mineral oil to chicken and gull eggs. *Wildl. Soc. Bull.*, 17, 435–441, 1989.

Bomford, M., A role for fertility control in wildlife management? Australian Government Publishing Service Bulletin No. 7, Canberra, Australia, 1990.

Bomford, M. and P. O'Brien, Potential use of contraception for managing wildlife pests in Australia. P. 205–214 in T. J. Kreeger, Ed., *Contraception in Wildlife Management.* U.S. Government Printing Office, Washington, D.C., 1997.

Bradley, M. P., Immunocontraceptive vaccines for control of fertility in the European red fox (*Vulpes vulpes*). P. 195–204 in T. J. Kreeger, Ed., *Contraception in Wildlife Management.* U.S. Government Printing Office, Washington, D.C., 1997.

Bray, O. E., J. J. Kennelly, and J. L. Guarino, Fertility of eggs produced on territories of vasectomized red-winged blackbirds. *Wilson Bull.*, 87, 187–195, 1975.

Brooks, J. E. and A. M. Bowerman, Estrogenic steroid used to inhibit reproduction in Norway rats. *J. Wildl. Manage.*, 35, 444–449, 1971.

Brooks, J. L., M. W. Fleming, and J. J. Kennelly, Beaver colony responses to fertility control: Evaluating a concept. *J. Wildl. Manage.*, 44, 568–575, 1980.

Butler, V., Elephants: trimming the herd. *Bioscience*, 48, 76–81, 1998.

Christens, E. and H. Blokpoel, Operational spraying of white mineral oil to prevent hatching of gull eggs. *Wildl. Soc. Bull.*, 19, 423–430, 1991.

Conn, P. M. and W. F. Crowley, Jr., Gonadotropin-releasing hormone and its analogs. *New England J. Med.*, 324, 93–103, 1991.

Converse, K. A. and J. J. Kennelly, Evaluation of Canada goose sterilization for population control. *Wildl. Soc. Bull.*, 22, 265–269, 1994.

Daels, P. F. and J. P. Hughes, Fertility control using intrauterine devices: an alternative for population control in wild horses. *Theriogenology,* 44, 629–639, 1995.

DeLiberto, T. J., E. M. Gese, F. F. Knowlton, J. R. Mason, M. R. Conover, L. Miller, R. H. Schmidt, and M. K. Holland, Fertility control in coyotes: is it a potential management tool? In *Proc. Vert. Pest Conf.*, 18, 144–149, 1998.

DeNicola, A. J., D. J. Kesler, and R. K. Swihart, Remotely delivered prostaglandin $F_{2\alpha}$ implants terminate pregnancy in white-tailed deer. *Wildl. Soc. Bull.*, 25, 527–531, 1997.

Dolbeer, R. A., Population dynamics: the foundation of wildlife damage management for the 21st century. In *Proc. Vert. Pest Conf.*, 18, 2–11, 1998.

Eagle, T. C., C. S. Asa, R. A. Garrott, E. D. Plotka, D. B. Siniff, and J. R. Tester, Efficacy of dominant male sterilization to reduce reproduction in feral horses. *Wildl. Soc. Bull.*, 21, 116–121, 1993.

Elder, W. H., Chemical inhibition of ovulation in the pigeon. *J. Wildl. Manage.*, 28, 556–575, 1964.

Fayrer-Hosken, R. A., P. Brooks, H. J. Bertschinger, J. F. Kirkpatrick, J. W. Turner, and I. K. M. Liu, Management of African elephant populations by immunocontraception. *Wildl. Soc. Bull.*, 25, 18–21, 1997.

Gardner, H. M., W. D. Hueston, and E. F. Donovan, Use of mibolerone in wolves and in three *Panthera* species. *J. Am. Vet. Med. Assoc.*, 187, 1193–1194, 1985.

Garrett, M. G. and W. L. Franklin. Diethylstilbestrol as a temporary chemosterilant to control black-tailed prairie dog populations. *J. Range Manage.*, 36, 753–756, 1983.

Garrott, R. A. and D. B. Siniff, Limitations of male-oriented contraception for controlling feral horse populations. *J. Wildl. Manage.*, 56, 456–464, 1992.

Greer, K. R., W. H. Hawkins, and J. E. Catlin, Experimental studies of controlled reproduction in elk (wapiti). *J. Wildl. Manage.*, 32, 368–376, 1968.

Guynn, D. C., Contraception in wildlife management: reality or illusion? P. 241–246 in T. J. Kreeger, Ed., *Contraception in Wildlife Management*. U.S. Government Printing Office, Washington, D.C., 1997.

Haight, R. G. and L. D. Mech, Computer simulation of vasectomy for wolf control. *J. Wildl. Manage.*, 61, 1023–1031, 1997.

Harder, J. D. and T. J. Peterle, Effects of diethylstilbestrol on reproductive performance in white-tailed deer. *J. Wildl. Manage.*, 38, 183–196, 1974.

Heilmann, T. J., R. A. Garrott, L. L. Cadwell, and B. L. Tiller, Behavioral response of free-ranging elk treated with an immunocontraceptive vaccine. *J. Wildl. Manage.*, 62, 243–250, 1998.

Hinds, L. A. and C. H. Tyndale-Biscoe, The effects of bromocriptine on lactation and subsequent reproduction in grey kangaroos, *Macropus fuliginosus* and *Macropus giganteus*. *Reprod. Fertil. Dev.,* 6, 705–711, 1994.

Hoffman, R. A. and B. B. Moorhead, Fertility control research on non-native mountain goats in Olympic National Park, Washington. P. 285–301 in P. N. Cohn, E. D. Plotka, and U. S. Seal, Eds., *Contraception in Wildlife.* Edwin Mellen Press, Lewiston, NY, 1996.

Hunter, A. and A. P. Byers, Immunological intervention in reproduction — potential for wildlife contraception. P. 101–118 in P. N. Cohn, E. D. Plotka, and U. S. Seal, Eds., *Contraception in Wildlife*. Edwin Mellen Press, Lewiston, NY, 1996.

Jacobsen, N. K., D. A. Jessup, and D. J. Kesler, Contraception in captive black-tailed deer by remotely delivered norgestomet ballistic implants. *Wildl. Soc. Bull.*, 23, 718–722, 1995.

Jolly, S. E., P. E. Cowan, and J. A. Duckworth, Research to develop contraceptive control of brushtail possums in New Zealand. P. 215–222 in T. J. Kreeger, Ed., *Contraception in Wildlife Management*. U.S. Government Printing Office, Washington, D.C., 1997.

Kennelly, J. J. and K. A. Converse, Surgical sterilization: an underutilized procedure for evaluating the merits of induced sterility. P. 21–28 in T. J. Kreeger, Ed., *Contraception in Wildlife Management.* U.S. Government Printing Office, Washington, D.C., 1997.

Kennelly, J. J. and P. J. Lyons, Evaluation of induced sterility for beaver (*Castor canadensis*) management problems. *East. Wildl. Damage Control Conf.*, 1, 169–175, 1983.

Kesler, D. J., Remotely delivered contraception with needle-less norgestomet implants. P. 171–184 in T. J. Kreeger, Ed., *Contraception in Wildlife Management*. U.S. Government Printing Office, Washington, D.C., 1997.

Kirkpatrick, J. F., I. K. M. Liu, and J. W. Turner, Jr., Remotely-delivered immunocontraception in feral horses. *Wildl. Soc. Bull.*, 18, 326–330, 1990.

Kirkpatrick, J. F. and J. W. Turner, Jr., Chemical fertility control and wildlife management. *Bioscience,* 35, 485–491, 1985.

Kirkpatrick, J. F., J. W. Turner, Jr, and I. K. M. Liu, Contraception of wild and feral equids. P. 161–170 in T. J. Kreeger, Ed., *Contraception in Wildlife Management*. U.S. Government Printing Office, Washington, D.C., 1997a.

Kirkpatrick, J. F., J. W. Turner, Jr., I. K. M. Liu, R. Fayrer-Hosken, and A. T. Rutberg, Case studies in wildlife immunocontraception: wild and feral equids and white-tailed deer. *Reprod. Fertil. Dev.,* 9, 105–110, 1997b.

Kreeger, T. J., A review of delivery systems for the administration of contraceptives to wildlife. P. 29–48 in T. J. Kreeger, Ed., *Contraception in Wildlife Management*. U.S. Government Printing Office, Washington, D.C., 1997.

Linhart, S. B. and R. K. Enders, Some effects of diethylstilbestrol in captive red foxes. *J. Wildl. Manage.*, 28, 358–363, 1964.

Marsh, R. E. and W. E. Howard, Evaluation of mestranol as a reproductive inhibitor of Norway rats in garbage dumps. *J. Wildl. Manage.*, 33, 133–138, 1969.

Matschke, G. H., Microencapsulated diethylstilbestrol as an oral contraceptive in white-tailed deer. *J. Wildl. Manage.*, 41, 87–91, 1977.

McCullough, D., Demography and management of wild populations by reproductive intervention. P. 119–132 in P. N. Cohn, E. D. Plotka, and U. S. Seal, Eds., *Contraception in Wildlife*. Edwin Mellen Press, Lewiston, New York, 1996.

McShea, W. J., S. L. Monfort, S. Hakim, J. Kirkpatrick, I. Liu, J. W. Turner, L. Chassy, and L. Munson, The effect of immunocontraception on the behavior and reproduction of white-tailed deer. *J. Wildl. Manage.*, 61, 560–569, 1997.

Mech, L. D., S. H. Fritts, and M. E. Nelson, Wolf management in the 21st century, from public input to sterilization. *J. Wildl. Res.*, 1, 195–198, 1996.

Menkhorst, P., D. Middleton, and B. Walters, Managing over-abundant koalas (*Phascolarctos cinereus*) in Victoria. *Occasional Papers of the Marsupial CRC No. 1*, Canberra, Australia, 1998.

Miller, L. A., Delivery of immunocontraceptive vaccines for wildlife management. P. 49–58 in T. J. Kreeger, Ed., *Contraception in Wildlife Management.* U.S. Government Printing Office, Washington, D.C., 1997.

Miller, L. A., B. E. Johns, and D. J. Elias, Immunocontraception as a wildlife management tool: some perspectives. *Wildl. Soc. Bull.*, 26, 237–243, 1998.

Morell, V., Australian pest control by virus causes concern. *Science,* 261, 683–684, 1993.

Nave, C. D., G. Shaw, R. V. Short, and M. B. Renfree, Contraceptive effects of levonorgestrel implants in a marsupial. *Reprod. Fertil. Dev.*, 12, 81–86, 2000.

Nelson, K. J., Sterilization of dominant males will not limit feral horse populations. *U.S.D.A. Forest Service Research Paper RM-226*, Rocky Mountain Forest and Range Experiment Station, Fort Collins, CO, 1980.

Nettles, V. F., Potential consequences and problems with wildlife contraceptives. *Reprod. Fertil. Dev.*, 9, 137–143, 1997.

Neville, P. F. and J. Remfry, Effect of neutering on two groups of feral cats. *Vet. Rec.*, 114, 447–450, 1984.

Oogjes, G., Ethical aspects and dilemmas of fertility control of unwanted wildlife: an animal welfarist's perspective. *Reprod. Fertil. Dev.*, 9, 163–167, 1997.

Orford, H. J. L., Hormonal contraception in free-ranging lions (*Panthera leo* L.) at the Etosha National Park. P. 303–320 in P. N. Cohn, E. D. Plotka, and U. S. Seal, Eds., *Contraception in Wildlife*. Edwin Mellen Press, Lewiston, NY, 1996.

Plotka, E. D., D. N. Eagle, D. N. Vevea, J. R. Tester, and D. B. Siniff, Development of an implantable contraceptive for feral horses. P. 209–228 in P. N. Cohn, E. D. Plotka, and U. S. Seal, Eds., *Contraception in Wildlife*. Edwin Mellen Press, Lewiston, NY, 1996.

Plotka, E. D. and U. S. Seal, Problems associated with fertility control in female white-tailed deer. P. 185–190 in P. N. Cohn, E. D. Plotka, and U. S. Seal, Eds., *Contraception in Wildlife*. Edwin Mellen Press, Lewiston, NY, 1996.

Pochop, P. A., J. L. Cummings, C. A. Yoder, and J. E. Steuber, Comparison of white mineral oil and corn oil to reduce hatchability in ring-billed gull eggs. In *Proc. Vert. Pest Conf.*, 18, 411–413, 1998.

Remfry, J., Control of feral cat populations by long term administration of megestrol acetate. *Vet. Rec.*, 103, 403–404, 1978.

Roughton, R. D., Effects of oral melengestrol acetate of reproduction in captive white-tailed deer. *J. Wildl. Manage.*, 43, 428–436, 1979.

Smith, A. E., S. R. Craven, and P. D. Curtis, Managing Canada geese in urban environments. *Jack Berryman Institute Publication 16*. Cornell University Cooperative Extension, Ithaca, NY, 1999.

Storm, G. L. and G. C. Sanderson, Effect of medroxyprogesterone acetate (Provera) on productivity in captive foxes. *J. Mammal.*, 50, 147–149, 1969.

Sturtevant, J. and B. C. Wentworth, Effect on acceptability and fecundity to pigeons of coating SC 12937 bait with zein or ethocel. *J. Wildl. Manage.*, 34, 777–782, 1970.

Till, J. A. and F. F. Knowlton, Efficacy of denning in alleviating coyote depredations upon domestic sheep. *J. Wildl. Manage.*, 47, 1018–1025, 1983.

Turner, J. W. and J. F. Kirkpatrick, New methods for selective contraception of wild animals. P. 191–208 in P. N. Cohn, E. D. Plotka, and U. S. Seal, Eds., *Contraception in Wildlife*. Edwin Mellen Press, Lewiston, NY, 1996.

Turner, J. W., I. K. M. Liu, and J. F. Kirkpatrick, Remotely delivered immunocontraception in captive white-tailed deer. *J. Wildl. Manage.*, 56, 154–157, 1992.

Turner, J. W., I. K. M. Liu, A. T. Rutberg, and J. F. Kirkpatrick, Immunocontraception limits foal production in free-roaming feral horses in Nevada. *J. Wildl. Manage.*, 61, 873–880, 1997.

Tyndale-Biscoe, C. H., Virus-vectored immunocontraception of feral mammals. *Reprod. Fertil. Dev.*, 6, 281–287, 1994.

Tyndale-Biscoe, C. H., Immunosterilization for wild rabbits: the options. P. 223–234 in T. J. Kreeger, Ed., *Contraception in Wildlife Management*. U.S. Government Printing Office, Washington, D.C., 1997.

Vandenbergh, J. G. and D. E. Davis, Gametocidal effects of triethylenemelamine on a breeding population of red-winged blackbirds. *J. Wildl. Manage.*, 26, 366–371, 1962.

Vickery, B. H., G. I. McRae, J. C. Goodpasture, and L. M. Sanders, Analog of LHRH: a new (anti) hormonal approach to contraception. P. 73–99 in P. N. Cohn, E. D. Plotka, and U. S. Seal, Eds., *Contraception in Wildlife*. Edwin Mellen Press, Lewiston, NY, 1996.

Waddell, R. B., Movements of an urban deer population during breeding and fawning, and the potential for controlling fertility using contragestion. M.S. thesis, University of Georgia, Athens, 2000.

Warren, R. J., R. A. Fayrer-Hosken, L. I. Muller, L. P. Willis, and R. B. Goodloe, Research and field applications of contraceptives in white-tailed deer, feral horses, and mountain goats. P. 133–145 in T. J. Kreeger, Ed., *Contraception in Wildlife Management.* U.S. Government Printing Office, Washington, D.C., 1997.

White, L. M., R. J. Warren, and R. A. Fayrer-Hosken, Levonorgestrel implants as a contraceptive in captive white-tailed deer. *J. Wildl. Dis.*, 30, 241–246, 1994.

Wofford, J. E. and W. H. Elder, Field trials of the chemosterilant, SC-12937, in feral pigeon control. *J. Wildl. Manage.*, 31, 507–515, 1967.

CHAPTER 9

Wildlife Translocation

"Toto, I don't think we're in Kansas anymore."

Dorothy
on being translocated to Oz by a tornado

Wildlife translocation, or relocation, consists of transporting live captured animals to a location different from their capture site and releasing them. Translocation has been used for various purposes, including the reintroduction of threatened or endangered species into portions of their former range and the stocking of popular game species. In wildlife damage management, translocation has been used either to remove individual animals responsible for depredations (culprits) or to reduce populations in specific areas (subpopulations) by removing large numbers of animals. Although this chapter discusses translocation in the context of wildlife damage management, data pertaining to other types of translocations will also be presented when relevant.

Many people regard lethal control methods as needlessly cruel. Translocation, on the other hand, appears to be the perfect solution: the offending animal is removed and given another chance to "live happily ever after" elsewhere, in its natural environment. Unfortunately, the biological realities of translocation are quite different and this technique is rather controversial, as we will see.

EXAMPLES OF THE USE OF TRANSLOCATION TO RESOLVE WILDLIFE CONFLICTS

Nuisance wildlife control operators report that live-trapping animals and releasing them off-site is their primary method of controlling raccoons, squirrels, skunks, and woodchucks, particularly in urban and suburban areas, because most of their customers do not want these animals harmed (Barnes 1993, 1995; Curtis et al. 1993). However, the effectiveness of this method is rarely monitored and the fate of translocated animals is unknown.

Translocation has often been used to manage problem bears at garbage dumps or campgrounds and also to alleviate problems caused by urban and suburban deer populations. Managing deer in suburban areas has proved particularly challenging because hunting may not be an option if local laws prohibit the discharge of firearms or if local residents are opposed to it. Surveys of suburban residents in New York and Virginia indicated that the management technique preferred by the highest number of respondents was live-trapping and translocating deer (Green et al. 1997; Stout et al. 1997). This technique was also perceived as the most effective short-term deer management technique by respondents in New York (Stout et al. 1997).

In order to evaluate wildlife translocation as a tool to alleviate wildlife damage, we need to address the following questions: 1. Do translocated animals return to their capture site? 2. Do translocated animals cause similar problems in their new homes? 3. Does translocating animals solve the original problem or do new animals just replace the old? 4. What is the survival rate of translocated animals, and what effect do relocated individuals have on the animals that are already living at the release site? 5. What factors influence the probability that translocation will satisfactorily resolve a wildlife problem? 6. Is translocation a cost-effective method to alleviate wildlife damage? These questions are explored in this chapter.

DO TRANSLOCATED ANIMALS RETURN TO THE SITE WHERE THEY WERE CAPTURED?

Some wildlife species have a strong homing instinct. Upon release, they tend to travel in the direction of their capture site and can cover extensive distances in an attempt to reach it. Bears are a classic example. Rogers (1986) summarized data from various studies reporting movements of black bears after translocation. In total, 81% of bears returned to their capture area when translocated less than 64 km, 48% returned from 64 to 120 km, 33% from 120 to 220 km, and 20% from over 220 km. Other studies, summarized by Linnell et al. (1997), showed return rates between 45% and 86% for black bears moved less than 64 km.

Grizzly and brown bears also demonstrate strong homing tendencies. Of 85 grizzly bears translocated less than 125 km in Yellowstone National Park, 59% returned to their capture site (Brannon 1987). In Alaska, Miller and Ballard (1982) translocated 20 radio-collared brown bears over distances of 150 to 300 km. At least 12 (60%) traveled back to their capture area, from an average distance of 198 km. None of the bears remained at their release site.

Less is known about the homing tendencies of canids and felids. In Alaska, five pen-reared gray wolves were translocated 282 km. One of them returned home and two others were killed after moving 140 to 160 km towards home (Henshaw and Stephenson 1974). In Iowa, 9 female red foxes and 171 pups were translocated 3 to 171 km. One female returned from 14 km, and a litter of four pups, translocated with their mother, returned from 3 km. The remaining foxes showed no homing tendencies (Andrews et al. 1973). In New Mexico, 14 cougars were experimentally translocated 342 to 510 km. Two males were able to return to their capture site, from 465 to 490 km away (Ruth et al. 1993, cited by Linnell et al. 1997).

In general, large herbivores stay relatively close to their release sites, although a few individuals may travel long distances. Rogers (1988) reviewed a series of papers reporting white-tailed deer movements after translocation and found that most animals remained within 30 km of their release sites. In other studies, translocated white-tailed deer moved an average of 3.3 km in Illinois (Hawkins and Montgomery 1969), between 2 and 5 km in another part of Illinois (Jones and Witham 1990), and 23 km in New York (Jones et al. 1997). Cromwell et al. (1999) captured 19 white-tailed deer, released 9 at their capture sites, and relocated the other 10. They found that half of the translocated deer dispersed from their release sites, while none of the deer released at their capture sites did so. Translocated black-tailed deer moved an average of 9 km in California (O'Bryan and McCullough 1985). However, individual deer occasionally find their way home over much longer distances. In Texas, three white-tailed deer successfully returned to their capture site from distances of 530 to 560 km, which represent the longest homing distances ever recorded for North American mammals. The same pattern was observed in moose translocated 700 km from Ontario (Canada) to Michigan: 22 of 29 individuals settled within 20 km of their release site, but one animal traveled 290 km in the direction of its capture site (Rogers 1988).

Birds typically have a better homing instinct than mammals. Because of this, translocation is not commonly used to solve bird problems. For example, 12 of 14 resident adult golden eagles captured in Wyoming and released 416 to 470 km away returned to their former territories (Phillips et al. 1991). Some eagles came back within a few weeks, whereas others took over five months to return (Figure 9.1).

Several variables influence the propensity of translocated animals to return home. As indicated in previous examples, homing tendencies decrease as distances between the capture and release sites increase. For example, 8 of 14 wolves returned from 50 to 64 km in Minnesota, whereas none of 20 wolves translocated more than 64 km returned (Fritts et al. 1984). Because translocating animals great distances is costly and time consuming, most people want to release the animals at the shortest distance possible while still achieving the purpose of the translocation. For most mammals, this distance is typically 5 to 10 times the width of the animal's home range, so that the animal is released into unfamiliar territory. For instance, if a skunk has a home range of 2 km in diameter, it would probably not return to its capture site if released 10 to 20 km away. The presence of physical barriers, such as rivers or mountain ranges, further impedes the ability of a translocated animal to return to its capture site. For example, factors reducing the homing success of black bears included the number of ridges, the elevation gain, and the presence of physiographic barriers (e.g., rivers) between capture and release sites (McArthur 1981).

The age of the translocated animal is also a factor in its propensity to return to its capture area. In general, adults are more likely to return than juveniles or subadults, in part because an adult may have been taken away from an established territory, a mate, or young. These provide strong incentives for translocated adults to return. In contrast, many subadults lack territories and may have been in the process of dispersing when captured. Thus, these animals have fewer reasons to return to the capture site. For example, translocated bear cubs, yearlings, and subadults return less frequently than adults (Rogers 1986). Also,

Figure 9.1 When golden eagles are translocated, many return to their capture site.

attempts to resolve urban Canada goose problems by translocating adults to rural areas have largely been abandoned because these birds are too likely to return home or to cause nuisance problems in other areas (Figure 9.2). In contrast, goslings that are captured just before they are able to fly and translocated to rural areas show little inclination to return home and are likely to remain in rural areas (Cooper and Keefe 1997).

Figure 9.2 Use of translocation to alleviate problems with urban Canada geese is becoming less common because the translocated geese are quickly replaced by others.

One way to reduce the tendency for translocated animals to return home is to keep the animals in large pens at the release site for several weeks. This method is called a "gentle," "soft," or "slow" release, whereas the immediate release of animals is called a "hard" or "quick" release. When wolves were reintroduced in Yellowstone National Park, a gentle release was used to increase the chances that the animals would stay in the park. Translocated wolves were kept in pens for two months before being released (Bangs and Fritts 1996). After some exploratory movements, these wolves established home ranges in the vicinity of their release site. Conversely, wolves reintroduced in Idaho were given a quick release and scattered over a much wider area.

Sidebar 9.1 Translocating Bears to Resolve a Short-Term Predation Problem

Even in the case of animals possessing a strong homing instinct, like bears, translocation can sometimes be used successfully to resolve a wildlife conflict, if that conflict is limited in space and time. In northeastern Oregon, for instance, translocation of black bears was used successfully to reduce predation on sheep (Armistead et al. 1994). In this case, several migratory sheep bands tended to travel along the same routes year after year. Records showed that predation by bears kept reoccurring at specific locations along those routes. Rerouting the sheep would have brought them in contact with other bears, so U.S.D.A. Wildlife Services initiated a preventive bear translocation program in which bears from problem sites were captured immediately prior to the arrival of sheep. These bears were then moved less than 33 km and released in areas with no sheep. The translocated bears were expected to return, but by that time the sheep would have left. This technique was successful in reducing sheep losses and the cost per bear was similar to that of lethal control. Most of the general public was supportive of this technique and so were sheep operators, who preferred preventive action rather than waiting for the bears to kill sheep before removing problem individuals.

DO NEW ANIMALS REPLACE THE TRANSLOCATED ONES SO THAT THE PROBLEM PERSISTS?

The answer to this question will depend on the species and circumstances, but very little information is available on the subject in scientific literature. A study of white-tailed deer movements in a New York forest revealed that the removal of small social groups of deer created a low-density area that persisted for several years (McNulty et al. 1997). Deer from adjacent areas did not tend to move into the low-density area, nor did they shift their ranges closer to it. The authors concluded that localized management of white-tailed deer was possible in some circumstances, owing to the deer's philopatry (tendency to remain near a particular site) to their home ranges. However, deer living in agricultural environments exhibit less philopatry than deer living in forested areas.

Conversely, after an experimental removal of resident golden eagles in Wyoming, all open territories were quickly reoccupied by new eagles originating from a nonbreeding, nonterritorial subpopulation (Phillips et al. 1991). Seven of the fourteen translocated eagles came back and were able to reclaim their territories, either during the same breeding season or during the next one. Three others were displaced from their territories and two more were found dead near their original nest tree after fighting with replacement eagles. The fate of the remaining two eagles was unknown.

In the case of eagles, replacement of translocated birds would be desirable from a damage management perspective because most eagles do not prey upon lambs or calves. Phillips et al. (1996) reported two cases of eagle predation on calves; the removal of a single eagle in one case, and of a pair in the other case, put an end to all depredations.

However, when nuisance behavior is exhibited by most individuals in a population, the problem is likely to reoccur as soon as the translocated animal is replaced by another. For example, translocating urban raccoons without removing the attractant will probably result in further depredations by other raccoons (Riley 1989). Similarly, when urban Canada geese are captured at a nuisance site and removed, new geese reoccupy the site within a few weeks or months, creating the same nuisance. For this reason, geese will have to be removed over and over from the same site. In these situations, translocation provides only a temporary solution to a wildlife damage problem. For example, Minnesota started translocating urban Canada geese during 1982, but this did not end the problem. Instead, the number of geese requiring translocation increased yearly until more than 6000 urban geese were being moved annually in the 1990's. The program was abandoned in 1997 because no more sites were willing to accept Minnesota's geese. In total, over 40,000 urban Canada geese were moved, making this one of the largest translocation programs in the world (Cooper and Keefe 1997).

DO TRANSLOCATED ANIMALS CREATE THE SAME PROBLEM ELSEWHERE?

Some individuals causing problems in one area will probably cause the same type of nuisance elsewhere if given the opportunity. Because wolves and bears can travel long distances when translocated, a certain proportion of these animals are bound to find their way to human settlements or livestock operations and cause further problems. In Glacier National Park, 11% of translocated black bears caused similar problems elsewhere (McArthur 1981). In Pennsylvania, McLaughlin et al. (1981, cited by Rogers 1986) found that 15% of translocated black bears had to be recaptured after causing problems in their new surroundings. In Yellowstone National Park, 43% of translocated grizzly bears were involved in further depredation incidents (Brannon 1987). In Minnesota, 13% of 107 translocated wolves captured at farms following depredation or harassment of livestock were recaptured at least once more for causing similar problems (Fritts et al. 1985). One of three cougars translocated in Alberta (Ross and Jalkotzy 1995) and one of two jaguars translocated in Belize (Rabinowitz 1986) resumed depredation in their new location. In Illinois, raccoons removed from urban or rural areas tended to establish new territories in proximity to human residences (Mosillo et al. 1999).

In the worst-case scenario, animals translocated because of inappropriate behavior may teach that behavior to other individuals through social facilitation. The inappropriate behavior may then quickly spread through the population. For instance, when a few black bears in Yosemite National Park learned how to break into cars to obtain any food that might be inside, the bears were relocated to other parts of the park rather than being euthanized. This practice compounded the problem by spreading the learned behavior throughout the bear population in the park.

WHAT HAPPENS TO TRANSLOCATED ANIMALS?

The general public usually views translocation as much more humane than the use of lethal means. Unfortunately, this is not necessarily the case. To translocate an animal, it first needs to be captured. A varying percentage of individuals may be injured or killed during capture, depending on the species, capture technique, experience of the capture crew, season, meteorological conditions, and condition of the animal prior to capture. Some capture mortality is usually inevitable. For instance, 5% of nuisance birds captured with the immobilizing drug alpha-chloralose died, even when this technique was used by trained biologists (Belant et al. 1999).

In white-tailed deer, mortality rates from capture ranged from 0 to 30% (Table 9.1). Usually, the highest mortality rates were observed for animals captured with dart guns or rocket nets (Hawkins et al. 1967; Palmer et al. 1980; Ishmael and Rongstad 1984). However, Diehl (1988) successfully darted 47 deer without a single fatality. Drive-netting and drop-netting resulted in 1 to 3% mortality rates (Conner et al. 1987; Sullivan et al. 1991) and the use of box or clover traps had a 0 to 8% mortality rate (Beringer et al. 1996). Capturing animals by running them down or tiring them out is particularly stressful and can result in high mortality. With African big game species, this method has occasionally resulted in a 100% mortality rate (Harthoorn 1975).

Individuals may die several weeks later as a result of the stress and trauma suffered during capture (Chalmers and Barrett 1982; Beringer et al. 1996). These deaths are usually a consequence of "capture myopathy," a condition associated with the intense muscle activity, fear, and stress experienced during pursuit and capture. It has been reported in ungulates, rhinoceros, elephants, baboons, flamingos, and geese (Chalmers and Barrett 1982). Capture myopathy occurs when blood flow to tissues is reduced, resulting in an oxygen deficiency in muscles and an increased production of lactic acid. This can produce muscle lesions, damage to internal organs, paralysis, and death. For example, when the Missouri Department of Conservation captured white-tailed deer using nets and translocated them, 19% died from capture myopathy (Low 1999). Capture myopathy was also a problem in Georgia when white-tailed deer were translocated (Cromwell et al. 1999).

Animals that survive capture are faced with more challenges once released in an unfamiliar environment. Often, they have difficulty finding food and shelter or avoiding predators and other dangers. This is particularly true for animals captured in urban and suburban areas and released in rural environments. A classic example of these problems is the translocation of black-tailed deer from Angel Island, in San Francisco

Table 9.1 Capture-Related Mortality of White-Tailed Deer Reported for Various Capture Techniques (unless indicated, these figures do not include deaths from capture myopathy following release)

Method	n	Mortality rate (%)	Reference
Box trap	47	0.0	Hawkins et al. 1967
	2035	2.1	Palmer et al. 1980
	85	7.6	Peery 1969
Clover trap	2	0.0	Ishmael and Rongstad 1984
	260	4.6	Fuller 1990
	115	5.1[1]	Beringer et al. 1996
Corral trap	302	13.9	Hawkins et al. 1967
Crossbow	83	15.7	Hawkins et al. 1967
Dart gun	47	0.0	Diehl 1988
	44	13.6	Palmer et al. 1980
	75	20.0	Hawkins et al. 1967
	6	33.3	Ishmael and Rongstad 1984
Drive net	5	0.0	Ishmael and Rongstad 1984
	430	<1.4[2]	Sullivan et al. 1991
	668	0.9	David F. Pac, Mont. Dep. of Fish, Wildl., and Parks, Bozeman, pers. commun. (Sullivan et al. 1991)
Drop net	175	≤7.4[3]	Conner et al. 1987
Longbow	63	33.3	Hawkins et al. 1967
Rocket net	300/187	2.6 + 11.2[4]	Beringer et al. 1996
	33	6.1	Hawkins et al. 1968
	17	23.5	Palmer et al. 1980
Tranimul (tranquilizer sprinkled on bait)	36	22.2	Hawkins et al. 1967
Unknown type of trap	27	30	DeNicola et al. 1997

[1] Deaths from capture myopathy = 0.
[2] 0.9% from capture myopathy and possibly 0.5% from injuries sustained during capture (the fate of these injured animals is unknown).
[3] 0.6% from capture accidents and 3.4 to 6.8% from capture myopathy (6 animals known dead, 6 disappeared).
[4] 2.6% of 300 deer died from capture accidents; 11.2% of 187 radio-tagged deer died from capture myopathy.

Bay, to the Cow Mountain Recreation Area, in the Mayacmas Mountains of California. Of 13 radio-collared deer whose fate is known, only 2 (15%) were still alive a year after translocation (O'Bryan and McCullough 1985). In comparison, the annual survival rate for indigenous black-tailed deer populations in chaparral habitat in the northern coastal range of California was 72% (Taber and Dasmann 1958, cited by O'Bryan and McCullough 1985). Unfortunately, the translocated deer were in poor condition when captured, and several of them died of malnutrition within two weeks of release. Subsequently, the main cause of death was collisions with vehicles. Predators and hunters also took their toll. The deer failed to recognize dangers that they had not been exposed to prior to translocation, such as fast-moving vehicles, predators, hunters, and poachers, all of which were absent from Angel Island.

In another translocation of black-tailed deer in California, 27 deer were moved from Ardenwood Regional Park in the city of Fremont to a more remote area. Three months later, 23 of these deer were known to be dead, 3 could not be located, and only 1 was known to be alive (McCullough et al. 1997).

Urban white-tailed deer translocated to a rural area in Illinois also had higher mortality than resident deer, particularly during the first winter and the first hunting season after translocation. The estimated annual mortality rate was 66% for translocated does, vs. 27% for resident does (Jones and Witham 1990). The major causes of mortality for translocated deer were collisions with vehicles, capture-related stress, and hunting. Of 25 translocated bucks, half were known to have died within a year after translocation, mostly due to hunting. In contrast, only 25% of translocated does in Kentucky died within 248 days of being released, and there was no difference in mortality rates between translocated and resident deer (Pais 1987, cited by McCall et al. 1988). In Wisconsin, 5 of 11 translocated deer whose fate was known died within a year (Diehl 1988). Mortality was again due to hunting and car accidents. In the Adirondack Mountains of New York, 47% of translocated deer perished in 1994 to 1995, vs. 12% of resident deer during the same time period (Jones et al. 1997).

Data available on post-translocation survival of raccoons are contradictory. In Illinois, Mosillo et al. (1999) found no difference in survival rates between urban raccoons translocated to rural areas, rural raccoons translocated to other rural sites, and resident raccoons released at their capture sites. In all cases, survival averaged 75 to 80% for the first two months following release. In contrast, Rosatte and MacInnes (1989) reported at least 50% mortality within three months for 24 urban raccoons translocated from Toronto either to rural areas or close to towns.

Translocation does not appear to increase mortality of black bears (Rogers 1986), except perhaps for cubs moved with their mother over long distances (more than 70 km). However, translocation does reduce survival of grizzly bears. In Yellowstone National Park, Blanchard and Knight (1995) compared survival of radio-collared grizzly bears which were either captured in management actions and translocated, or captured for research purposes and released at their capture site. They found that translocated bears had lower survival rates than nontranslocated bears (71 to 80% vs. 87 to 94%, respectively). In Alaska, Miller and Ballard (1982) observed a high mortality of brown bear cubs and yearlings translocated with their mothers. They also reported that none of the translocated females produced young the year after they were moved. Blanchard and Knight (1995), on the other hand, did not observe a decrease in the reproduction of translocated female grizzlies.

The annual mortality rate (40%) for translocated gray wolves in Minnesota was similar to that for nontranslocated wolves in other parts of the state (Fritts et al. 1985). The behavior of translocated wolves and their ability to find mates and reproduce were similar to those of naturally dispersing wolves.

For social species, survival of translocated animals may increase if animals of the same social group are released together (Figure 9.3). However, this is not always true. Jones et al. (1997) found no difference in the survival of female white-tailed deer translocated with other members of their social group or with unrelated individuals. Deer did not remain together in either case.

Figure 9.3 For some species, survival of translocated individuals is enhanced if they are moved as a group. (Photo courtesy of Mark McClure.)

WHAT ARE THE CONSEQUENCES OF TRANSLOCATION ON RESIDENT WILDLIFE POPULATIONS?

Competitive Interactions

Any given area can only support a limited number of animals; this is the concept of biological carrying capacity. If translocated animals are released in an area where the population is already at its carrying capacity, some animals must die or leave. Usually, the translocated animals will be the ones to perish because of their inability to find food or avoid danger as successfully as resident animals.

In territorial species, translocating animals may cause an increase in territorial fights. Territory owners usually win these fights because a territory has much more value to an individual possessing all the information about the resources available on that territory. Hence, the territory owners will be more determined and more likely to win territorial disputes. This is called site dominance. Although resident animals usually win these disputes, they also suffer from them because they must spend both time and energy to intimidate, chase, or fight off these intruders. Doing so reduces their own chances of survival.

Disease and Parasite Transmission

Transmission of diseases and parasites by translocated wildlife is a serious concern; infected individuals translocated to an area where a certain disease is absent could potentially start an epidemic. For example, a major rabies outbreak in the mid-Atlantic states in 1982 was attributed to the translocation of raccoons from Florida to Virginia (Nettles et al. 1979; Jenkins and Winkler 1987). Similarly, during the late 1970s, a rabies outbreak in skunks in Ontario was traced to the translocation of nuisance animals (Rosatte and MacInnes 1989). Likewise, the parasite *Plasmodium kempi* probably spread into new areas through the translocation of infected turkeys (Castle and Christensen 1990). This blood parasite does not pose much of a threat to turkeys but can severely affect other species (i.e., quail, pheasant, and endangered species, such as the lesser prairie chicken).

Some wildlife diseases also can be transmitted to pets, livestock, or humans. Raccoons, for instance, can carry rabies, pseudo-rabies, canine parvovirus, canine distemper, canine adenovirus, feline panleukopenia, and the parasite *Baylisascaris procyonis* (Rosatte and MacInnes 1989). Additionally, epidemics can have indirect effects on species that are not infected by the disease. For instance, a plague epidemic in prairie dogs was partly responsible for the population decline of the black-footed ferret (Hutchins et al. 1991, cited by Cunningham 1996).

The opposite problem can also occur when translocated individuals that have never been exposed to a certain disease are released into an area where that disease is endemic. In this case, the outcome of translocation could be the death of the translocated animals. This is especially a concern in translocations of threatened or endangered species when every individual is valuable.

Reproduction and Population Genetics

In some cases, individuals have been translocated out of the range of their subspecies and into the range of a different subspecies. This practice can cause serious problems to local populations. Subspecies are adapted to the particular environment in which they evolved. An influx of genetic material from a different subspecies can result in individuals that are not as well adapted to local conditions, causing a decrease in fitness in future generations. This is termed "outbreeding depression." A classic example of a disaster that can result when a subspecies is translocated out of its range and breeds with the local subspecies is that of the alpine ibex. Following extinction of this species in the former Czechoslovakia, some individuals were successfully reintroduced from Austria and a small population became established. To augment this population, animals from different subspecies were introduced from Turkey and Sinai a few years later, and these animals bred with the established individuals. The resulting hybrids were fertile but bred in the fall instead of winter. Consequently, their offspring were born in the cold of winter and did not survive, which caused the eventual extinction of the population.

Another example of problems resulting from the translocation of a subspecies out of its range can be found in Mexico. In 1952, this country adopted a law to curb the decline of deer populations by prohibiting the harvest of antlerless deer. As a result, white-tailed deer populations increased on many ranches in northern Mexico, which eventually led to problems of overabundance. Because the 1952 law precluded doe hunting, landowners and government agencies translocated does to reduce populations. Between 1992 and 1997, over 2000 deer were captured and moved to ranches that wanted more deer (Martinez et al. 1997). Although expensive, this technique was successful as long as there were local ranches willing to accept the deer. However, their willingness to do so declined over time. By 1997, 65% of the captured deer had to be translocated outside the range of the subspecies. This practice may have detrimental consequences on the reproduction and genetics of local deer populations at the release sites (Galindo-Leal and Weber 1994). The translocated deer subspecies (*Odocoileus virginianus texanus*) is larger than other subspecies of white-tailed deer found in Mexico, such as the Coues deer (*O. v. couesi*). During mating season, the male offspring of the large translocated does may have a competitive advantage over the smaller resident bucks and obtain a disproportionate number of matings (Galindo-Leal and Weber 1994). In addition to the alteration of population genetics of the local deer, more immediate problems can occur. Small does carrying offspring sired by large males may experience difficulties during parturition (dystocia), to the point that both mother and fawn may die (Galindo-Leal and Weber 1994).

IS TRANSLOCATION COST-EFFECTIVE?

Translocation costs vary widely. Boyer and Brown (1988) surveyed state wildlife agencies in the U.S. about their translocation activities in 1985. For nuisance white-tailed deer, the reported translocation cost was $100 to $125 per animal; for black bears, it was $100 to $350. These data, however, are incomplete, as some states did not provide cost estimates for their translocations. In Minnesota, it cost the government $10 to translocate an urban Canada goose, vs. $0 to have it shot by sports hunters, and $24 to capture and process it for food.

Other authors report costs for nuisance white-tailed deer capture and translocation. In New Hampshire, as mentioned earlier, the Fish and Game Department sponsored a translocation program in 1983, during which 27 deer were removed from Long Island in Lake Winnipesaukee, at a cost of $800 per deer and with a capture mortality of 30%. As a result, the program was canceled after one year (DeNicola et al. 1997). In Ohio, the Columbus and Franklin County Park District reported that it cost $133 to translocate a deer, $207 to kill it using a sharpshooter, and $45 to kill it using hunters (Peck and Stahl 1997). The Missouri Department of Conservation translocated suburban deer at a cost of $350 per deer (Low 1999).

Ishmael and Rongstad (1984) compared the costs of various deer removal techniques at the University of Wisconsin Arboretum during 1982 and 1983. Shooting over bait was the least expensive technique at $74 per deer (for $n = 34$ deer removed), followed by dart gunning at $179 per deer ($n = 6$), drive netting at $523 per deer ($n = 5$), and using a clover trap at $570 per deer ($n = 2$). Capture with rocket nets was also attempted but not a single animal was caught. Other authors had success trapping deer using the Stephenson box trap, spending as little as 1.3 man-hours per deer captured (Hawkins et al. 1967). Palmer et al. (1980) reported values of 1.8, 2.8, 3.3, 4.1, and 6.9 man-hours per deer for public hunting, box trapping, shooting by biologists, dart-gunning, and rocket netting, respectively, to remove deer from an enclosed area in Ohio. Similarly, Conner et al. (1987) reported values of 2.2 man-hours and $28 per deer captured by drop-netting in Maryland ($n = 175$). Translocating mule deer from Angel Island cost $431 per deer removed, but so many deer died after being released that the cost per surviving deer was $2876 (O'Bryan and McCullough 1985).

Most of the cost of translocating animals is in capturing them (Figure 9.4). In general, the cost to capture a single animal will be lowest when 1. densities are high, 2. the animals have not been trapped previously, and 3. wildlife biologists are not trying

Figure 9.4 Capturing and relocating animals can become very expensive, especially when aircraft are required. (Photo courtesy of Jessica Pettee.)

to capture a particular animal. In any trapping operation, it is always much easier and less expensive to capture the first 10% of a population than the last 10%, which may well be impossible to remove by any means. Of course, capturing animals is only one cost of translocation. Another cost is the transportation of animals to the release site, which can be reduced if the animals are not taken far from the capture site. However, this will increase the chances of the animal returning to the capture site. Likewise, the cost of releasing an animal will be lowest if the animal is released immediately (hard or quick release). However, the chances of the translocated animal surviving and remaining near the release site will increase if the animal is maintained in a large pen for several weeks prior to release. Because this gentle release is much more expensive than a hard release, it is used mainly for highly valuable animals, such as wolves.

To summarize, translocation can be very expensive per animal when a small number of animals are captured, but cost-efficiency increases considerably when large numbers of animals are captured. In some circumstances, capturing deer can be less expensive and/or more time-effective than sharpshooting (e.g., Palmer et al. 1980; Peck and Stahl 1997); however, it is usually more expensive and labor intensive than public hunting.

WHAT ARE GOVERNMENTAL POLICIES CONCERNING THE TRANSLOCATION OF NUISANCE ANIMALS?

Regulations concerning translocation vary among states. In a national survey, Craven and Nosek (1992, cited by Craven et al. 1998) found that 47 states allowed translocation of nuisance wildlife (California, Massachusetts, and Alaska did not). Some states required a permit for wildlife translocations and/or had specific regulations restricting the translocation of some species due to disease concerns. For instance, Connecticut did not allow translocation of foxes, skunks, or raccoons because these species were potential rabies vectors.

WHEN IS TRANSLOCATION WARRANTED?

Translocation is not a panacea. Its effectiveness varies widely with the circumstances, as does its cost. The main benefit of translocation lies in public relations. Translocation is socially acceptable; people always prefer to "give the animal a second chance," even when knowing that the odds are against that animal surviving. For instance, in the case of the translocation of white-tailed deer from a small nature center in Wisconsin, Scott Diehl (1988, p. 246) stated that relocation was "a remarkably positive solution to a very sensitive public relations dilemma." Most people do not like to be responsible for causing an animal's death. When homeowners hire someone to catch a nuisance animal for them, they prefer to think that the animal will be translocated rather than killed. Most nuisance wildlife control operators (NWCOs) will comply with their clients' wishes, although the reality of translocation differs from the myth of animals living "happily ever after" in their new homes. Capturing and translocating wildlife can be extremely stressful to the animals and can result in high mortality rates, particularly for urban wildlife released in a rural habitat where they do not know how to survive. In such cases, it may be more humane to euthanize animals than to translocate them (Craven et al. 1998).

Essentially, the translocation of problem animals makes sense in three situations: 1. when the animal is so valuable that euthanasia is not an option, 2. when the population is below carrying capacity at the release site, and 3. when public relations are more important than other factors. These conditions would not apply for most urban wildlife captured by NWCOs. Yet, these people are responsible for translocating hundreds of thousands of animals annually (Craven et al. 1998). State wildlife agencies should consider requiring nuisance animals to be euthanized upon capture, rather than allowing them to be translocated. Wildlife agencies should then also require all NWCOs to inform their clients that the captured animals will be humanely euthanized. This way, clients will realize that one of the costs of removing the nuisance animal will be its death. Then they can make their own decision whether the problem they are experiencing is worth that cost.

Translocating nuisance animals creates two problems that need to be considered before a decision is made to use this technique. The first problem is a moral one. If an animal is likely to cause the same problem elsewhere, then translocating animals merely shifts the problem to another group of people. The second problem concerns liability. The person(s), organization, or agency translocating an animal may be held liable for any damage or injuries caused by that animal after translocation and, in some cases, by that animal's offspring. Hence, animals should be translocated only if there is a willingness to assume this liability, which may be in the millions of dollars (Sidebar 9.2). For this reason, species that may cause human injuries or excessive damage should not be translocated.

Sidebar 9.2 Knochel vs. the State of Arizona

The Arizona Department of Fish and Game had a policy of capturing nuisance bears and translocating them elsewhere. At a release site in a mountain range near Phoenix, one of these translocated bears severely mauled a young girl, who was camping at the time. The attack left her with permanent injuries. The girl and her family sued the state for damages, arguing that the state knew, or should have known, that it was creating a dangerous situation by translocating nuisance bears and that, therefore, the state should be held liable for the girl's injuries. On advice of its counsel, the state settled out of court for $4.5 million.

SUMMARY

The translocation of animals is an appealing method of resolving wildlife damage problems because the public perceives it as giving the problem animal a second chance in a new home. Unfortunately, reality is usually different. Translocated animals often return to their original homes or cause similar problems at their release sites. Most translocated animals have a short life expectancy and may adversely affect resident animals at the release sites. People, organizations, or agencies who translocate animals should be willing to assume liability for any damage or injuries caused by these animals. For this reason, animals posing a threat to human safety should not be translocated. In many cases, translocating animals will not resolve the original problem because new animals will replace them and cause the same problems. Hence, it is often more effective to make the site less attractive to wildlife (e.g., it is better to cap a chimney than constantly translocate raccoons that attempt to nest in it). Capturing animals alive and translocating them is usually expensive. However, while translocating animals is not a panacea, it is a viable option in some situations. Translocation will be most appropriate when: 1. the animals have high value (e.g., an endangered species), 2. the population of the species at the release site is below carrying capacity, which is most likely to occur for species which are hunted or trapped or after a population crash, and 3. public relation concerns outweigh other considerations.

LITERATURE CITED

Andrews, R. D., G. L. Storm, R. L. Phillips, and R. A. Bishop, Survival and movements of transplanted and adopted red fox pups. *J. Wildl. Manage.*, 37, 69–72, 1973.

Armistead, R. A., K. Mitchell, and G. E. Connolly, Bear relocations to avoid bear/sheep conflicts. In *Proc. Vert. Pest Conf.*, 16, 31–35, 1994.

Bangs, E. E. and S. H. Fritts, Reintroducing the gray wolf to central Idaho and Yellowstone National Park. *Wildl. Soc. Bull.*, 24, 402–413, 1996.

Barnes, T. G., A survey comparison of pest control and nuisance wildlife control operators in Kentucky. In *Proc. East. Wildl. Damage Control Conf.*, 6, 39–48, 1993.

Barnes, T. G., Survey of the nuisance wildlife control industry with notes on their attitudes and opinions. In *Proc. Great Plains Wildl. Damage Control Workshop,* 12, 104–108, 1995.

Belant, J. L., L. A. Tyson, and T. W. Seamons, Use of alpha-chloralose by the Wildlife Services program to capture nuisance birds. *Wildl. Soc. Bull.*, 27, 938–942, 1999.

Beringer, J., L. P. Hansen, W. Wilding, J. Fischer, and S. L. Sheriff, Factors affecting capture myopathy in white-tailed deer. *J. Wildl. Manage.*, 60, 373–380, 1996.

Blanchard, B. M. and R. R. Knight, Biological consequences of relocating grizzly bears in the Yellowstone ecosystem. *J. Wildl. Manage.*, 59, 560–565, 1995.

Boyer, D. A. and R. D. Brown, A survey of translocations of mammals in the United States 1985. P. 1–11 in L. Nielsen and R. D. Brown, Eds., *Translocation of Wild Animals.* Wisconsin Humane Society, Milwaukee, WI and Caesar Kleberg Wildlife Research Institute, Kingsville, TX, 1988.

Brannon, R. D., Nuisance grizzly bear, *Ursus arctos*, translocations in the Greater Yellowstone area. *Can. Field-Naturalist*, 101, 569–575, 1987.

Castle, M. D. and B. M. Christensen, Hematozoa of wild turkeys from the midwestern United States: translocation of wild turkeys and its potential role in the introduction of *Plasmodium kempi. J. Wildl. Dis.*, 26, 180–185, 1990.

Chalmers, G. A. and M. W. Barrett, Capture myopathy. P. 84–94 in G. L. Hoff and J. W. Davis, Eds., *Noninfectious Diseases of Wildlife*. Iowa State University Press, Ames, 1982.

Conner, M. C., E. C. Soutiere, and R. A. Lancia, Drop-netting deer: costs and incidence of capture myopathy. *Wildl. Soc. Bull.*, 15, 434–438, 1987.

Cooper, J. A. and T. Keefe, Urban Canada goose management: policies and procedures. *Trans. North Am. Wildl. Nat. Res. Conf.*, 62, 412–430, 1997.

Craven, S., T. Barnes, and G. Kania, Toward a professional position on the translocation of problem wildlife. *Wildl. Soc. Bull.*, 26, 171–177, 1998.

Craven, S. R. and J. A. Nosek, Final report to the NPCA: summary of a survey on translocation of urban wildlife. University of Wisconsin, Department of Wildlife Ecology, Madison, 1992.

Cromwell, J. A., R. J. Warren, and D. W. Henderson, Live-capture and small-scale relocation of urban deer on Hilton Head Island, South Carolina. *Wildl. Soc. Bull.*, 27, 1025–1031, 1999.

Cunningham, A. A., Disease risks of wildlife translocations. *Conserv. Biol.*, 10, 349–353, 1996.

Curtis, P. D., M. L. Richmond, P. A. Wellner, and B. Tullar, Characteristics of the private nuisance wildlife control industry in New York. In *East. Wildl. Damage Control Conf.*, 6, 49–57, 1993.

DeNicola, A. J., S. J. Weber, C. A. Bridges, and J. L. Stokes, Nontraditional techniques for management of overabundant deer populations. *Wildl. Soc. Bull.*, 25, 496–499, 1997.

Diehl, S. R., The translocation of urban white-tailed deer. P. 239–249 in L. Nielsen and R. D. Brown, Eds., *Translocation of Wild Animals*. Wisconsin Humane Society, Milwaukee, WI and Caesar Kleberg Wildlife Research Institute, Kingsville, TX, 1988.

Fritts, S. H., W. J. Paul, and L. D. Mech, Movements of translocated wolves in Minnesota. *J. Wildl. Manage.*, 48, 709–721, 1984.

Fritts, S. H., W. J. Paul, and L. D. Mech, Can relocated wolves survive? *Wildl. Soc. Bull.*, 13, 459–463, 1985.

Fuller, T. K., Dynamics of a declining white-tailed deer population in north-central Minnesota. *Wildl. Monographs* 110, 1–37, 1990.

Galindo-Leal, C. and M. Weber, Translocation of deer subspecies: reproductive implications. *Wildl. Soc. Bull.*, 22, 117–120, 1994.

Green, D., G. R. Askins, and P. D. West, Public opinion: obstacle or aid to sound deer management? *Wildl. Soc. Bull.*, 25, 367–370, 1997.

Harthoorn, A. M., *The Chemical Capture of Animals. A Guide to the Chemical Restraint of Wild and Captive Animals.* Ralph Curtis Books, Hollywood, FL, 1975.

Hawkins, R. E., D. C. Autry, and W. D. Klimstra, Comparison of methods used to capture white-tailed deer. *J. Wildl. Manage.*, 31, 460–464, 1967.

Hawkins, R. E. and G. G. Montgomery, Movements of translocated deer as determined by telemetry. *J. Wildl. Manage.*, 33, 196–203, 1969.

Hawkins, R. E., L. D. Montoglio, and G. G. Montgomery, Cannon-netting deer. *J. Wildl. Manage.*, 32, 191–195, 1968.

Henshaw, R. E. and R. O. Stephenson, Homing in the gray wolf (*Canis lupus*). *J. Mammal.,* 55, 234–237, 1974.

Hutchins, M., T. Foose, and U. S. Seal, The role of veterinary medicine in endangered species conservation. *J. Zoo Wildl. Med.*, 22, 277–281, 1991.

Ishmael, W. E. and O. J. Rongstad, Economics of an urban deer-removal program. *Wildl. Soc. Bull.*, 12, 394–398, 1984.

Jenkins, S. R. and W. G. Winkler, Descriptive epidemiology from an epizootic of raccoon rabies in the middle Atlantic states, 1982–1983. *Am. J. Epidemiol.*, 126, 429–437, 1987.

Jones, J. M. and J. H. Witham, Post-translocation survival and movements of metropolitan white-tailed deer. *Wildl. Soc. Bull.*, 18, 434–441, 1990.

Jones, M. L., N. E. Mathews, and W. F. Porter, Influence of social organization on dispersal and survival of translocated female white-tailed deer. *Wildl. Soc. Bull.*, 25, 272–278, 1997.

Linnell, J. D. C., R. Aanes, J. E. Swenson, J. Odden, and M. E. Smith, Translocation of carnivores as a method for managing problem animals: a review. *Biodiversity Conserv.,* 6, 1245–1257, 1997.

Low, J., Mixed success. *Missouri Conserv.*, 60(4), 28, 1999.

Martinez, A., D. G. Hewitt, and M. Cotera Correa, Managing overabundant white-tailed deer in northern Mexico. *Wildl. Soc. Bull.*, 25, 430–432, 1997.

McArthur, K. L., Factors contributing to effectiveness of black bear transplants. *J. Wildl. Manage.*, 45, 102–110, 1981.

McCall, T. C., R. D. Brown, and C. A. DeYoung, Mortality of pen-raised and wild white-tailed deer bucks. *Wildl. Soc. Bull.*, 16, 380–384, 1988.

McCullough D. R., K. W. Jennings, N. B. Gates, B. G. Elliott, and J. E. DiDonato, Overabundant deer populations in California. *Wildl. Soc. Bull.*, 25, 478–483, 1997.

McLaughlin, C. R., C. J. Baker, A. Sallade, and J. Tamblyn, Characteristics and movements of translocated nuisance black bears in north central Pennsylvania. Pennsylvania Game Commission Report, Harrisburg, PA, 1981.

McNulty, S. A., W. F. Porter, N. E. Mathews, and J. A. Hill, Localized management for reducing white-tailed deer populations. *Wildl. Soc. Bull.*, 25, 265–271, 1997.

Miller, S. D., and W. B. Ballard, Homing of transplanted Alaskan brown bears. *J. Wildl. Manage.*, 46:869–876, 1982.

Mosillo, M., E. J. Heske, and J. D. Thompson, Survival and movements of translocated raccoons in north central Illinois. *J. Wildl. Manage.*, 63, 278–286, 1999.

Nettles, V. F., J. H. Shaddock, and R. K. Sikes, Rabies in translocated raccoons. *Am. J. Pub. Health,* 69, 601–602, 1979.

O'Bryan, M. K. and D. R. McCullough, Survival of black-tailed deer following relocation in California. *J. Wildl. Manage.*, 49, 115–119, 1985.

Pais, R. C., Mortality, dispersal, and habitat use of resident and translocated white-tailed deer does in the Cumberland Plateau of eastern Kentucky. M.S. thesis, University of Kentucky, Lexington, 1987.

Palmer, D. T., D. A. Andrews, R. O. Winters, and J. W. Francis, Removal techniques to control an enclosed deer herd. *Wildl. Soc. Bull.*, 8, 29–33, 1980.

Peck, L. J. and J. E. Stahl, Deer management techniques employed by the Columbus and Franklin County Park District, Ohio. *Wildl. Soc. Bull.*, 25, 440–442, 1997.

Peery, C. H., III, The economics of Virginia's deer transplantation program. In *Proc. Southeastern Assoc. Game Fish Commissioners*, 22, 142–144, 1969.

Phillips, R. L., J. L. Cummings, and J. D. Berry, Responses of breeding golden eagles to relocation. *Wildl. Soc. Bull.*, 19, 430–434, 1991.

Phillips, R. L., J. L. Cummings, G. Notah, and C. Mullis, Golden eagle predation on domestic calves. *Wildl. Soc. Bull.*, 24, 468–470, 1996.

Rabinowitz, A. R., Jaguar predation on domestic livestock in Belize. *Wildl. Soc. Bull.*, 14, 170–174, 1986.

Riley, D. G., Controlling raccoon damage in urban areas. *Great Plains Wildl. Damage Control Workshop,* 9, 85–86, 1989.

Rogers, L. L., Effect of translocation distance on frequency of return by adult black bears. *Wildl. Soc. Bull.*, 14, 76–80, 1986.

Rogers, L. L., Homing tendencies of large mammals: a review. P. 76–92 in L. Nielsen and R. D. Brown, Eds., *Translocation of Wild Animals*. Wisconsin Humane Society, Milwaukee, WI and Caesar Kleberg Wildlife Research Institute, Kingsville, TX, 1988.

Rosatte, R. C. and C. D. MacInnes, Relocation of city raccoons. *Great Plains Wildl. Damage Control Workshop*, 9, 87–92, 1989.

Ross, P. I. and M. G. Jalkotzy, Fates of translocated cougars, *Felis concolor*, in Alberta. *Can. Field-Naturalist,* 109, 475–476, 1995.

Ruth, T. K., K. A. Logan, L. L. Sweanor, J. F. Smith, and L. J. Temple, *Evaluating Mountain Lion Translocation: Final Report*. Hornocker Wildlife Research Institute, Moscow, ID, 1993.

Stout, R. J., B. A. Knuth, and P. D. Curtis, Preferences of suburban landowners for deer management techniques: a step towards better communication. *Wildl. Soc. Bull.*, 25, 348–359, 1997.

Sullivan, J. B., C. A. DeYoung, S. L. Beasom, J. R. Heffelfinger, S. P. Coughlin, and M. W. Hellickson, Drive-netting deer: incidence of mortality. *Wildl. Soc. Bull.*, 19, 393–396, 1991.

Taber, R. D., and R. F. Dasmann, The black-tailed deer of the chaparral: its life history and management in the North Coast Range of California. Game Bulletin 8. California Department of Fish and Game, Sacramento, CA, 1958.

CHAPTER 10

Fear-Provoking Stimuli

"All we have to fear is fear itself."

President Franklin Roosevelt

"As James Thurber said, 'You can fool too many of the people too much of the time.' But, unfortunately, animals are a lot smarter."

Adapted from William Fitzwater

"Fear-provoking stimuli can be classified scientifically into three categories – those that don't work, those that break, and those that get lost."

Adapted from Russell Baker

The optimal foraging theory predicts that animals should forage in a way that maximizes their nutritional benefits. Foraging, however, can be dangerous as animals venture out from cover and into areas that they would otherwise avoid. Hence, foraging decisions are often tempered by the need to avoid predation. That is, animals have to ignore some good foraging opportunities because it is too risky to exploit them. Because animals need to find food and to avoid predation, they often exhibit risk-averse foraging. For instance, marmots and prairie dogs remain close to the safety of their burrow even though better foraging opportunities are located elsewhere.

Wildlife managers can reduce wildlife damage to agricultural crops or predation on livestock by exploiting an animal's tendency to avoid foraging in risky areas. This is usually accomplished through the use of fear-provoking stimuli that increase the animal's fear of areas where crops or livestock are located. Fear-provoking stimuli are any objects that increase an animal's wariness or fear. They can be visual (e.g., predator models or scarecrows), auditory (e.g., distress calls, loud noises), or olfactory stimuli (e.g., predator urines). An animal's fear of an area can also be increased by making habitat modifications, such as removing hiding cover, which makes the animal more vulnerable to predators. Habitat modification, however, is discussed in Chapter 14 rather than in this chapter.

VISUAL STIMULI

Several visual stimuli can be used as fear-provoking stimuli. They are designed primarily to repel birds, which rely more on vision than on their other senses to avoid danger. Most visual stimuli designed to repel birds are models of predators (Figure 10.1). The use of predator models is based on the fact that foraging is more dangerous when predators are present. Hence, predator models can be placed in areas with human–wildlife conflicts to convince birds that it is safer to forage elsewhere.

Some predator models are designed to mimic hawks or owls (Craven and Lev 1985), but the most common predator model is the scarecrow, which is a model of a human. Scarecrows have been used for centuries to reduce bird predation on agricultural crops (Figure 10.2). They are effective because many wildlife species view humans as dangerous predators. For instance, Canada geese, which are hunted in North America, stopped visiting agricultural fields once scarecrows were erected in them (Heinrich and Craven 1990). Scarecrows, especially those capable of some movement, have provided at least some protection from birds at fish farms (Lagler 1939; Stickley et al. 1995), containment ponds (Boag and Lewin 1980), grain fields (Knittle and Porter 1988) and sunflower fields (Cummings et al. 1986).

Figure 10.1 Shiny pie pans and plastic owl models can be used to keep birds from loafing and defecating on sailing boats, but without much success in this case (note the loafing birds which are ignoring the pie pans and owl model).

Figure 10.2 Scarecrows have been used for hundreds of years to scare birds away from agricultural crops. (Figure from Vertebrate Pest Conference. With permission.)

Some birds may avoid an area where there are several dead conspecifics. Naef-Daenzer (1983) tested this idea in Switzerland but found that carrion crows were not repelled from grain fields by the presence of dead crows suspended above them.

Most animals are neophobic or fearful of novel objects, and many fear-provoking stimuli exploit this behavior. Examples of novel objects used to scare birds include flags, streamers, and flashing objects. Aluminum pie pans are often hung on fruit trees or in gardens and are somewhat effective in deterring birds for short periods of time (Marsh et al. 1991). Spotlights, flashing lights, and strobe lights are often used to repel nocturnal animals. Unfortunately, these lights are often ineffective or work for only short periods of time (Koehler et al. 1990; Andelt et al. 1997). Mylar® is a shiny tape which can be strung above a field or garden. Mylar tape twists in a breeze, causing it to flash in the sunlight. Dolbeer et al. (1986) strung Mylar tape above fields of sunflower, millet, and corn and reduced damage caused by blackbirds and house sparrows. Heinrich and Craven (1990) kept Canada geese out of agricultural crops using Mylar flags. However, Mylar was ineffective in deterring American robins, house finches, northern mockingbirds, and catbirds from blueberry fields

(Tobin et al. 1988). These findings led Dolbeer et al. (1986) to speculate that Mylar® may be effective against species that forage in large flocks but not against those that forage by themselves or in small groups.

AUDITORY STIMULI

Exploders and Bangers

Several fear-provoking stimuli are based on auditory stimuli (sound). One of the most common is the use of a loud bang to scare animals from areas where they are unwanted. There are many ways to produce a loud bang, including the use of firecrackers or firearms. A number of projectiles, including cracker shells, bird bombs, and whistle bombs, have been designed to scare birds when fired from a starter pistol or shotgun (Figure 10.3; Marsh et al. 1991). Crackers shells are fired from a shotgun and explode with a bang after traveling 100 to 150 m. Hence, cracker shells produce two loud blasts: the gun charge which propels the cracker shell through the air and then the cracker shell when it explodes. The latter is especially alarming to animals because the explosion occurs much closer than expected. For this reason, cracker shells are particularly useful in scaring birds from aquaculture facilities where people are along the edge of a pond and the birds are in the middle of the pond. Bird bombs are fired from a modified starter pistol and travel 20 to 30 m before exploding. Whistle bombs, racket bombs, and screamer shells make a whistling or screaming sound as they travel through the air. They are usually less effective than cracker shells, but are useful in that they generate a different sound (Marsh et al. 1991). Cracker shells, whistle bombs, and noise bombs have been used to disperse birds from airports,

Figure 10.3 Pyrotechnic devices used in wildlife damage management. (Photo courtesy of U.S.D.A. Wildlife Services.)

Figure 10.4 A propane exploder. (Photo courtesy of U.S.D.A. Wildlife Services.)

roosts, fish farms, agricultural fields, urban parks, and recreation areas (Mott 1980; Long 1982; Aguilera et al. 1991; Marsh et al. 1991; Andelt et al. 1997).

A disadvantage of shooting a firearm, starter pistol, or bird banger is that someone has to be present to fire it. Hence, they are labor intensive. To overcome this problem, several firecrackers can be attached to a slow-burning fuse so they ignite individually over a long period of time. Rope firecrackers can be strung over a pond or water bucket to lessen the risk of a wild fire. Rope firecrackers have been used to reduce bird predation at fish farms, protect rice and grain fields from blackbirds, and disperse starlings from vineyards (Marsh et al. 1991).

Due to the danger of fire, firecrackers have largely been replaced by propane exploders. Propane exploders function by releasing a measured amount of propane into a firing chamber where a spark ignites the gas (Figure 10.4). The explosions are loud and a timer can be used so that the cannon fires at preselected intervals ranging from a minute to an hour. The advantage of propane exploders is that they are safe and can be used when no one is around. Their disadvantage is that they can disturb the neighbors and therefore should not be used if homes or people are nearby. Also, because they fire at regular intervals, animals can habituate to them over the course of a few days. Propane exploders have been used to reduce bird damage in grain fields (Stephen 1961; Stickley et al. 1972), citrus groves (Hobbs and Leon 1987), and aquaculture facilities (Parkhurst et al. 1987). The effectiveness of propane exploders increases when paired with scarecrows (Figure 10.5).

Novel Sounds

Animals are wary of unfamiliar sounds, and some auditory devices designed to repel wildlife are based on this principle. The Av-Alarm® is a noise generator used to keep birds out of agricultural fields. It produces sounds believed to be irritating to birds or to "jam" the birds' own calls, but there is little evidence to support this claim (Marsh et al. 1991). Av-Alarms® have been used successfully to reduce the number of Canada geese foraging in agricultural fields (Heinrich and Craven 1990). A similar device developed by the U.S.D.A. National Wildlife Research Center

Figure 10.5 This pairing of a propane exploder with a scarecrow is used to protect a sunflower field from blackbirds. (Photo courtesy of John Cummings and U.S.D.A. Wildlife Services. With permission.)

combines a warbling siren with a strobe light and has had some success in scaring coyotes away from sheep flocks for short periods of time (Curnow 1991). However, noise generators in general have been largely unsuccessful in reducing wildlife damage (Craven and Lev 1985; Woronecki 1988; Bomford 1990).

Devices capable of emitting ultrasonic frequencies are often advertised to repel nuisance birds. Because humans cannot hear these high frequencies, these devices are supposed to bother birds but not people. However, Brand and Kellogg (1939a, b) documented that birds are unable to detect ultrasonic frequencies. Hence, it is not surprising that birds are not repelled or bothered by these devices (Marsh et al. 1991). While some mammals can hear ultrasonic frequencies, there is little evidence that ultrasound is useful in reducing problems caused by rodents or other mammals (Howard and Marsh 1985; Bomford and O'Brien 1990). Ultrasonic devices did not keep deer away from feeding sites (Curtis et al. 1997; Belant et al. 1998a). Ultrasonic "deer whistles" designed to be mounted on automobiles are ineffective despite the claim that they frighten deer away from moving vehicles (Romin and Dalton 1992).

Distress Calls and Alarm Calls

When a bird has been captured by a predator, it may emit a loud distress call that sounds more like a scream than a normal bird call. The call functions either by startling the predator into accidently dropping the bird or by attracting so much attention that the predator gives up and lets the bird go (Conover 1994a; Wise et al. 1999). When other birds hear a distress call, they know that a bird has been captured but often approach the source of a distress call before fleeing. Approaching a predator may seem maladaptive, but by doing so, birds can learn valuable information about predators in the area and how to avoid them.

Alarm calls are given when a bird detects a predator. This is a soft call and sounds like "seeet, seeet." Birds often respond to alarm calls by gathering in tree branches where they can watch the predator. During the breeding season, birds with nests or young nearby may dive at a predator or mob it to drive it away.

Recordings of both distress and alarm calls are often broadcast in an effort to scare birds from areas where they are unwanted. Recordings of these calls have been effective in scaring starlings from airports and vineyards, gulls from airports and garbage dumps, finches from grain fields, herons from fish farms, and flying-fox bats from orchards (Bomford and O'Brien 1990; Andelt and Hooper 1996). Some airports even have special vehicles equipped with mounted speakers that are used to broadcast distress and alarm calls at avian culprits. Distress call recordings have also been used to disperse jackdaws and American crows from roosts (Frings et al. 1958; Gorenzel and Salmon 1993).

The main drawback with distress and alarm calls is that birds ultimately learn they are not real and habituate to them. However, it takes birds longer to habituate to these calls than to loud noises that have no biological meaning to them. For instance, Spanier (1980) compared the effectiveness of distress calls to a loud bang in repelling black-crowned night herons from a fish farm. He reported that the loud bang lost its effectiveness within a week, while the distress calls remained effective for several months. Still, it is best to broadcast these calls sparingly to slow the rate of habituation. Recordings of distress and alarm calls are most effective if made from the same species that one is trying to scare. That is, when scaring crows from a tomato field, it is better to use a playback of distress and alarm calls given by a crow than those of another species.

Because distress calls are given when a bird is being held by a predator and alarm calls are given when there is a predator in the neighborhood, birds probably expect to see a predator whenever they hear one of these calls. If they do not, they may realize that something is not right and habituate more rapidly to the distress and alarm calls. For this reason, it is wise to pair the broadcast of these calls with a predator model (Nakamura 1997). Conover (1985) designed an owl model which held a model of a crow in its talons. This model was paired with the playback of crow distress calls. In this case, the visual predator model reinforced the distress call playback because crows that flew closer to investigate saw what they expected: a crow captured by a predator. Crows habituated more slowly to this model than to an owl model placed out by itself.

OLFACTORY STIMULI

Olfactory stimuli (odors) can also be used to repel animals. Many herbivorous animals are repelled by the odor of predators. By not foraging in areas where the scent of predators is strong, herbivores can avoid placing themselves in dangerous situations. Wildlife biologists have exploited this avoidance behavior by spreading predator odors in crops and orchards to keep herbivores away.

The effectiveness of predator odors is greatly enhanced when the odor is sprayed directly on the resource needing protection (Sullivan et al. 1985; Swihart et al. 1991). One drawback to this approach, however, is that predator odor may quickly evaporate. To prevent this from happening, predator odors can be placed in slow-release devices (Epple et al. 1995) or encapsulated within a semipermeable material (Burwash et al. 1998).

Many different types of predator odors have been effective in repelling herbivores. Chemicals from the anal gland secretions of the short-tailed weasel and mink have been used to reduce feeding by snowshoe hares (Sullivan and Crump 1984). Synthetic stoat odor and fox fecal odor reduced depredation on chickpeas by house mice (Coulston et al. 1993), weasel odor reduced vole gnawing on trees (Sullivan et al. 1988a), and bobcat urine reduced woodchuck damage to fruit trees (Swihart 1991; Belant et al. 1998b). Predator urine or feces discouraged white-tailed deer and elk from browsing (Melchiors and Leslie 1985; Swihart et al. 1991). Elk foraged less in areas which smelled of gray wolves (Andelt et al. 1992). Odors from minks, bobcats, coyotes, and domestic dogs have been used to reduce foraging by mountain beavers (Nolte et al. 1993, 1994; Epple et al. 1995).

Several studies have investigated which components of predator odors may be responsible for repelling animals. Identifying the active ingredients would make it possible to produce a synthetic spray and eliminate the need to collect predator urine. Sulfur-containing compounds in predator urines provoke an avoidance response or inhibit browsing by snowshoe hares (Sullivan and Crump 1984), European rabbits (Sullivan et al. 1988b), pocket gopher (Sullivan et al. 1988c), rats (Tobin et al. 1995; Burwash et al. 1998), mice (Epple et al. 1995), deer (Melchiors and Leslie 1985; Belant et al. 1998b), and domestic sheep (Epple et al. 1995).

CHEMICAL STIMULI

One chemical, 4-aminopyridine, seems to act as a fear-provoking stimuli and has been registered for this use under the trademark Avitrol®. When red-winged blackbirds ingest a piece of corn or other material sprayed with a small amount of this chemical, they go through a progression of behavioral changes beginning with erratic flight, convulsions, and distress calls, ultimately ending with the bird's death. Other blackbirds that witness these intoxicated individuals often become alarmed and leave the area. Reportedly less than 1% of a flock has to ingest the material and respond with these distress symptoms to cause the flock to abandon a field (De Grazio et al. 1972). Field tests evaluating the effectiveness of Avitrol® in protecting fields of ripening corn and sunflowers have produced mixed, but generally favorable, results (De Grazio et al. 1971, 1972; Stickley et al. 1972, 1976; Besser and Guarino 1976; Woronecki et al. 1979). However, other investigators have noted that 4-aminopyridine kills birds that consume it and have questioned whether the chemical's effectiveness is because it scares birds from fields, or kills them (Jaeger et al. 1983; Conover 1994b).

THE PROBLEM OF HABITUATION

The main problem with most fear-provoking stimuli is that animals soon learn that they pose no real threat and then ignore them. Habituation is the main factor that limits the effectiveness of fear-provoking stimuli as a method to resolve human–wildlife conflicts. Unless the fear-provoking stimuli are actually capable of killing animals, habituation is inevitable, but it takes time, maybe a few days or weeks, before it occurs. In the meantime, fear-provoking stimuli can be used to reduce wildlife damage. Hence, fear-provoking stimuli are most suitable for problems lasting only a few days. For example, when fields are planted in the spring, birds often go down the rows digging up the newly planted seeds. Fear-provoking stimuli may be useful in this situation because, a few days after the seeds sprout, birds lose interest in them. Fear-provoking stimuli may be useful to reduce raccoon damage to sweet corn fields because raccoons only eat ripe corn ears. Hence, protection is needed only for a few days before the corn is harvested. Fear-provoking stimuli can also be used to reduce bird damage in a ripening blueberry field, because there may only be a few days between when the berries are ripe enough to attract birds and when they are harvested. Unfortunately, the most common usage of fear-provoking stimuli is to place owl models in barns or other buildings, where they are supposed to repel pigeons and starlings. They have little chance of being successful because protection is needed for months or years.

CAN HABITUATION TO FEAR-PROVOKING STIMULI BE DELAYED?

Several methods can be used to retard an animal's habituation to fear-provoking stimuli. One method is to use fear-provoking stimuli that mimic real predators. Hawk models work best against those species that are common prey of hawks and in areas where raptors are abundant. In this case, an animal's experience with real hawks helps make the animal wary of the predator model. Interestingly, birds were more wary of a hawk model that appeared to be grasping a starling with its talons, than of a similar model not holding a starling (Conover and Perito 1981). Distress and alarm calls are effective fear-provoking stimuli because birds hear these calls occasionally during their normal activities. These experiences reinforce a bird's fear when it hears a distress or alarm call and delay the bird's habituation to broadcasted calls. Likewise, scarecrows are most effective against species, such as ducks and geese, that are often hunted by humans. They are also more effective during hunting seasons and in areas where hunting is popular. Scarecrows are a poor choice for keeping songbirds out of an orchard because these birds have no reason to fear humans.

Lack of movement is an obvious cue to animals that something is not alive. Hence, predator models that are capable of movement are more effective than stationary ones. Hawk models with mobile wings were more effective than hawk models without them (Conover 1985). Another example of a visual model that moves is the pop-up scarecrow: the Scarey Man®. This is an inflatable model of a person attached to an air pump. At timed intervals, the air pump inflates the scarecrow, making it appear to be standing up (Cummings et al. 1986; Stickley et al. 1995). These scarecrows often are used in conjunction with an exploder or siren and have been used in keeping birds out of sunflower fields, aquaculture facilities, and airports (Cummings et al. 1986; Stickley et al. 1995). Andelt et al. (1997) found that the Scarey Man had little effect on black-crowned night herons and great blue herons.

Figure 10.6 Hawk-kites are suspended below helium balloons, and their swaying in the wind makes them appear more life-like than a stationary predator model.

Another example of predator models capable of movement is hawk kites (Figure 10.6). These are clear plastic kites upon which a picture of a soaring hawk is printed. The movements of the kite as it soars in the wind are similar to those of actual hawks. These hawk kites have reduced bird damage to blueberry orchards and corn fields (Conover 1982, 1984) and often are suspended beneath helium balloons that keep them airborne when there is no wind. However, these models are labor intensive and prone to breakage (Conover 1982, 1984). Model airplanes also have been used to chase birds away from an area where they are unwanted.

The more animals are exposed to a fear-provoking stimuli, the faster they will habituate it. For this reason, bangers and shell-crackers should be used sparingly and propane cannons should be set so that they fire only a couple of times per hour.

Ideally, a propane cannon or other auditory stimuli should fire only when wildlife are close to it. One way to accomplish this is to have the noisemaker remotely controlled by someone who fires it only when an animal is nearby. This, however, is too labor intensive to be practical for most wildlife damage problems. One alternative is to connect a noisemaker to a motion detector, auditory sensor, or infrared sensor that activates the noisemaker whenever an animal is detected (Heinrich and Craven 1989). For instance, Belant et al. (1996) found that deer habituated more slowly to motion-activated gas exploders than to those that fired at timed intervals. Fear-provoking stimuli also should be moved to new locations every few days to delay habituation.

USING LIVE PREDATORS AS FEAR-PROVOKING STIMULI

The problem of habituation is lessened if live predators are used as fear-provoking stimuli, rather than innocuous models of them. For instance, JFK International Airport in New York uses falconry to reduce the danger of bird–aircraft strikes (Garber 1996). Peregrine falcons, goshawks, and gyrfalcons are most frequently used for this purpose (Blokpoel 1976). These raptors are a threat to most avian species, causing them to leave an area once a hawk or falcon is seen. However, the usefulness of falconers is limited by their small number. Trained hawks and falcons cannot be left unattended; hence, the method also is very labor intensive (Blokpoel 1976). Hawks and falcons also cannot be flown at night or during rain or fog. Because of these limitations, most facilities use trained raptors to supplement other fear-provoking stimuli, such as fire crackers, distress call playbacks, and human patrols (Marsh et al. 1991).

USING GUARD DOGS AS FEAR-PROVOKING STIMULI

Live dogs can be used as fear-provoking stimuli to scare wildlife. Dogs have been used to keep Canada geese away from sites where they are unwanted (Castelli and Sleggs 2000) and deer out of apple orchards and plant nurseries (Beringer et al. 1994). One way to use dogs as fear-provoking stimuli is simply to keep the dog under the control of its handler. Another approach is to enclose the dog within a fence on the property. Dog-proof fences are economically feasible and easier to build than wildlife-proof fences. Dogs can also be kept from wandering by having the dog wear a training collar that shocks the dog whenever it comes within a few meters of a wire that gives off a weak radio signal of the right frequency. The wire can either be buried or left on the ground around the area needing protection. Dogs quickly learn that if they try to cross the wire, they will get a shock and cease doing so. This method can be quite cost effective when the dogs behave as expected (Beringer et al. 1994). However, when guard dogs are left alone, some quickly tire of chasing animals that they cannot catch and start to ignore them.

Sidebar 10.1 Using Electronic Training Collars to Change Coyote Behavior

An electronic training collar provides an electric shock to the animal wearing it when the collar receives a specific radio signal. These collars were developed to train dogs in the field. They allow the dog's handler to punish the dog instantly for misbehaving by using a radio transmitter. Andelt et al. (1999) tested whether such a system could be used to teach coyotes to stop killing sheep. They placed training collars on captive coyotes and shocked them whenever they attacked a lamb or got within 2 m of one. Coyotes quickly associated the electric shocks with the lamb, stopped trying to kill them, and kept their distance from them.

Using training collars on free-ranging wildlife might not seem very practical. After all, it would be difficult to catch predators and place training collars on them. However training collars may be useful for resolving some very specific human–wildlife conflicts (Andelt et al. 1999). For instance, bears that raid campgrounds are usually relocated. Too often, these bears simple repeat their camp-raiding behavior in their new surroundings and have to be euthanized. One possible way to stop these bears from repeating their destructive behavior would be to place a training collar on them before they are released. If campgrounds were equipped with a radio transmitter that was constantly on and had a limited range, the bear would receive a shock whenever it approached the campground. Alternatively, park rangers could have a radio transmitter and signal the collar to shock the bear whenever they see the bear misbehaving. A similar system could be used to keep newly released wolves from attacking livestock or California condors from straying into dangerous areas.

Guard dogs have also been used to reduce depredation on livestock, especially sheep. Guard dogs, which have been raised with sheep since they were pups and have bonded with them, stay with the sheep and aggressively repel any predators. Use of guard dogs generally results in a reduction in the number of ewes or lambs killed by coyotes (Andelt 1992).

Guard dogs are a low-impact method to deal with predation, and many people view them as a humane and acceptable method of predator control (Green et al. 1984). Use of guard dogs, however, is not a panacea. Some guard dogs harass, injure, or even kill the livestock that they are supposed to protect, some stray from the flock and others attack wildlife or people (Green et al. 1984). Predators also may learn to "gang up" on a guard dog by having one predator distract the guard dog while another kills livestock.

Considerable time, effort, and money must be invested in guard dogs for them to be effective protectors (Green et al. 1984). One company spent over $9000 to implement a program using border collies to control nuisance Canada geese and $2000 annually to maintain the program. This investment provides no guarantee that a guard dog will be effective. Guard dogs are subject to early death and injury (Lorenz et al. 1986). For these reasons, donkeys (Green 1989; Walton and Feild 1989) and llamas (Meadows and Knowlton 2000) also are used to guard sheep (Figure 10.7). Whether it is cost effective to use a guard animal depends on many factors, such as the annual rate of predation, the ability and longevity of the animal, and the costs to purchase and maintain it (Green et al. 1984).

Figure 10.7 Use of llama to guard a sheep flock from predators. (Photo courtesy of Laurie Meadows. With permission.)

HAZING OR HARASSMENT

Most wildlife allow a person or predator to approach to a certain distance before fleeing. When an animal is constantly chased away from a site, it will stop returning to that site, especially if it has to travel some distance to get there. This approach, called harassment, has been used successfully to keep wood-pigeons from agricultural fields (Kenward 1978), grey herons or cormorants from fish farms (Draulans and van Vessem 1985; Moerbeek et al. 1987), and Canada geese off golf courses (Castelli and Sleggs 2000). Chasing wildlife away from crops was the most common method used by farmers living near India's Sariska Tiger Reserve to reduce wildlife damage (Sekhar 1998). Harassment and nest destruction have also been used to keep birds from nesting in unwanted areas (Ickes et al. 1998). Drawbacks with hazing and harassment are that they are very labor intensive, because the animals have to be chased away as soon as they arrive, and many chases may be required before the animals stop returning.

Dogs can add an extra frightening dimension to human patrols (Marsh et al. 1991). Border collies are most often used, and these dogs can be specially trained to chase geese (Castelli and Sleggs 2000). Use of border collies to harass geese is effective, but also labor intensive because the dogs must be under human supervision to be successful (Smith et al. 1999).

Wildlife have been hazed using human-piloted aircraft, including fixed-wing airplanes, ultralight recreational aircraft, and helicopters (Marsh et al. 1991). For instance, fixed-winged airplanes have been used to scare large flocks of blackbirds from sunflower fields in North Dakota (Figure 10.8). Helicopters have been used to keep bison from wandering outside of Yellowstone National Park (Meagher 1989). However, the use of aircraft to harass wildlife can be dangerous.

Small radio-controlled aircraft have proven to be an effective tool to disperse birds. Their use is both cheaper and safer than using a normal-sized aircraft. They

Figure 10.8 Fixed-wing airplane used to harass flocks of blackbirds from sunflower fields. (Photo courtesy of David Bergman and U.S.D.A. Wildlife Services. With permission.)

have been successful in dispersing dunlins, robins, gulls, pigeons, Canada geese, lapwings, and starlings. Though effective, use of model aircraft is labor intensive, and requires a skilled operator (Blokpoel 1976). Furthermore, birds often return to the area soon after the model airplane lands. For these reasons, model aircraft are best employed as one part of an integrated hazing program (Marsh et al. 1991).

Airboats and outboard motorboats have been used to haze birds on large ponds and lakes where the birds are too far offshore to be repelled by someone standing on the bank (Batten 1977). For instance, airboats were used to disperse geese from wildlife refuges (Craven et al. 1986; Taylor and Kirby 1990). Boats also are expensive and labor intensive, and their noise can be an annoyance to neighbors. Furthermore, birds often return after the boat leaves the area or simply move to places inaccessible to boats.

SUMMARY

Fear-provoking stimuli function by increasing the animal's fear of the area where they are deployed. They may be either visual (e.g., scarecrows, predator models), auditory (e.g., firecrackers, exploders, distress calls), or olfactory stimuli (e.g., predator odors). Most fear-provoking stimuli are effective for a few days, but animals quickly habituate to them. Thus, these stimuli are useful in alleviating wildlife problems which are of short duration. Most fear-provoking stimuli are actually innocuous and pose no danger to wildlife but are designed to deceive an animal into thinking that they pose a danger to it. Habituation to fear-provoking stimuli can be delayed by using lifelike stimuli that are capable of movement, by switching among different stimuli, and by changing the location of the stimuli. Some fear-provoking stimuli, such as hazing, harassment by humans, trained falcons and hawks, and guard dogs, pose a real threat to wildlife and, therefore, are effective for much longer periods of time than fear-provoking stimuli that are based on deception.

LITERATURE CITED

Aguilera, E., R. L. Knight, and J. L. Cummings, An evaluation of two hazing methods for urban Canada geese. *Wildl. Soc. Bull.*, 19, 32–35, 1991.

Andelt, W. F., Effectiveness of livestock guarding dogs for reducing predation on domestic sheep. *Wildl. Soc. Bull.*, 20, 55–62, 1992.

Andelt, W. F., D. L. Baker, and K. P. Burnham, Relative preference of captive cow elk for repellent-treated diets. *J. Wildl. Manage.*, 56, 164–173, 1992.

Andelt, W. F. and S. N. Hooper, Effectiveness of alarm-distress calls for frightening herons from a fish rearing facility. *Prog. Fish-Culturist,* 58, 258–262, 1996.

Andelt, W. F., R. L. Phillips, K. S. Gruver, and J. W. Guthrie, Coyote predation on domestic sheep deterred with electronic dog-training collar. *Wildl. Soc. Bull.*, 27, 12–18, 1999.

Andelt, W. F., T. P. Woolley, and S. N. Hopper, Effectiveness of barriers, pyrotechnics, flashing lights, and Scarey Man® for deterring heron predation on fish. *Wildl. Soc. Bull.*, 25, 686–694, 1997.

Batten, L. A., Sailing on reservoirs and its effect on water birds. *Biol. Conserv.*, 11, 49–58, 1977.

Belant, J. L., T. W. Seamans, and C. P. Dwyer, Evaluation of propane exploders as white-tailed deer deterrents. *Crop Prot.,* 15, 575–578, 1996.

Belant, J. L., T. W. Seamans, and L. A. Tyson, Evaluation of electronic frightening devices as white-tailed deer deterrents. In *Proc. Vert. Pest Conf.*, 18, 107–110, 1998a.

Belant, J. L., T. W. Seamans, and L. A. Tyson, Predator urines as chemical barriers to white-tailed deer. In *Proc. Vert. Pest Conf.*, 18, 359–362, 1998b.

Beringer, J., L. P. Hansen, R. A. Heinen, and N. F. Giessman, Use of dogs to reduce damage to a white pine plantation. *Wildl. Soc. Bull.*, 22, 627–632, 1994.

Besser, J. F., and J. L. Guarino, Protection of ripening sunflowers from blackbird damage by baiting with Avitrol® FC Corn Chops-99s. In *Proc. Bird Control Sem.*, 7, 200–203, 1976.

Blokpoel, H., *Bird Hazards to Aircraft.* Clarke, Irwin and Co., Ltd., Ottawa, Canada, 1976.

Boag, D. A. and V. Lewin, Effectiveness of three waterfowl deterrents on natural and polluted ponds. *J. Wildl. Manage.*, 44, 145–154, 1980.

Bomford, M., Ineffectiveness of a sonic device for deterring starlings. *Wildl. Soc. Bull.*, 18, 151–156, 1990.

Bomford, M. and P. H. O'Brien, Sonic deterrents in animal damage control: a review of device tests and effectiveness. *Wildl. Soc. Bull.*, 18, 411–422, 1990.

Brand, A. R. and P. P. Kellogg, Auditory responses of starlings, English sparrows, and domestic pigeons. *Wilson Bull.*, 51, 38–41, 1939a.

Brand, A. R. and P. P. Kellogg, The range of hearing of canaries. *Science,* 90, 354, 1939b.

Burwash, M. D., M. E. Tobin, A. D. Woodhouse, and T. P. Sullivan, Field testing synthetic predator odors for roof rats (*Rattus rattus*) in Hawaiian macadamia nut orchards. *J. Chem. Ecol.,* 24, 603–630, 1998.

Castelli, P. M. and S. E. Sleggs, Efficacy of border collies to control nuisance Canada geese. *Wildl. Soc. Bull.*, 28, 385–392, 2000.

Conover, M. R., Comparison of two behavioral techniques to reduce bird damage to blueberries: Methiocarb and a hawk-kite predator model. *Wildl. Soc. Bull.*, 10, 211–216, 1982.

Conover, M. R., Comparative effectiveness of Avitrol®, exploders, and hawk-kites to reduce blackbird damage to corn. *J. Wildl. Manage.*, 48, 109–116, 1984.

Conover, M. R., Protecting vegetables from crows using an animated crow-killing owl model. *J. Wildl. Manage.*, 49, 631–636, 1985.

Conover, M. R., Stimuli eliciting distress calls in adult passerines and response of predators and birds to their broadcast. *Behaviour,* 131, 19–38, 1994a.

Conover, M. R., Behavior of red-winged blackbirds during and after viewing a conspecific distressed by ingestion of 4-aminopyridine. *Pest. Sci.,* 41, 13–19, 1994b.

Conover, M. R. and J. J. Perito, Response of starlings to distress calls and predator models holding conspecific prey. *Z. für Tierpsychologie,* 57, 163–172, 1981.

Coulston, S., D. M. Stoddart, and D. R. Crump, Use of predator odors to protect chick-peas from predation by laboratory and wild mice. *J. Chem. Ecol.*, 19, 607–612, 1993.

Craven, S. R. and E. Lev, Double-crested cormorant damage to a commercial fishery in the Apostle Islands, Wisconsin. In *Proc. East. Wildl. Damage Control Conf.*, 2, 14–24, 1985.

Craven, S. R., G. A. Bartelt, D. H. Rusch, and R. E. Trost, Distribution and movement of Canada geese in response to management changes in east central Wisconsin, 1975–1981. *Wisconsin Department of Natural Resources Technical Bulletin 158*, Madison, WI, 1986.

Cummings, J. L., C. E. Knittle, and J. L. Guarino, Evaluating a pop-up scarecrow coupled with a propane exploder for reducing blackbird damage to ripening sunflower. In *Proc. Vert. Pest Conf.*, 12, 286–291, 1986.

Curnow, R. D., The Denver Wildlife Research Center: an update. In *Proc. Great Plains Wildl. Damage Conf.*, 10, 160–164, 1991.

Curtis, P. D., C. Fitzgerald, and M. E. Richmond, Evaluation of the Yard Gard ultrasonic yard protector for repelling white-tailed deer. In *Proc. Eastern Wildl. Damage Manage. Conf.*, 7, 172–176, 1997.

De Grazio, J. W., J. F. Besser, T. J. DeCino, J. L. Guarino, and R. I. Starr, Use of 4-aminopyridine to protect ripening corn from blackbirds. *J. Wildl. Manage.*, 35, 565–569, 1971.

De Grazio, J. W., J. F. Besser, T. J. DeCino, J. L. Guarino, and E. W. Shafer, Protecting ripening corn from blackbirds by broadcasting 4-aminopyridine baits. *J. Wildl. Manage.*, 36, 1316–1320, 1972.

Dolbeer, R. A., P. P. Woronecki, and R. L. Bruggers, Reflecting tapes repel blackbirds from millet sunflowers, and sweet corn. *Wildl. Soc. Bull.*, 14, 418–425, 1986.

Draulans, D. and J. van Vessem, The effect of disturbance on nocturnal abundance and behaviour of grey herons (*Ardea cinerea*) at a fish-farm in winter. *J. App. Ecol.*, 22, 19–27, 1985.

Epple, G., J. R. Mason, E. Arnov, D.L. Nolte, R. L. Hartz, R. Kaloostian, D. Campbell, and A. B. Smith, Feeding responses to predator-based repellents in the mountain beaver. *Ecol. Appl.*, 5, 1163–1170, 1995.

Frings, H., M. Frings, J. Jumber, R. Busnel, J. Giban, and P. Gramet, Reactions of American and French species of *Corvus* and *Larus* to recorded communication signals tested reciprocally. *Ecology,* 39, 127–131, 1958.

Garber, S. D., Effectiveness of falconry in reducing risk of birds strikes under study at JFK International. *J. Int. Civ. Aviation Org.*, 51, 5–7, 1996.

Gorenzel, W. P. and T. P. Salmon, Tape-recorded calls disperse American crows from urban roosts. *Wildl. Soc. Bull.*, 21, 334–338, 1993.

Green, J. S., Donkeys for predation control. In *Proc. East. Wildl. Damage Control Conf.*, 4, 83–86, 1989.

Green, J. S., R. A. Woodruff, and T. T. Tuller, Livestock-guarding dogs for predator control: costs, benefits, and practicality. *Wildl. Soc. Bull.*, 12, 44–50, 1984.

Heinrich, J. W. and S. R. Craven, Evaluation of a Canada goose call-activated switch for crop damage abatement. In *Proc. East. Wildl. Damage Control Conf.*, 4, 65–69, 1989.

Heinrich, J. W. and S. R. Craven, Evaluation of three damage abatement techniques for Canada geese. *Wildl. Soc. Bull.*, 18, 405–410, 1990.

Hobbs, J. and F. G. Leon, III, Great-tailed grackle predation on south Texas citrus. In *Proc. East. Wildl. Damage Control Conf.*, 3, 143–148, 1987.

Howard, W. E. and R. E. Marsh, Ultrasonics and electromagnetic control of rodents. *Acta Zoo. Fennica*, 173, 187–189, 1985.

Ickes, S. K., J. L. Belant, and R. A. Dolbeer, Nest disturbance techniques to control nesting by gulls. *Wildl. Soc. Bull.*, 26, 269–273, 1998.

Jaeger, M. M., J. L. Cummings, D. L. Otis, J. L. Guarino, and C. E. Knittle, Effect of Avitrol® baiting on bird damage to ripening sunflower within a 144-section block of North Dakota. In *Proc. Bird Control Sem.*, 9:247–254, 1983.

Kenward, R. E., The influence of human and goshawk *Accipiter gentilis* activity on wood-pigeons *Columba palumbus* at brassica feeding sites. *Ann. Appl. Biol.*, 89, 277– 286, 1978.

Knittle, C. E. and R. D. Porter, Waterfowl damage and control methods in ripening grain: an overview. *U. S. Fish and Wildlife Service Technical Report 14*. Washington, D.C., 1988.

Koehler, A. E., R. E. Marsh, and T. P. Salmon, Frightening methods and devices/stimuli to prevent mammal damage — a review. In *Proc. Vert. Pest Conf.*, 14, 168–173, 1990.

Lagler, K. F., The control of fish predators at hatcheries and rearing stations. *J. Wildl. Manage.*, 3, 169–179, 1939.

Long, G. L., Pyrotechnics for bird control. In *Proc. Great Plains Wildl. Damage Control Workshop*, 5, 278–282, 1982.

Lorenz, J. R., R. P. Coppinger, and M. R. Sutherland, Causes and economic effects of mortality in livestock guarding dogs. *J. Range Manage.*, 39, 293–295, 1986.

Marsh, R. E., W. A. Erickson, and T. P. Salmon, *Bird Hazing and Frightening Techniques*. University of California, Davis, 1991.

Meadows, L. E. and F. F. Knowlton, Efficacy of guard llamas to reduce canine predation on domestic sheep. *Wildl. Soc. Bull.*, 28, 614–622, 2000.

Meagher, M., Evaluation of boundary control for bison of Yellowstone National Park. *Wildl. Soc. Bull.*, 17, 15–19, 1989.

Melchiors, M. A. and C. A. Leslie, Effectiveness of predator fecal odors as black-tailed deer repellents. *J. Wildl. Manage.*, 49, 358–362, 1985.

Moerbeek, D. J., W. H. van Dobben, E. R. Osieck, G. C. Boere, and C. M. Bungenberg de Jong, Cormorant damage prevention at a fish farm in the Netherlands. *Biol. Conserv.*, 39, 23–38, 1987.

Mott, D. F., Dispersing blackbirds and starlings from objectionable roost sites. In *Proc. Vert. Pest Conf.*, 9, 38–42, 1980.

Naef-Daenzer, L., Scaring of carrion crows (*Corvus corone corone*) by species-specific distress calls and suspended bodies of dead crows. In *Proc. Bird Control Sem.*, 9, 91–95, 1983.

Nakamura, K., Estimation of effective area of bird scarers. *J. Wildl. Manage.*, 61, 925–934, 1997.

Nolte, D. L., J. P. Farley, D. L. Campbell, J. R. Mason, and G. Epple, Potential repellents to prevent mountain beaver damage. *Crop Prot.*, 12, 624–626, 1993.

Nolte, D. L., J. R. Mason, G. Epple, E. Arnov, and D. L. Campbell, Why are predator urines aversive to prey? *J. Chem. Ecol.,* 20, 1505–1516, 1994.

Parkhurst, J. A., R. P. Brooks, and D. E. Arnold, A survey of wildlife depredation and control techniques at fish-rearing facilities. *Wildl. Soc. Bull.*, 15, 386–394, 1987.

Romin, L. A. and L. B. Dalton, Lack of response by mule deer to wildlife warning whistles. *Wildl. Soc. Bull.*, 20, 382–384, 1992.

Sekhar, N. V., Crop and livestock depradation caused by wild animals in protected areas: the case of Sariska Tiger Reserve, Rajasthan, India. *Environ. Conserv.,* 25, 160–171, 1998.

Smith, A. E., S. R. Craven, and P. D. Curtis, Managing Canada geese in urban environments. *Jack Berryman Institute Publication 16*, Logan, UT, 1999.

Spanier, E., The use of distress calls to repel night herons (*Nycticorax nycticorax*) from fish ponds. *J. App. Ecol.*, 17, 287–294, 1980.

Stephen, W. J. D., Experimental use of acetylene exploders to control duck damage. *Trans. North Am. Wildl. Conf.,* 26, 98–111, 1961.

Stickley, A. R., R. T. Mitchell, R. G. Heath, C. R. Ingram, and E. L. Bradley, A method for appraising the bird repellency of 4-aminopyridine. *J. Wildl. Manage.*, 36, 1313–1316, 1972.

Stickley, A. R., R. T. Mitchell, J. L. Seubert, C. R. Ingram, and M. I. Dyer, Large-scale evaluation of blackbird frightening agent 4-aminopyridine in corn. *J. Wildl. Manage.*, 40, 126–131, 1976.

Stickley, A. R., D. F. Mott, and J. O. King, Short-term effects of an inflatable effigy on cormorants at catfish farms. *Wildl. Soc. Bull.*, 23, 73–77, 1995.

Sullivan, T. P. and D. R. Crump, Influence of mustelid scent gland compounds on suppression of feeding by snowshoe hares (*Lepus americanus*). *J. Chem. Ecol.*, 10, 1809–1821, 1984.

Sullivan, T. P., D. R. Crump, and D. S. Sullivan, Use of predator odors as repellents to reduce feeding damage by herbivores. III. Montane and meadow voles (*Microtus montanus* and *Microtus pennsylvanicus*). *J. Chem. Ecol.*, 14, 363–377, 1988a.

Sullivan, T. P., D. R. Crump, and D. S. Sullivan, Use of predator odors as repellents to reduce feeding damage by herbivores. IV. Northern pocket gophers (*Thomomys talpoides*). *J. Chem. Ecol.*, 14, 379–389, 1988c.

Sullivan, T. P., L. O. Nordstrom, and D. S. Sullivan, Use of predator odors as repellents to reduce feeding damage by herbivores. II. Black-tailed deer (*Odocoileus hemionus columbianus*). *J. Chem. Ecol.*, 11, 921–935, 1985.

Sullivan, T. P., D. S. Sullivan, D. R. Crump, H. Weiser and E. A. Dixon, Predator odors and their potential role in managing pest rodents and rabbits. In *Proc. Vert. Pest Conf.*, 13, 145–150, 1988b.

Swihart, R. K., Modifying scent-marking behavior to reduce woodchuck damage to fruit trees. *Ecol. Appl.*, 1, 98–103, 1991.

Swihart, R. K., J. J. Pignatello, and M. J. I. Mattina, Aversive responses of white-tailed deer, *Odocoileus virginianus*, to predator urines. *J. Chem. Ecol.*, 17, 767–775, 1991.

Taylor, J. P., and R. E. Kirby, Experimental dispersal of wintering snow and Ross' geese. *Wildl. Soc. Bull.*, 18, 312–319, 1990.

Tobin, M. E., R. M. Engeman, and R. T. Sugihara, Effects of mongoose odors on rat capture success. *J. Chem. Ecol.*, 21, 635–639, 1995.

Tobin, M. E., P. P. Woronecki, R. A. Dolbeer, and R. L. Bruggers, Reflecting tape fails to protect ripening blueberries from bird damage. *Wildl. Soc. Bull.*, 16, 300–303, 1988.

Walton, M. T., and C. A. Feild, Use of donkeys to guard sheep and goats in Texas. In *Proc. East. Wildl. Damage Control Conf.*, 4, 87–94, 1989.

Wise, K. K., F. K. Knowlton, and M. R. Conover, Response of captive coyotes to the starling distress call: testing the startle-predator and predator-attraction hypotheses. *Behaviour*, 136, 935–949, 1999.

Woronecki, P. P., Effect of ultrasonic, visual, sonic devices on pigeon numbers in a vacant building. In *Proc. Bird Control Sem.*, 13, 266–272, 1988.

Woronecki, P. P., R. A. Dolbeer, C. R. Ingram, and A. R. Stickley, Jr., 4-aminopyridine effectiveness reevaluated for reducing blackbird damage to corn. *J. Wildl. Manage.*, 43, 184–191, 1979.

CHAPTER 11

Chemical Repellents

"Organic chemistry is the study of carbon compounds. Biochemistry is the study of carbon compounds that crawl."

Mike Adams

"The rabbits in my garden only eat a light meal – they start at first light and don't stop until evening."

Adapted from Henny Youngman

Many plant species are protected from herbivores because they contain toxins that make them poisonous. Unfortunately, most agricultural crops lack chemical defenses against insect or vertebrate herbivores. This is particularly true of plants grown for human consumption, as toxins would also make them unpalatable or poisonous to humans. Therefore, farmers often apply chemicals to crops to give them protection that they would otherwise lack. For insect pests, lethal chemicals are applied, but because vertebrates are so valuable, pesticides are rarely applied with the intention of killing them. Rather, we apply repellents in an effort to dissuade wildlife from eating plants without causing the animals serious harm. Efficacious use of repellents depends on a knowledge of how plants protect themselves from herbivores and why animals have preferences for certain tastes and foods.

HOW PLANTS USE CHEMICALS TO DEFEND THEMSELVES FROM HERBIVORES

Plants have contended with herbivores for millions of years. Over this period, plants have evolved mechanisms, such as thorns and toxins, to discourage herbivory. Many plants use nutrient dilution to discourage herbivores by packing their tissues with hard-to-digest fibers, such as lignin and cellulose. Likewise, some grasses contain high concentrations of silica, making them difficult to digest. Because herbivores seek nutrient-rich plants, they avoid those containing large concentrations of fiber or silica. Humans refer to plants that are well defended with fibers as "tough" or "stringy" and generally avoid eating them, preferring more "tender" plants.

Many plants contain high levels of tannins, which also cause nutrient dilution, but in a different way. Tannins reduce an animal's ability to absorb proteins and carbohydrates in the food by binding to them so they are excreted, rather than absorbed (Robbins 1983). For instance, condensed tannins from blackbrush and bitterbrush inhibit food digestion by snowshoe hares (Clausen et al. 1990).

Alkaloids and other plant secondary compounds also discourage herbivory. These chemicals are often toxic even in very low concentrations. Yet, some animals have developed methods of detoxifying these chemicals. However, the toxins take their toll because the detoxification and excretion processes are energetically expensive. For instance, coniferyl benzoate is a phenylpropanoid ester produced by quaking aspen. It is repellent to most birds. Ruffed grouse have the ability to detoxify it (Jakubas et al. 1997), and this species forages heavily on aspen buds during winter and aspen flowers during spring (Svoboda and Gullion 1972; Doerr et al. 1974; Jakubas and Gullion 1991). Yet, ruffed grouse limit their foraging to aspen buds with low concentrations of coniferyl benzoate because the energy and nitrogen costs of detoxification are substantial (Jakubas et al. 1997). As another example, snowshoe hares select green alders to browse based on their concentrations of pinosylvin and pinosylvin methyl ether (Bryant et al. 1983), and they select birches to browse based on concentrations of papyriferic acid (Reichardt et al. 1984; Bryant et al. 1994).

BIOLOGICAL BASIS OF FOOD PREFERENCES

To survive, every animal must extract enough energy from its diet to meet its energetic needs. It also must consume certain chemicals that it needs to survive and is unable to produce itself, such as salts, minerals, amino acids, and vitamins. In trying to meet their daily metabolic needs, animals face a wide array of consumable items. Some choices are nutritious and contain a high concentration of energy, some are poisonous, and others contain such low levels of nutrients and energy that an animal cannot afford to eat them. Animals must be able to sift through these food items and select only those that are able to meet the animal's nutritional needs. How an animal makes these decisions involves a complex array of innate preferences for certain tastes and learned preferences from post-ingestive feedback and early experience.

Role of Olfaction in Shaping Food Preferences

Olfaction serves many functions. It provides information about the environment and assists in locating and identifying foods. In some species, it plays a role in sexual, maternal, and aggressive behavior. Given its importance, it is not surprising that animals have hundreds or thousands of different olfactory receptors (Beauchamp 1997). Olfactory ability and preferences differ among species and provide reasons why diets differ among animals. For instance, many herbivores are repelled by sulfur compounds common in meat, while carnivores are attracted to them.

Role of Taste in Shaping Food Preferences

Mammal have hundreds of different olfactory receptors, but there are only a small number of specific taste receptors. The most important ones are sweet, sour, salty, and bitter. Taste receptors probably evolved for the purpose of regulating food intake to ensure the rejection of toxins and the acceptance of nutrient-rich foods. Although taste preferences are modified by experience and learning, an animal's initial response to tastes is innately determined (Beauchamp 1997). For example, acidic foods produce a "sour" taste which animals often avoid, perhaps because those foods may injure the lining of an animal's mouth or throat. Many animals exhibit a craving for salt, especially when deprived of salt. In particular, herbivores often exhibit a strong preference for salt, due to its low concentration in plants. Herbivores and omnivores also exhibit a preference for "sweet" foods (i.e., plants containing large concentrations of simple carbohydrates). This preference for sweet foods may help animals identify energy-rich foods. Yet, some predators, such as cats, do not prefer carbohydrate sweeteners and may not even be able to taste them (Beauchamp et al. 1977).

Some foods contain poisonous alkaloids and taste "bitter." The ability to detect these chemicals helps protect animals from poisonous foods. Carnivores exhibit a strong dislike of alkaloids, which is not surprising because their normal diet rarely contains these chemicals. In contrast, herbivores exhibit a graded response to alkaloids. Given a choice, they will select a diet low in alkaloids but will consume bitter foods in larger quantities when there is no alternative. By doing so, they can keep their intake of alkaloids within tolerable levels. Furthermore, there is only a rough correlation between toxicity and bitterness (Beauchamp 1997). By sampling bitter plants in small quantities, herbivores learn which plants can be eaten safely and which need to be avoided. Likewise, they can identify plant parts, such as flowers, fruit, or young leaves, containing lower levels of alkaloids. These findings suggest that bitter chemicals are more likely to be effective repellents for use against carnivores and omnivores rather than herbivores (Beauchamp 1997).

Role of Tactile Stimuli in Shaping Food Preferences

Some repellents function by changing the texture or tactile features of food (Mason and Clark 1994). For instance, beaver do not like eating gritty foods and can be dissuaded from gnawing on trees by painting the bark with a latex paint mixed with sand. Likewise, birds can be repelled from food contaminated with dolomitic lime, activated charcoal, or sand (Belant et al. 1997).

Role of Irritants in Shaping Food Preferences

The trigeminal nerve is the fifth cranial nerve and runs from the brain to the facial skin, eyes, mouth, throat, and nasal passages. Some chemicals, such as capsaicin in hot peppers, stimulate the trigeminal nerve. Sometimes, people seek low levels of stimulation of the trigeminal nerve (e.g., people enjoy eating spicy food) but, usually, stimulation of the trigeminal nerve causes pain.

Chemicals that irritate the trigeminal nerve may serve as good repellents because they are innately unpleasant. Furthermore, the sensation of pain often increases with repeated stimulation of the trigeminal nerve (Green 1991). In contrast, animals habituate to most odor and taste stimuli after repeated exposure due to nerve desensitization. When capsaicin is consumed, it stimulates the trigeminal nerve, causing irritation in mammals but apparently not bothering birds due to neurological differences between these groups of animals (Mason and Maruniak 1983). As a result, capsaicin can be used to repel mammals but not birds. Such a repellent could be used, for instance, to keep squirrels from eating birdseed from feeders (Fitzgerald et al. 1997).

Role of Post-Ingestion Feedback in Shaping Food Preferences

Some chemicals inhibit food consumption by causing illness, nausea, or gastrointestinal distress. These chemicals function by a post-ingestion feedback mechanism (Provenza 1995). That is, when humans eat something and become nauseated, we associate the sickness with the food just eaten, and develop an aversion to that food. We may rationalize that we are allergic to the food or that it was ill-prepared. The end result, however, is that we become wary of that particular food, no longer like its taste, and time will pass before we eat it again. We make this association between nausea and the consumed food even when an illness or seasickness is the source of the nausea. Human brains have evolved to make this association innately because doing so helps us avoid poisonous foods. Other species have evolved the same response.

Often animals eat various foods in one meal. How then do they decide what caused the illness? When nausea follows the consumption of many items, animals develop an aversion to the food that had an unusual taste or odor. When animals become ill after eating a mix of familiar and novel food, they develop an aversion to the novel food (Kalat 1974).

Likewise, animals may feel more energetic after consuming an item high in energy or other essential nutrients. As a result, the animal may develop a preference for the item and increase future consumption of it (Mehiel and Bolles 1988). Horses become more active after eating oats, which contain a high concentration of carbohydrates (this is the source of the phrase, "feeling its oats," which means being frisky). Horses develop a preference for oats as a result of their enhanced vigor. Animals can develop a preference for items which alleviate illnesses. For example, healthy lambs prefer plain water to water containing sodium bicarbonate, but lambs suffering from acidosis prefer the latter (Provenza 1997).

Food preferences are shaped by nutritional needs. Humans, for instance, prefer high-carbohydrate foods when hungry, low-carbohydrate foods when satiated, and salty foods when suffering from a sodium deficiency (Booth and Toase 1983). Rats develop a preference for foods containing threonine when suffering from a deficiency of this amino acid but not at other times (Gietzen 1993). Protein and carbohydrate preferences depend on an animal's nutritional state (Gibson and Booth 1986, 1989). Thus, food preferences constantly change depending on nutritional needs and generally decline when foods are eaten to satiety (Provenza 1997).

Role of Early-Life Experiences in Shaping Food Preferences

An animal's experiences early in life can have a dramatic and long-lasting effect on its food preferences (Figure 11.1). A young animal learns what foods to eat based on what its parents feed it, being escorted by its parents to places where certain foods can be obtained, and watching what its parents eat (Mirza and Provenza 1992). Young mammals can acquire taste preferences associated with their mother's milk (Galef and Henderson 1972; Nolte and Provenza 1991). Even prenatal experiences with chemicals can influence later taste preferences (Nolte and Mason 1995). As they mature, animals become more reluctant to eat novel foods or familiar foods whose flavors have changed (Birch and Marlin 1982; Provenza et al. 1993). This phenomenon is referred to as neophobia: a fear of the new.

Figure 11.1 Goslings learn what to eat by following their parents and watching what they eat.

TYPES OF REPELLENTS

Area Repellents

Much research has been conducted on olfactory or "area repellents" used to repel animals from specific areas by imparting a particular odor to the area. Area repellents are used to keep deer out of orchards by hanging rags dipped in bone tar oil, open-mesh bags containing human hair, and bars of soap from the trees (Figure 11.2). Although some area repellents provide some protection, most are ineffective (Hygnstrom and Craven 1988; El Hani and Conover 1997).

Some area repellents (e.g., moth balls) may function by producing an odor which animals find so offensive or noxious that they are willing to leave the area to avoid it. Others function by increasing an animal's fear of an area. The fear-provoking

Figure 11.2 Bone tar oil can be used as an area repellent to discourage deer herbivory when poured on a rag and tied to a plant.

repellents are more successful at keeping animals out of a protected area. Many predators have unique odors which other animals can use to detect and avoid them. Predator urine has demonstrated good efficacy against a broad range of herbivores when used as an area repellent (Sullivan et al. 1988). The repellency is associated with the odor of sulfur compounds in the urine, which result from the digestion of meat (Nolte et al. 1994). Research has shown that urine of coyotes maintained on a meat diet repelled herbivores while the urine of coyotes on a vegetarian diet did not (Nolte et al. 1994). Some species, such as skunks, also emit odorous chemicals when threatened by predators. These odors may serve as effective area repellents if animals respond to them by avoiding the area.

Some sick, parasitized, or dying animals have a unique odor (Beauchamp 1997). By avoiding areas containing these odors, conspecifics reduce their chances of becoming ill or parasitized themselves. Hence, these odors have a potential for use as area repellents.

Many species use chemicals (pheromones) to mark their territories. These chemicals may serve as area repellents if animals, especially juveniles and subadults, avoid treated areas because they think they are entering an adult's defended territory (Lindgren et al. 1997).

Several problems limit the usefulness of area repellents in reducing wildlife damage. One limitation is that animals lose their ability to detect an odor after repeated or continuous exposure. For instance, exposure to an odor over several weeks leads to a decline in the ability to detect that odor. This loss in sensitivity can last for weeks (Beauchamp 1997). Another limitation is that, if animals are starving, they will enter a site containing food regardless of how badly it smells. Hence, area repellents fail when needed most (i.e., when alternate food supplies are lacking).

Contact Repellents

Contact repellents are chemicals designed to be applied directly to food items needing protection, so that the repellents are ingested along with the food (these are also called food repellents). The idea is that, if an animal cannot eat the food without also consuming the repellent, it will stop eating that food. Hopefully, application of a repellent will cause the animal to shift its food preference from the treated food to an alternate food. Several contact repellents have been developed to stop deer from eating plants (Sidebar 11.1).

Sidebar 11.1 Effectiveness of Repellents to Protect Plants from Deer

White-tailed deer and mule deer cause more damage to North American crops than any other wildlife species. Their browsing can be particularly destructive to orchards, tree farms, reforestation areas, and plant nurseries. Numerous odor and taste repellents have been developed to reduce deer damage, including BGR® (a putrefied egg extract), blood meal, bobcat excrement, chicken eggs, coyote urine, feather meal, Hinder®, Hot Sauce (contains capsaicin), human hair, Magic Circle®, meat meal, Ropel®, soap, and thiram. There have been numerous studies to test the effectiveness of these repellents, often producing conflicting results. The variance in findings probably results from differences in location, deer species, plant species, time of year, test duration, or criteria of success. Despite these problems, BGR was found to be the most effective repellent in a number of field tests. Predator odors also have shown promise as repellents; however, more field tests are needed, and they are not yet registered by the EPA for use as repellents. No repellent, including BGR, has eliminated deer damage entirely, and, in fact, none has reduced browsing by more than 50%. Hence, growers using repellents should expect some browsing damage. Repellents differ considerably in cost, and cost effectiveness should also be considered when selecting repellents (El Hani and Conover 1997).

Contact repellents function by imparting a noxious taste or odor to the food (taste repellents), by producing illness when consumed (aversive conditioning chemicals), or by a combination of both (dual-action chemicals). An advantage of taste repellents is that they have an immediate effect. In contrast, aversive conditioning chemicals may not inhibit feeding the first time an animal consumes them. Because aversive conditioning chemicals produce a learned aversion, it takes time for an animal to become ill and to learn what made it sick. This time lag could be a problem when low levels of damage cannot be tolerated. For instance, taste repellents would be a more effective method to prevent rodents from gnawing the insulation of electric wires. Taste repellents would also work better when attempting to reduce crop damage near a roost containing a large number of blackbirds. If each bird must become ill before being conditioned to avoid the food, there may be little left by the time all of the birds have become averted to it. The disadvantage of taste repellents is that they change only the taste of the food and not the nutritional value. Animals may ignore the repellent if they are hungry enough. Therefore, taste repellents work best when alternate foods are available and may not work at all when most needed (e.g., during winter). This disadvantage can be minimized by using dual-action food repellents, which have both a noxious taste and produce illness when consumed. The advantage of dual-action repellents is that, if an animal ignores the noxious taste and consumes the food anyway, the chemical causes illness. Hence, animals learn the error of ignoring the taste through postingestive feedback.

One bird repellent that functions with these dual modes of action is methiocarb. This chemical is an astringent, which produces a puckering sensation that birds find unpleasant. If, however, the taste is not sufficient to deter birds from eating food treated with methiocarb, they become ill. As a result of its dual action, methiocarb is very effective at stopping birds from eating treated seed (Stickley and Guarino 1972) or fruit (Guarino et al. 1974; Stone et al. 1974).

One of the biggest drawbacks associated with contact repellents is that they usually wear off or decompose in a matter of days and, consequently, may only provide protection for a limited period of time. This is not an issue for short-term problems, but it is for long-term ones. For instance, tree seedlings or fruit trees need protection from deer browsing for the entire winter, and most repellents applied in November will have washed off or broken down by March. The severity of this problem can be reduced by mixing the repellent with another chemical (called a sticker) which helps the repellent adhere more tightly to plants.

Another problem with applying contact repellents is that plants grow constantly. Any new leaves or shoots produced after the repellent was applied will be unprotected (Figure 11.3). For instance, consider the issue of trying to reduce goose grazing on turf. Because grass grows constantly and is mowed often, a repellent would have to be reapplied every few weeks to be effective.

Systemic Repellents

One way to avoid the problem of contact repellents either washing off or decomposing is to use a chemical contained within the plant (i.e., a systemic repellent). Selenium, which is the chemical that gives onions their pungent smell, eye-watering effect, and "hot" taste, repels many animals. Plants absorb selenium through their roots and transport it throughout their system. Selenium pellets have been used to protect conifer tree seedlings from deer herbivory by placing them near the roots. Although selenium protected some trees from deer browsing, it killed others (Allan et al. 1984; Angradi and Tzilkowski 1987; Engeman et al. 1995). Thus, it was not an effective way to resolve this human–wildlife conflict.

Another approach to using systemic repellents is breeding plants to contain high concentrations of chemicals capable of warding off herbivores. For example, there are two types of grain sorghum: a yellow type, which birds prefer, and a purple type, which they do not. The latter contains a high concentration of tannins, which inhibit herbivory.

Figure 11.3 One problem with spraying a repellent on turf is that the grass is constantly growing; new leaves that grow after the repellent has been applied will not be treated with the repellent.

Some plants, such as tall fescue grass, do not have the ability to produce chemicals to protect themselves from herbivory. However, tall fescue obtains alkaloids, notably ergovaline, through a unique mutualistic relationship with a fungus (Clay 1988, 1993). The fungus grows within the plant and benefits from the relationship by living off the plant's nutrients. In fact, the fate of the fungus is completely tied to that of its plant host. The fungus cannot produce spores or live outside the plant. It is passed from one plant to another through the plant's seeds. The only way to get an infected plant is to have an infected parent (Clay 1988, 1993). The plant also benefits from this relationship because the fungus produces alkaloids, which sicken both avian and mammalian herbivores. After consuming fungus-infected grass, herbivores develop an aversion for it and stop eating it (Aldrich et al. 1993; Schmidt and Osborn 1993; Conover and Messmer 1996; Conover 1998). For this reason, fungus-infected plants may make an ideal ground cover in areas where vertebrate herbivores are causing damage. They could be planted along dikes and canals where burrowing animals are a threat or in parks and golf courses where Canada geese are a nuisance.

Biotechnology offers another approach to developing systemic repellents. Geneticists have protected cotton and corn plants from insect herbivores by inserting a piece of DNA that produces an insect toxin into the cotton or corn genome. The genetically engineered plants produce the toxin and suffer less herbivory from corn rootworms and cotton bollworms. Such an approach could also be used to produce plants containing a systemic repellent effective against vertebrate pests. However, the costs of creating such a plant makes it unlikely that one will be bio-engineered just to resist vertebrate pests. Also, the chemical must not reduce the quality of the crop for human use.

CONDITIONED FOOD AVERSIONS BASED ON DECEPTION

There are a plethora of problems caused by foraging wildlife for which we cannot apply a contact repellent to protect the resource from predation. For instance, protecting pecans from squirrels is not an easy task. Simply spraying a repellent on pecans is ineffective because squirrels do not eat the shell, but rather crack it open and eat the seed. As another example, consider the problem of trying to protect a duck nest from predation (Figure 11.4). Spraying eggs with repellent is ineffective because most predators eat only egg contents and not the shell. Injecting a repellent into the egg would solve this problem, but we lack the technology to do so without killing the embryo. Even if this problem were overcome, there would still be the logistical task of finding duck nests so that we could inject them. Nest searches also are counterproductive; nests visited by humans are more likely to be depredated than unvisited nests because predators often follow human trails.

Aversive conditioning can still be used to protect those resources that cannot be directly treated with a repellent by resorting to a technique called deception-

Figure 11.4 Contact repellents are ineffective in protecting the eggs of nesting birds because the nests are difficult to locate and most predators do not eat egg shells where the repellent is located. (Photo courtesy of M. Kirkland and U.S. Bear River Wildlife Refuge. With permission.)

based food aversions or DBFA. With this approach, the resource needing protection is not treated. Rather, the aversive chemical is applied to a bait that mimics the resource. DBFA is effective when a depredating animal consumes the treated mimic, becomes ill, develops an aversion to the mimic, and, finally, generalizes its aversion to the resource.

In the last two decades, there have been several attempts to use DBFA to solve wildlife problems. Gustavson et al. (1974) tried to use DBFA to teach coyotes not to kill sheep. They hypothesized that, by distributing sheep carcasses and bait packages laced with lithium chloride, coyotes would consume them, become ill, develop an aversion to the taste of mutton, and stop killing sheep and lambs. While some initial tests yielded positive results (Gustavson et al 1974, 1976; Ellins et al. 1977), other tests did not (Burns 1980, 1983a, b; Burns and Connolly 1980). Large-scale field tests involving replicated samples also produced mixed results (Bourne and Dorrance 1982; Gustavson et al. 1982; Jelinski et al. 1983; Conover and Kessler 1994).

Other applications of DBFA, however, have been more successful. For instance, Nicolaus et al. (1983) used DBFA to train American crows not to eat eggs. Woodchucks generalized an aversion to untreated tomatoes from treated ones (Swihart and Conover 1991), and livestock learned not to graze certain plant species (Burritt and Provenza 1989, 1990; Lane et al. 1990). Nonetheless, Ratnaswamy et al. (1997) were unable to stop raccoons from depredating the nests of sea turtles by using DBFA.

SIMILARITIES BETWEEN BATESIAN MIMICRY AND DBFA

Many animals that have evolved chemical defenses against predation, especially insects and amphibians, also have evolved conspicuous color patterns or behaviors to enhance a predator's ability to recognize them (Jaervi et al. 1981; Sillen-Tullberg et al. 1982). Other species lacking chemical defenses have evolved color patterns and behaviors that mimic those of poisonous species. This phenomenon is known as Batesian mimicry (Brower 1969; Brower and Moffitt 1974; Fink and Brower 1981). Batesian mimicry functions because these palatable prey obtain some degree of protection from predators through their mimicry (Brower 1958a, b, c; Duncan and Sheppard 1965; Morell and Turner 1970).

DBFA can be viewed as a reverse form of Batesian mimicry because, in DBFA, the resource is innocuous and the mimics (the chemically treated baits) are poisonous. Nevertheless, the general ecological and behavioral principles that apply to Batesian mimicry should also apply to DBFA. Two of these principles involve the precision of the mimicry and the cost to benefit ratio.

Precision of Mimicry

Some Batesian mimics are discovered and preyed upon because predators learn to detect subtle differences between them and the poisonous models (Sillen-Tullberg et al. 1982; Brower and Fink 1985). Likewise, some attempts to use DBFA have

Figure 11.5 Deception-based aversion conditioning can be used to teach animals not to accept food from humans. (Figure courtesy of Vertebrate Pest Council. With permission.)

failed because depredating animals could distinguish between the chemically treated mimics and the untreated resource. Attempts to use DBFA to teach coyotes not to kill sheep failed because chemically treated mimics are not similar to live sheep. Sheep mimics were created by injecting chemicals into mutton baits wrapped in sheep fur or into sheep carcasses. Coyotes had little difficulty differentiating between chemically treated baits and live sheep; many coyotes developed an aversion to mutton baits but continued killing sheep.

Fortunately, a closer mimicry is possible for other human–wildlife conflicts. Bears can destroy beehives, and it is easy to create a chemically treated beehive which mimics untreated beehives (Colvin 1975; Gilbert and Roy 1977; Polson 1983). It is also possible to mimic closely food handouts from humans (Figure 11.5; Cornell and Cornely 1979; Conover 1999) or nests (Nicolaus 1987; Conover 1990; Semel and Nicolaus 1992), and attempts to use DBFA to address these problems have been more successful. Likewise, precise mimics of vegetable crops (Swihart and Conover 1991) or seeded fields (Avery 1989) can be created by treating part of a field while leaving the rest untreated.

Costs to Benefits Ratio

If treated baits (mimics) can be produced that are identical to the resource, then animals will not be able to distinguish between the two and will face a dilemma. Consider, for instance, an attempt to protect turtle eggs from raccoons by distributing treated mimics. If a raccoon finds a nest and eats the eggs, it runs the risk that the eggs are mimics and that sickness will ensue. If it does not eat any, it runs the risk of missing a nutritious meal. Which decision the animal makes will depend upon which mistake it can most afford to make. That, in turn, depends on the severity of the potential illness and the nutritive value of the resource.

If the risk of making an error and eating a treated mimic is low, the animal probably will take that risk. However, if the illness caused by eating a mimic is severe and long-lasting, the animal is less likely to take the chance of getting ill. Predation rates on monarch butterfly populations, for example, vary with their toxicity (Calvert et al. 1979; Brower and Fink 1985). Aversive conditioning chemicals also vary in the severity of the illness they produce. Nicolaus and Nellis (1987) reported that 9 of 32 mongooses that were fed carbachol-treated eggs were killed by the drug, while the survivors averted from eggs. In contrast, consumption of lithium chloride is rarely fatal and its effects last only for a few hours; not surprisingly, most attempts to avert predators from consuming eggs using this chemical have failed (Hopkins and Murphy 1982; Sheaffer and Drobney 1986; Conover 1989).

Animals are more likely to consume a novel food and risk becoming ill when they are nutritionally deficient and the food contains the needed nutrients. Birds forage on toxic monarch butterflies and their Batesian mimics when alternate food supplies are lacking (Fink and Brower 1981; Brower 1985). If nutrients in the resource are irreplaceable, DBFA cannot be expected to change the animals' behavior. For example, raccoons generalized an aversion to untreated food from treated sources of the same food, as long as alternate foods were available, but they did not do so when alternate foods were unavailable (Conover 1989).

FACTORS INFLUENCING REPELLENT EFFECTIVENESS TO REDUCE WILDLIFE DAMAGE

Weather

Rain can drastically reduce the effectiveness of contact repellents. Andelt et al. (1991) reported that several repellents were effective in reducing browsing of apple twigs by mule deer, but when the twigs were sprinkled with water to simulate rainfall, their repellency decreased. Sullivan et al. (1985) found that Big Game Repellent® (BGR) or the feces of coyotes, cougars, and wolves completely suppressed feeding by captive mule deer on salal, Douglas-fir, and western red cedar for 20 days in dry weather. But after just one day of heavy rain, the repellents became ineffective.

Repellent Concentration

Bullard et al. (1978a, b) showed that the repellency of synthetic fermented egg could be enhanced considerably by increasing its concentration. In a study evaluating the effectiveness of predator fecal odor on mule deer, Melchiors and Leslie (1985) found that repellency of fecal extracts from five predators was correlated with concentration. Repellent concentration also was a determining factor in decreasing browsing damage by elk (Andelt et al. 1992).

Duration of the Problem

Many authors have noted that repellent effectiveness declined over time. Hence, repellents may have to be reapplied repeatedly to retain their effectiveness. Reapplying repellents, however, is not always successful in lowering browsing rates. Conover and Kania (1988) found that even a midwinter reapplication of BGR® was not sufficient to stop deer browsing on young apple trees. Clearly, repellents are more appropriate for short-term problems rather than long-term ones.

Availability of Alternate Food Supplies

Repellents normally function by reducing the palatability of the treated plant to a level lower than that of other available plants. Hence, repellent effectiveness depends upon the availability of alternate forage. Andelt et al. (1991, 1992) reported that repellents reduced deer and elk browsing when alternate foods were available but not when the treated food was presented to hungry animals lacking alternative forage.

Relative Plant Palatability

Repellents should be more effective when used on a less palatable plant species than on those which are highly palatable. Adding a repellent to a less palatable food is more likely to reduce the food's palatability to below that of alternate foods. Support for this hypothesis comes from several studies that evaluated the same repellent on numerous plant species. Dietz and Tigner (1968) found that repellents were more effective at protecting aspen than the more palatable chokecherry. Sodium selenite was more effective at protecting white ash seedlings than the more palatable seedlings of black cherry (Angradi and Teilkowski 1987). Swihart et al (1991) found that bobcat and coyote urine were more effective when used on eastern hemlock than on the more palatable Japanese yews.

LAWS GOVERNING THE USE OF VERTEBRATE REPELLENTS

All repellents sold in the U.S. and in most other countries must be approved by the government. In the U.S., the Federal Insecticide, Fungicide, and Rodenticide Act (FIFRA) of 1947 gave the federal government the right to ban the sale of any repellent or pesticide when the risks to human health or the environment outweigh the benefits. Since 1970, the U.S. Environmental Protection Agency (EPA) has had the responsibility for making these decisions. This is done by requiring that all pesticides and repellents be registered with and approved by the EPA for the intended use before being sold. The EPA makes its decision based on evaluations of scientific tests of product chemistry, toxicity to target and nontarget species, potential hazards to human health, environmental fate, and chemical longevity in the environment. Chemicals intended for use on food crops face an even more rigorous standard (Fagerstone et al. 1990).

It is the responsibility of the company wishing to register a pesticide or repellent to demonstrate that it meets the requirements of registration. Before the EPA will approve a repellent, the manufacturer of the chemical must submit the results of several medical and environmental tests to verify its safety. These tests can cost more that a million dollars (Fagerstone et al. 1990).

Additionally, repellents or pesticides must be registered with and approved by state governments before they can be used in each state. Usually, the states accept the decisions of the EPA and do not require additional testing. However, some states, such as California, Florida, and New York, require their own review of the data and sometimes require additional testing. Dealing with the different state requirements can be laborious for chemical companies because requirements differ among states, and each state requires a separate application and application fee (Hushon 1997). These requirements provide disincentives to companies or individuals wishing to register a repellent.

Other countries have their own laws governing the use of repellents within their jurisdiction. In Canada, repellents and pesticides fall under the Pest Control Products Acts and must be approved before they can be used. Canada's registration requirements for repellents are even more laborious than those in the U.S.; it takes approximately two years for a potential repellent to be reviewed in Canada (Hushon 1997). In the European Union, repellents are regulated under the Biocides Directive and require rigorous testing before they can be approved for use (Hushon 1997).

One potential way to reduce costs associated with registering a new repellent is to use pesticides already registered for other uses on food. The advantage of registering these chemicals is that much of the required testing has already been conducted (Reidinger 1997). For instance, the bird repellent methiocarb was first used as an insecticide.

Unfortunately, the cost to register a repellent is so prohibitive that chemical companies are not willing to spend the necessary funds because the market for vertebrate repellents is relatively small compared to the use of insecticides on major crops. Even when a repellent is registered for use by the EPA, the company must recover the money it spent conducting the required tests. Hence, a repellent that may cost only a few dollars per kilogram to manufacture may sell for several hundred dollars per kilogram so that the company can recover the money spent registering the chemical. Due to their high cost, the demand for repellents is limited because few people can afford to use them. Hence, there are very few chemical repellents available for use throughout the world.

SUMMARY

Through evolution, many plants have developed chemicals that protect them from herbivores. Chemical repellents are designed to provide protection to plants by exploiting an animal's need to avoid poisonous foods and concentrate its foraging on highly nutritious foods. Animals recognize nutritious food through taste and olfactory cues, learning through postingestive feedback, and early-life experiences. Repellents can be used as area repellents designed to keep wildlife out of an area, contact or food repellents sprayed on or attached to the food item, or systemic repellents incorporated within the plant. DBFA can also be used to protect food items by tricking animals into believing that a safe food is actually poisonous and should be avoided. Repellents are most effective when used during periods of good weather, in high concentrations, in small areas, and for problems with a short duration. Because repellents are designed to persuade animals to eat something else rather than the treated food, they function best when alternate food supplies are readily available and when used on plants of low palatability. The biggest factor limiting the use of repellents in wildlife damage management is that all vertebrate repellents in the U.S. must be approved for that use by the federal government. This is necessary to ensure that repellents are effective and pose no threat to the environment or human health. Although no one would disagree with these goals, the necessary tests to verify that a repellent meets these criteria are so expensive that chemical companies have little interest in developing vertebrate repellents. Thus, few repellents can be used legally in the U.S. or in many other countries.

LITERATURE CITED

Aldrich, C. G., J. A. Paterson, J. L. Tate, and M. S. Kelley, The effects of untufted-infected tall fescue on diet utilization and thermal regulation in cattle. *J. Anim. Sci.*, 71, 164–170, 1993.

Allan, G. G., D. I. Gustafson, R. A. Mikels, J. M. Miller, and S. Neogi, Reduction of deer browsing of Douglas-fir (*Pseudotsuga menziesii*) seedlings by quadrivalent selenium. *For. Ecol. Manage.*, 7, 163–181, 1984.

Andelt, W. F., D. L. Baker, and K. P. Burnham, Relative preference of captive cow elk for repellent-treated diets. *J. Wildl. Manage.*, 56, 164–173, 1992.

Andelt, W. F., K. P. Burnham, and J. A. Manning, Relative effectiveness of repellents for reducing mule deer damage. *J. Wildl. Manage.*, 55, 341–347, 1991.

Angradi, T. R., and W. M. Tzilkowski, Preliminary testing of a selenium-based systemic deer browse repellent. In *Proc. East. Wildl. Damage Control Conf.*, 3, 102–107, 1987.

Avery, M. L., Experimental evaluation of partial repellent treatment for reducing bird damage to crops. *J. App. Ecol.*, 26, 433–439, 1989.

Beauchamp, G. K., Chemical signals and repellency: problems and prognosis. P. 1–10 in J. R. Mason, Ed., *Repellents in Wildlife Management: Proceedings of a Symposium.* National Wildlife Research Center, Fort Collins, CO, 1997.

Beauchamp, G. K., O. Maller, and J. G. Rogers, Flavor preference in cats (*Felis catus* and *Panthera* sp.). *J. Comp. Physiol. Psychol.*, 91, 1118–1127, 1977.

Belant, J. L., S. K. Ickes, L. A. Tyson, and T. W. Seamans, Comparison of four particulate substances as wildlife feeding repellents. *Crop Protect.*, 16, 439–447, 1997.

Birch, L. L. and D. W. Marlin, I don't like it; I never tried it: effects of exposure on two-year-old children's food preferences. *Appetite,* 3, 353–360, 1982.

Booth, D. A. and A. M. Toase, Conditioning of hunger/satiety signals as well as favor cues in dieters. *Appetite,* 4, 235–236, 1983.

Bourne, J. and M. J. Dorrance, A field test of lithium chloride aversion to reduce coyote predation on domestic sheep. *J. Wildl. Manage.*, 46, 235–239, 1982.

Brower, J. V. Z., Experimental studies of mimicry in some North American butterflies. Part I. The monarch, *Danaus plexippus* and the viceroy, *Limenitis archippus archippus*. *Evolution,* 12, 32–47, 1958a.

Brower, J. V. Z., Experimental studies of mimicry in some North American butterflies. Part II. *Battus philenar* and *Papilio trailus, P. polyxenes*, and *P. glaucus*. *Evolution,* 12, 123–136, 1958b.

Brower, J. V. Z., Experimental studies of mimicry in some North American butterflies. Part III. *Danaus gilippus herenice* and *Limenitis archippus floridensis*. *Evolution,* 12, 273–285, 1958c.

Brower, L. P., Ecological chemistry. *Sci. Am.*, 220, 22–29, 1969.

Brower, L. P., Foraging dynamics of bird predators on overwintering monarch butterflies in Mexico. *Evolution*, 39, 852–868, 1985.

Brower, L. P. and L. S. Fink, A natural toxic defense system: cardenolides in butterflies versus birds. P. 171–188 in N. S. Braveman and P. Braveman, Eds., *Experimental Assessments and Clinical Applications of Conditioned Food Aversions*. New York Academy of Science, New York, 1985.

Brower, L. P. and C. M. Moffitt, Palatability dynamics of cardenolides in the monarch butterfly. *Nature*, 249, 280–283, 1974.

Bryant, J. P., R. K. Swihart, P. B. Reichardt, and L. Newton, Biogeography of woody plant chemical defense against snowshoe hare browsing: comparison of Alaska and eastern North America. *Oikos,* 70, 385–395, 1994.

Bryant, J. P., G. D. Wieland, P. B. Reichardt, V. E. Lewis, and M. C. McCarthy, Pinosylvin methyl ether deters snowshoe hare feeding on green alder. *Science,* 222, 1023–1025, 1983.

Bullard, R. W., T. J. Leiker, J. E. Peterson, and S. R. Kilburn, Volatile components of fermented egg, an animal attractant and repellent. *J. Agric. Food Chem.*, 26, 155–159, 1978a.

Bullard, R. W., S. A. Shumake, D. L. Campbell, and F. J. Turkowski, Preparation and evaluation of a synthetic fermented egg coyote attractant and deer repellent. *J. Agric. Food Chem.*, 26, 160–163, 1978b.

Burns, R. J., Evaluation of conditioned predation aversion for controlling coyote predation. *J. Wildl. Manage.*, 44, 938–942, 1980.

Burns, R. J., Coyote predation aversion with lithium chloride: management implications and comments. *Wildl. Soc. Bull.*, 11, 128–133, 1983a.

Burns, R. J., Micro encapsulated lithium chloride bait aversion did not stop coyote predation on sheep. *J. Wildl. Manage.*, 47, 1010–1017, 1983b.

Burns, R. J. and G. E. Connolly, Lithium chloride bait aversion did not influence prey killing by coyotes. In *Proc. Vert. Pest Conf.*, 9, 200–204, 1980.

Burritt, E. A. and F. D. Provenza, Food aversion learning: conditioning lambs to avoid a palatable shrub (*Cercocarpus montanus*). *J. Anim. Sci.*, 67, 650–653, 1989.

Burritt, E. A. and F. D. Provenza, Food aversion learning in sheep: persistence of conditioned taste aversion to palatable shrubs (*Cercocarpus montanus* and *Amelanchier alnifolia*). *J. Anim. Sci.*, 68, 1003–1007, 1990.

Calvert, W. H., L. E. Hedricks, and L. P. Brower, Mortality of monarch butterfly (*Danaus plexippus* L.): avian predation at five overwintering sites in Mexico. *Science*, 204, 847–851, 1979.

Clausen, T. P., F. D. Provenza, E. A. Burritt, P. B. Reichardt, and J. P. Bryant, Ecological implications of condensed tannin structure: a case study. *J. Chem. Ecol.*, 16, 2381–2392, 1990.

Clay, K., Fungal endophytes of grasses: a defensive mutualism between plants and fungi. *Ecology,* 69, 10–16, 1988.

Clay, K., The ecology and evolution of endophytes. *Agric. Ecosystems Environ.*, 44, 39–64, 1993.

Colvin, T. R., Aversive conditioning of black bear to honey utilizing lithium chloride. In *Proc. Ann. Conf. Southeastern Assoc. Game Fish Commissioners,* 29, 450–453, 1975.

Conover, M. R., Potential compounds for establishing conditioned food aversions in raccoons. *Wildl. Soc. Bull.*, 17, 430–435, 1989.

Conover, M. R., Reducing mammalian predation on eggs by using a conditioned taste aversion to deceive predators. *J. Wildl. Manage.*, 54, 360–365, 1990.

Conover, M. R. Impact of consuming tall fescue leaves with the endophytic fungus, *Acremonium coenophialum* on meadow voles. *J. Mammal.,* 79, 457–463, 1998.

Conover, M. R., Can waterfowl be taught to avoid food handouts through conditioned food aversions? *Wildl. Soc. Bull.*, 27, 160–166, 1999.

Conover, M. R. and G. S. Kania, Effectiveness of human hair, BGR, and a mixture of blood meal and peppercorns in reducing deer damage to young apple trees. In *East. Wildl. Damage Control Conf.*, 3, 97–101, 1988.

Conover, M. R. and K. K. Kessler, Diminished producer participation in an aversive conditioning program to reduce coyote predation on sheep. *Wildl. Soc. Bull.*, 22, 229–233, 1994.

Conover, M. R. and T. A. Messmer, Feeding preferences and changes in mass of Canada geese grazing untufted-infected tall fescue. *Condor*, 98, 859–862, 1996.

Cornell, D. and J. E. Cornely, Aversive conditioning of campground coyotes in Joshua Tree National Monument. *Wildl. Soc. Bull.*, 7, 129–131, 1979.

Dietz, D. R. and J. R. Tigner, Evaluation of two mammal repellents applied to browse species in the Black Hills. *J. Wildl. Manage.*, 32, 109–114, 1968.

Doerr, P. D., L. B. Keith, D. H. Rusch, and C. A. Fischer, Characteristics of winter feeding aggregations of ruffed grouse in Alberta. *J. Wildl. Manage.*, 38, 601–615, 1974.

Duncan, C. L. and P. M. Sheppard, Sensory discrimination and its role in the evolution of Batesian mimicry. *Behaviour*, 24, 270–282, 1965.

El Hani, A. and M. R. Conover, Comparative analysis of deer repellents. P. 147–155. in J. R. Mason, Ed., *Repellents in Wildlife Management: Proceedings of a Symposium.* National Wildlife Research Center, Fort Collins, CO, 1997.

Ellins, S. R., S. M. Catalano, and S. A. Schechinger, Conditioned taste aversion: a field application to coyote predation on sheep. *Behav. Biol.*, 20, 91–95, 1977.

Engeman, R. M., D. L. Campbell, D. Nolte, and G. W. Witmer, Some recent results on nonlethal means of reducing animal damage to reforestation projects in the western United States. In *Aust. Vert. Pest Control Conf.*, 10, 150–154, 1995.

Fagerstone, K. A., R. W. Bullard, and C. A. Ramey, Politics and economics of maintaining pesticide registrations. In *Proc. Vert. Pest Conf.*, 14, 8–11, 1990.

Fink, L. S. and L. P. Brower, Birds can overcome the cardenolide defense of the monarch butterflies in Mexico. *Nature*, 291, 67–70, 1981.

Fitzgerald, C. S., P. D. Curtis, M. E. Richmond, and J. A. Dunn, Effectiveness of capsaicin as a repellent to birdseed consumption by gray squirrels. P. 169–183 in J. R. Mason, Ed., *Repellents in Wildlife Management: Proceedings of a Symposium.* National Wildlife Research Center, Fort Collins, CO, 1997.

Galef, D. G. and P. W. Henderson, Mother's milk: a determinant of the feeding preference of weaning rat pups. *J. Comp. Physiol. Psychol.*, 83, 374–378, 1972.

Gibson, E. L. and D. A. Booth, Acquired protein appetite in rats: dependence on a protein-specific need state. *Experientia,* 42, 1003–1004, 1986.

Gibson, E. L. and D. A. Booth, Dependence of carbohydrate-conditioned flavor preference on internal state in rats. *Learning Motivation,* 20, 36–47, 1989.

Gietzen, D. W., Neural mechanisms in the responses to amino acid deficiency. *J. Nutr.*, 123, 610–625, 1993.

Gilbert, B. K. and L. D. Roy, Prevention of black bear damage to bee yards using aversive conditioning. P. 93–102 in R. L. Phillips and C. Jonkel, Eds., *Proceedings of the 1975 Predator Symposium.* Montana Forest Conservation Experiment Station, University of Montana, Missoula, 1977.

Green, B. G., Temporal characteristics of capsaicin sensitization and desensitization on the tongue. *Physiol. Behavior*, 49, 501–505, 1991.

Guarino, J. L., W. F. Shake, and E. W. Schaefer, Jr., Reducing bird damage to ripening cherries with methiocarb. *J. Wildl. Manage.*, 38, 338–342, 1974.

Gustavson, C. R., J. Garcia, W. G. Hankies, and K. W. Rusiniak, Coyote predation control by aversive conditioning. *Science,* 184, 581–583, 1974.

Gustavson, C. R., J. R. Jowsey, and D. N. Milligan, A 3-year evaluation of taste aversion coyote control research in Saskatchewan. *J. Range Manage.*, 35, 57–59, 1982.

Gustavson, C. R., M. Sweeney, and J. Garcia, Prey-lithium aversions. I. Coyotes and wolves. *Behav. Biol.*, 17, 61–72, 1976.

Hopkins, S. R. and T. M. Murphy, Testing of lithium chloride aversion to mitigate raccoon depredation of loggerhead turtle nests. In *Proc. Ann. Conf. Southeastern Assoc. Fish Wildl. Agencies*, 36, 484–491, 1982.

Hushon, J. M., Review of regulatory-imposed marketing constraints to repellent development. P. 423–428 in J. R. Mason, Ed., *Repellents in Wildlife Management: Proceedings of a Symposium.* National Wildlife Research Center, Fort Collins, CO, 1997.

Hygnstrom, S. E. and S. R. Craven, Electric fences and commercial repellents for reducing deer damage in cornfields. *Wildl. Soc. Bull.*, 16, 291–296, 1988.

Jakubas, W. J., C. G. Guglielmo, and W. H. Karasov, Dilution and detoxification costs: relevance to avian herbivore food selection. P. 53–70 in J. R. Mason, Ed., *Repellents in Wildlife Management: Proceedings of a Symposium.* National Wildlife Research Center, Fort Collins, CO, 1997.

Jakubas, W. J. and G. W. Gullion, Use of quaking aspen flower buds by ruffed grouse: Its relationship to grouse densities and bud chemical composition. *Condor,* 93, 473–485, 1991.

Jaervi, T., G. Sillen-Tullberg, and C. Wiklund, The cost of being aposematic. an experimental study of predation on larvae of *Papilio machaon* by the great tit *Parus major. Oikos,* 36, 267–275, 1981.

Jelinski, D. E., R. C. Rounds, and J. R. Jowsey, Coyote predation on sheep, and control by aversive conditioning in Saskatchewan. *J. Range Manage.*, 36, 16–19, 1983.

Kalat, J. W., Taste salience depends on novelty, not concentration, in taste-aversion learning in rats. *J. Comp. Physiol. Psychol.*, 86, 47–50, 1974.

Lane, M. A., M. H. Ralphs, J. D. Olsen, F. D. Provenza, and J. A. Pfister, Conditioned taste aversion: potential for reducing cattle loss to larkspur. *J. Range Manage.*, 43, 127–131, 1990.

Lindgren, P. M. F., T. P. Sullivan, and D. R. Crump, Review of synthetic predator odor semiochemicals as repellents for wildlife management in the Pacific Northwest. P. 217–230 in J. R. Mason, Ed., *Repellents in Wildlife Management: Proceedings of a Symposium.* National Wildlife Research Center, Fort Collins, CO, 1997.

Mason, J. R. and L. Clark, Use of activated charcoal and other particulate substances as food additives to suppress bird feeding. *Crop Protect.,* 13, 219–224, 1994.

Mason, J. R. and J. A. Maruniak, Behavioral and physiological effects of capsaicin in red-winged blackbirds. *Pharmacol. Biochem. Behav.*, 19, 857–862, 1983.

Mehiel, R., and R. C. Bolles, Learned flavor preferences based on calories are independent of initial hedonic value. *Anim. Learn. Behav.,* 16, 383–387, 1988.

Melchiors, M. A. and C. A. Leslie, Effectiveness of predator fecal odors as black-tailed deer repellents. *J. Wildl. Manage*., 49, 358–362, 1985.

Mirza, S. N. and F. D. Provenza, Effects of age and conditions of exposure on maternally mediated food selection by lambs. *Appl. Anim. Behav. Sci.,* 33, 35–42, 1992.

Morell, G. M. and J. R. G. Turner, Experiments on mimicry: I. The response of wild birds to artificial prey. *Behaviour,* 36, 116–130, 1970.

Nicolaus, L. K., Conditioned aversions in a guild of egg predators: implications for aposematism and prey defense mimicry. *Am. Midland Nat*., 117, 405–419, 1987.

Nicolaus, L. K, J. F. Cassell, R. B. Carlson, and C. R. Gustavson, Taste-aversion conditioning of crows to control predation on eggs. *Science*, 220, 212–214, 1983.

Nicolaus, L. K. and D. W. Nellis, The first evaluation of the use of conditioned taste aversion to control predation by mongooses upon eggs. *Appl. Anim. Behav. Sci*., 17, 329–346, 1987.

Nolte, D. L. and J. R. Mason, Maternal ingestion of ortho-aminoacetophenone during gestation affects intake by offspring. *Physiol. Behav*., 58, 925–928, 1995.

Nolte, D. L., J. R. Mason, G. Epple, E. Aronov, and D. L. Campbell, Why are predator urines aversive to prey? *J. Chem. Ecol.,* 20, 1505–1516, 1994.

Nolte, D. L. and F. D. Provenza, Food preferences in lambs after exposure to flavors in milk. *Appl. Anim. Behav. Sci*., 32, 381–389, 1991.

Polson, J. E., Application of Aversion Techniques for the Reduction of Losses to Beehives by Black Bears in Northeast Saskatchewan. *Saskatchewan Research Council C-* 805–13-E-83, 1983.

Provenza, F. D., Postingestive feedback as an elemental determinant of food preference and intake in ruminants. *J. Range Manage*., 48, 2–17, 1995.

Provenza, F. D., Origins of food preference in herbivores. P. 81–90 in J. R. Mason, Ed., *Repellents in Wildlife Management: Proceedings of a Symposium*. National Wildlife Research Center, Fort Collins, CO, 1997.

Provenza, F. D., J. J. Lynch, and J. V. Nolan, The relative importance of mother and toxicosis in the selection of foods by lambs. *J. Chem. Ecol*., 19, 313–323, 1993.

Ratnaswamy, M. J., R. J. Warren, M. T. Kramer, and M. D. Adam, Comparisons of lethal and nonlethal techniques to reduce raccoon depredation of sea turtle nests. *Behaviour,* 61, 368–376, 1997.

Reichardt, P. B., J. P. Bryant, T. P. Clausen, and G. D. Wieland, Defense of winter dormant Alaska paper birch against snowshoe hares. *Oecologia (Berlin)*, 654, 58–69, 1984.

Reidinger, R. F., Jr., Recent studies on flavor aversion learning in wildlife damage management. P. 101–120 in J. R. Mason, Ed., *Repellents in Wildlife Management: Proceedings of a Symposium*. National Wildlife Research Center, Fort Collins, CO, 1997.

Robbins, C. T., *Wildlife Nutrition*. Academic Press, New York, 1983.

Schmidt, S. P. and T. G. Osborn, Effects of endophyte-infected tall fescue on animal performance. *Agric. Ecosystems Environ*., 44, 233–262, 1993.

Semel, B. and L. K. Nicolaus, Estrogen-based aversion to eggs among free-ranging raccoons. *Ecol. Appl*., 2, 439–449, 1992.

Sheaffer, S. E. and R. D. Drobney, Effectiveness of lithium chloride induced taste aversions in reducing waterfowl nest predation. In *Trans. Missouri Academy Sci*., 20, 59–63, 1986.

Sillen-Tullberg, B., C. Wiklund, and T. Jaervi, Aposematic coloration in adults and larvae of *Lygaeus equestris* and its bearing on Muellerian mimicry: an experimental study on predation on living bugs by the great tit *Parus major. Oikos*, 39, 131–136, 1982.

Stickley, A. R., Jr. and J. L. Guarino, A repellent for protecting corn seed from blackbirds and crows. *J. Wildl. Manage.*, 36, 150–152, 1972.

Stone, C. P., W. F. Shake, and D. J. Langowski, Reducing bird damage to highbush cranberries with a carbamate repellent. *Wildl. Soc. Bull.*, 2, 135–139, 1974.

Sullivan, T. P., D. R. Crump, and D. S. Sullivan, Use of predator odors as repellents to reduce feeding damage by herbivores, II. Black-tailed deer (*Odocoileus hemonius columbianus*). *J. Chem. Ecol.*, 11, 921–935, 1985.

Sullivan, T. P., D. S. Sullivan, D. R. Crump, H. Wiser, and E. A. Dixon, Predator odors and their potential role in managing pest rodents and rabbits. In *Proc. Vert. Pest Conf.*, 13, 145–150, 1988.

Svoboda, F. J. and G. W. Gullion, Preferential use of aspen by ruffed grouse in northern Minnesota. *J. Wildl. Manage.*, 36, 1166–1180, 1972.

Swihart, R. K. and M. R. Conover, Responses of woodchucks to potential garden crop repellents. *J. Wildl. Manage.*, 55, 177–181, 1991.

Swihart, R. K., J. J. Pignatello, and M. J. I. Mattina, Aversive responses of white-tailed deer, *Odocoileus virginianus*, to predator urine. *J. Chem. Ecol.*, 17, 767–777, 1991.

CHAPTER 12

Diversion

"If you cannot win using force, try kindness."

Anonymous

"One for the mouse, one for the crow, one to rot, and two to grow."

Doggerel from the American colonies reminding farmers why they need to plant extra corn seeds

One approach to reduce wildlife damage to agriculture is to provide an alternate source of food for the problem animal in the hope that it consumes the alternate food in lieu of the affected crop. This is one of the most philosophically appealing approaches to resolving a human–wildlife conflict because the animal voluntarily changes its behavior when offered a more attractive alternative.

Usually, diversionary feeding programs involve planting a crop in a separate field and encouraging wildlife to use it, or placing food at feeder stations on an "as needed" basis. Many different names have been used for this approach. Unharvested fields used to feed wildlife have been called lure fields, sacrificial fields, diversionary fields, supplemental fields, alternative fields, or decoy fields. In this chapter, I refer to them as diversionary fields and refer to the general approach of providing alternate food to wildlife as the use of diversion because this label seems clearest.

In the context of wildlife damage management, diversion is defined as a method to solve wildlife damage by providing an alternative which diverts the culprit from its damaging behavior or lures it away from the problem site. Private citizens and government agencies often distribute supplemental food for wildlife, especially in the winter, but this differs from diversion owing to human motivation. The purpose of supplemental feeding is to increase the health and survival of wildlife species, while the intention of diversion is to reduce wildlife damage. However, the response of wildlife to diversion and supplemental feeding is the same.

Humans have been familiar with the concept of diversion for centuries, although not in the context of wildlife damage. In human history, diversion was used to deal with other people. For example, a vanquished tribe could pay a tribute (often money, gold, or other valuable goods) to their stronger opponents to appease them and to avoid being attacked.

One of the earliest examples of diversion to reduce wildlife damage comes from the 1600s. It is the doggerel quoted at the beginning of this chapter, which advised American colonial farmers to plant extra corn seeds so there would be plenty of seeds to grow, even after birds and mammals ate their fill (Conover and Conover 1987). As early as 1943, the U.S. government used diversionary fields to divert ducks from commercial rice fields in California (Horn 1949). Before investigating the uses of diversion in wildlife damage management, we need to consider the optimal foraging theory and how animals use diversion as an antipredator behavior.

OPTIMAL FORAGING THEORY

Optimal foraging theory predicts that those animals which make optimal foraging decisions have a higher probability of survival and reproduction. Hence, the genes of wise decision makers will predominate in future generations. The diet a forager selects should depend upon those factors most important to its survival and reproductive success (i.e., to its fitness). In many species, individuals enhance their fitness by increasing their energy intake; these animals seek a high caloric diet, and we refer to them as "energy maximizers."

The foraging decisions of other species, especially those vulnerable to predation while foraging, are shaped by the need to forage as quickly as possible and return to a safe place. These animals are "time minimizers." Many birds are time minimizers; they can quickly stuff themselves with food and then retire to a safe place to digest the food. They are able to do this because many birds have an enlarged esophagus (a sack-like organ called the crop) that allows them to store food for later digestion.

Herbivores face their own foraging problems. For them, potential food (i.e., vegetation) is abundant although most of it is of low nutritional quality and is difficult to digest because it contains cellulose and lignin. Hence, herbivores may search for plants which are more nutritious and easier to digest. Acquiring sufficient protein is another difficult problem for herbivores, especially for pregnant or lactating females. Thus, the foraging decisions of herbivores during certain times of the year are based more on a need for protein than a need for energy.

Large herbivores have adapted to the problem of low nutritional forage by developing efficient digestive systems capable of handling large volumes of food. Such digestive systems allow the animals to survive on poor quality food. The optimal foraging strategy for these animals is one of quantity vs. quality to ensure their digestive systems are full by the end of the foraging bout. Other herbivores select a varied diet and will eat small amounts of many different plants, including toxic ones. This practice allows them to acquire needed nutrients without consuming enough of a poisonous plant to become ill.

Some herbivores, such as Canada geese, are selective foragers; their digestive systems are too small to handle large quantities of food. Their foraging strategy is to seek out those plants or parts of plants (such as shoots and new leaves) that are both nutritious and easily digested, because their food consumption is limited by how fast they can digest what they have already eaten (Conover 1991). Diversion programs must take into account the foraging strategy of the species involved.

LARGE GROUP FORMATION AS AN ANTIPREDATOR BEHAVIOR — A NATURAL FORM OF DIVERSION

Many animals avoid predation by congregating in large groups. This defensive strategy reduces the probability of any one individual falling victim to a predator. For example, thousands of gulls may nest in a single colony, often located on an island, where access by mammalian predators is limited. For those few predators able to reach the nesting colony, the large number of gulls and eggs on the island provides a line of defense: satiation of the predator by providing it with an abundance of food (Figure 12.1). This is the same strategy (satiate a predator with food) which is manifested in the use of diversion, except that in the case of nesting birds, their neighbors' eggs provide the diversionary food.

One interesting question is why all prey do not use the defensive strategy of congregating in large groups to avoid predation. The answer is that there are three major drawbacks with this approach: 1. being in a large group can interfere with foraging due to competition for food with other group members, 2. large groups are detrimental to those animals that rely on being cryptic as their primary anti-predator strategy, and 3. predators may respond to the aggregation of prey by congregating in that area and, if they do so, the prey's strategy of concentrating

Figure 12.1 Gulls nest in dense colonies where there are so many gull nests that, if a predator reaches the colony, the chances of its destroying the nest of a particular gull are reduced.

together loses its effectiveness (e.g., the main threat to schools of small fish are schools of predatory fish). This suggests that a major limitation with the use of diversion in resolving human–wildlife conflicts is that the problem animals may congregate in large numbers in response to the food and overwhelm the diversionary food supply. For this reason, diversion will be more successful for those species that are territorial or solitary and less unsuccessful for those species that congregate into large groups. For example, diversion may be an option to protect a blueberry field from territorial mockingbirds, but not from juvenile starlings, which flock in the hundreds or thousands.

EXAMPLES OF DIVERSION TO RESOLVE HUMAN–WILDLIFE CONFLICTS

Diversion has been used mostly by government agencies in response to complaints of crops eaten by wildlife populations that have increased as a result of government action. Many U.S. National Wildlife Refuges, which consist mainly of wetlands, were established in the 20th century to provide secure loafing and roosting areas for waterfowl. These refuges have been successful in increasing local waterfowl populations. However, with these increased populations come additional problems because many ducks and geese feed outside the refuges. Local farmers become angry when their crops are destroyed by such waterfowl, and the conflict between farmers and refuge managers can become quite heated. In the Prairie Potholes region where many of the ducks in North America nest, an antiduck sentiment developed in the 1950s, and farmers threatened to take matters into their own hands. The Souris Duck Control Association advocated the reduction of waterfowl populations, duck sterilization, the drainage of wetlands, and compensation for crop damage (Fairaizl and Pfeifer 1987). Because the mandate of the U.S. Fish and Wildlife Service was to increase wildlife populations, these complaints created an awkward situation for the agency. The U.S. Fish and Wildlife Service responded to farmers' complaints by trying to scare ducks from the farmers' unharvested grain fields. The effort was not very successful because alternative food sources were unavailable. Instead, ducks scared from one field just flew to another one (Fairaizl and Pfeifer 1987). Realizing the problem, the government set up a series of bait stations in the refuges and planted grain fields to serve as a diversion for ducks scared from agricultural fields. The initial results of this program were encouraging, and the program ultimately was expanded to all of North Dakota. It consisted of planting 12-ha fields of barley or wheat, which cost an average of $3587 per diversionary field in 1975 and produced an overall benefit to cost ratio of 2 to 1 (Fairaizl and Pfeifer 1987; Knittle and Porter 1988).

In addition to diverting ducks, some refuges in North Dakota planted sunflower fields on idle ground to reduce blackbird depredation in commercial sunflower fields. Blackbird use of these diversionary fields was heavy, and the diversion was successful in reducing damage in nearby sunflower fields (Cummings et al. 1987). In California, waterfowl damage in commercial rice fields near refuges was abated by the procurement of a rice field that was left to the birds as a diversion.

The other major use of diversion has been to reduce mammalian damage to the timber industry. One such problem occurs when forest clear-cuts are restocked with trees. Initially, restocking was accomplished by spreading seeds of desirable tree species over the tract, but this approach often failed because too many seeds were consumed by birds and small mammals. However, Sullivan and Sullivan (1982a) were able to increase lodgepole pine seed survival fourfold by spreading sunflower seeds with the pine seeds at a ratio of two sunflower seeds to each pine seed.

Another successful use of diversion involved trying to stop black bears from debarking 15- to 30-year-old Douglas fir trees in the spring to feed on the underlying sapwood. Trees often were killed outright or died as a result of fungus or disease entering through the wound. This problem was alleviated by providing the bears with alternative food in the spring. By 1993, the Washington Forest Protection Association was saving 300,000 ha of trees from 1200 bears by placing out 150,000 kgs of feed annually. Similar problems involving red squirrels girdling lodgepole pines in the spring (Sullivan and Klenner 1993) and voles girdling apple trees during the winter (Sullivan and Sullivan 1988) were avoided by providing the mammals with diversionary food.

Diversion has been used to reduce crop damage by ungulates (Johnson et al. 1985). In one imaginative program, the Utah Division of Wildlife Resources purchased Hardware Ranch. This site had historically been a source of friction between the division and local farmers who experienced severe elk depredation of their hay. The division began to manage Hardware Ranch for use by elk, and this ranch became a winter home for local elk. This reduced elk damage in the rest of the county and created a popular tourist attraction where people now come to view the large elk herd.

Diversion also has been tried, with mixed results, to solve problems other than crop damage. Crabtree and Wolfe (1988) reduced predation on duck nests in Utah by distributing dead fish during the duck nesting season. In contrast, Greenwood et al. (1998) were unable to reduce mammalian predation on duck nests in North Dakota by distributing more than 13,000 kg of diversionary food (chopped fish offal and sunflower seeds) annually at 12 food plots. Wood and Wolfe (1988) attempted to reduce deer–automobile collisions during winter by diverting deer away from roads by providing alfalfa hay, deer pellets, and apple mash at feeding stations located 0.4 to 1.2 km off the roads. In general, stretches of road with feeding stations had fewer deer–automobile collisions than those without them (Figure 12.2). Their findings, however, were ambiguous because of small sample sizes and inconsistent results.

Occasionally, diversion is used to lure nuisance animals out of buildings. For instance, homeowners erect bat houses in their backyards in the hope that bats will prefer roosting in them to roosting in attics. Some people build nest boxes hoping that raccoons will occupy them and not their chimneys. Diversion has been used to alleviate the problem of starlings displacing wood ducks from nest boxes erected for the latter (Sidebar 12.1).

Figure 12.2 The number of ungulate–vehicle collisions can be reduced if the resources that attract them to the road (e.g., forage or salt) are provided elsewhere.

Sidebar 12.1 A Novel Use of Diversion

Populations of cavity-nesting birds are often limited by a shortage of adequate nesting sites. Wildlife managers often install nest boxes to help specific species, but this approach can be ineffective if boxes are taken over by more aggressive species, such as starlings or squirrels. At Pungo National Wildlife Refuge in North Carolina, more wood duck boxes were occupied by starlings than by ducks. To resolve this problem, Grabill (1977) exploited the territorial behavior of starlings by attaching a smaller nest box to an existing wood duck nest box. Both nest boxes were placed on a single pole (Figure 12.3). Starlings, like most animals, are aggressive toward members

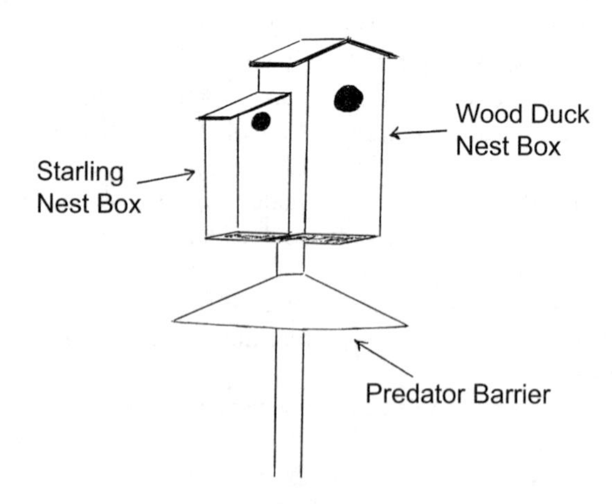

Figure 12.3 A wood duck nest box that has been altered by attaching to it a smaller nest box intended to divert starlings from occupying the duck nest box.

of their own species and exclude them from their territories, but are tolerant toward other species. In this case, starlings occupied over 95% of the smaller boxes but not the attached larger boxes, which were then used for nesting by wood ducks.

CREATING A FOOD DIVERSION THROUGH HABITAT MODIFICATION

An indirect approach to diversion is to increase the quantity or quality of native foods. One indication that such an approach may have merit comes from observations that wildlife damage often is more severe when natural food supplies are low. For instance, coyote predation of mule deer fawns (Hamlin et al. 1984) and sheep (Stoddart et al. 2001) was highest when alternate prey populations (rodents and jackrabbits) were low.

Damage by herbivores can be reduced by increasing the amount of native browse. Deer damage in plant nurseries usually does not begin until the mast crop in the surrounding woodlots has been consumed (Conover 1989). Hence, measures to increase the mast crop, such as fertilizing oak trees, might reduce wildlife damage. Campbell and Evans (1978) recommended seeding forest clear-cuts with palatable forbs so deer would eat the forbs and not the tree seedlings. In Wyoming, elk were diverted from foraging on private lands to public lands by increasing the quality of the native browse on public land through herbicide treatment, burning, and fertilization (Long 1989). In Alaska, bison started feeding in barley fields after the suppression of fires reduced vegetation quality on their normal winter range. A burning program was restated to lure the bison back to their winter range (Van Vuren 1998).

One problem with the use of native browse as a diversion from agricultural crops or nurseries is that plants receiving adequate light, water, and fertilizer have higher growth rates and are more nutritious than those that do not. Consequently, these plants are more likely to be browsed by wildlife (Giusti 1988; Nelson 1989). They are also more likely to occur in agricultural fields, plant nurseries, and managed tree plantations than in unmanaged fields and forests. Hence, these plants are usually the ones needing protection from wildlife, and it may be difficult to increase the palatability of native browse to a higher level than cultivated plants.

Some bird roosts are located in areas where they pose a health hazard or nuisance problem for local residents. For instance, a school playground is not an appropriate place for a large crow roost because the accumulation of fecal material poses a health hazard. Diversion could be used to alleviate this problem by establishing attractive roosting habitat away from human habitations, so the birds will move their roost site (Sidebar 12.2).

Sidebar 12.2 Using Diversion to Convince Blackbirds to Abandon Objectionable Urban Roosts

During the winter, thousands of blackbirds can roost in the same place night after night. When roost sites are located near houses, schools, or occupied buildings, the roosting birds cause nuisance problems for local residents due to their noise, accumulation of fecal material, and odor. Roosts also pose a health hazard due to the risk of histoplasmosis (Chapter 4). A common solution to the presence of objectionable roosts is to force the birds to abandon them through the use of fear-provoking stimuli or harassment (Chapter 10). However, little has been accomplished if the blackbirds simply move to another site where they cause the same problems. A method is needed to convince blackbirds to roost at a remote site where they will not bother anyone. Blackbirds prefer to roost in thick vegetation offering insulation from cold temperatures and protection from predators. Glahn et al. (1994) observed that bamboo stands provide this type of roosting habitat and that blackbirds often roost in such stands. By selecting sites where blackbird roosts would not cause problems and then planting them with bamboo, blackbirds could be lured from objectionable roost sites (Flynt and Glahn 1995).

RESPONSE OF WILDLIFE TO SUPPLEMENTAL FOOD

Animals respond to supplemental food in a variety of ways, all of which have the effect of increasing their density in the area where the food is provided. The most immediate response to supplemental food is a change in animal behavior. With their great mobility, birds often respond to the availability of supplemental food by congregating in large numbers, especially outside of the breeding season. White-tailed deer shift their home ranges toward sites with abundant food, and for this reason, deer numbers can increase within a few weeks to changes in localized food resources (Figure 12.4; VerCauteren and Hygnstrom 1998). In general, this congregation at supplemental food sites will be greatest for those species that have over-

Figure 12.4 White-tailed deer will shift their home ranges towards sites with abundant food, and for this reason, deer numbers can increase rapidly in response to diversionary food. (Photo courtesy of Lou Cornicelli.)

lapping home ranges or which travel to foraging sites from a central location, such as a roost site or nesting colony.

If food is provided for several weeks or months, population changes may begin to occur. During a five-year period when supplemental food was provided to white-tailed deer within a 250-ha enclosure, deer density increased sevenfold to 63 deer per km^2 (Ozoga and Verme 1982). During this period, the deer's dependence on supplemental food increased as native forage became overbrowsed. Compared to prefeeding years, the supplemental food accelerated body growth and physical maturation of deer and caused an increase in reproduction (Ozoga and Verme 1982).

Sullivan and Klenner (1993) showed that diversion could be used to stop red squirrels from debarking lodgepole pines. However, within two years, it also caused squirrel populations to increase three- to fivefold. Vessey (1987) summarized 15 supplemental feeding studies involving rodents and found that most studies reported an increase in the local population owing to higher reproduction, immigration, and/or survival rates. Boutin (1990) reviewed 138 studies in which terrestrial vertebrates received supplemental food and found that animals that received supplemental food usually had smaller home ranges, higher body weights, and more offspring that those not receiving supplemental food. These factors generally resulted in a two- to threefold increase in density, although the populations were still subject to periodic crashes. Most studies on birds have reported that an increase in food availability led to an earlier starting date for egg laying and an increased survival of young but rarely changed egg mass or clutch size (Hochachka and Boag 1987).

There can be an interaction between food availability and predation rates for many species. Ward and Kennedy (1996) found that providing supplemental food to goshawks during the breeding season increased survival rates of their young by reducing predation rates on the chicks. Apparently, when the parents did not have to spend as much time searching for food, they were able to devote more time to nest defense.

Despite its philosophical appeal and apparent success, diversion is rarely used in wildlife damage management. To understand why, we must examine the factors influencing its effectiveness and utility. In doing so, we can identify the conditions required for diversion to be a viable option.

RESPONSE OF WILDLIFE TO DIVERSION

As noted previously, animals tend to congregate near food supplies by making behavioral changes in home ranges and time budgets. These behavioral changes are welcomed because they must occur if the diversion is going to resolve a human–wildlife conflict elsewhere. Nonetheless, in the long run, increasing populations are a major drawback to the use of diversion to reduce wildlife damage. Because of this, diversion may initially reduce wildlife damage, but its effectiveness probably will wane over the years. Consequently, diversion is least suited for problems that occur during the season when natural food supplies are scarce. Providing food at that time will allow more animals to survive the year and guarantee that more and more animals will have to be fed each year (Figure 12.5).

Figure 12.5 One problem with providing diversionary food is that animal numbers will increase at the diversionary food site (even if it is in a backyard), and this may cause unanticipated problems.

The use of diversion is best suited for short-term problems or those that do not occur on a yearly or regular basis. For instance, diversion may be an excellent way to prevent birds from eating tree seeds that have been distributed to restock a forest clear-cut. In this situation, the seeds only need to be protected for a short period of time before germinating. Moreover, many years will pass before the new trees are harvested and the site has to be reseeded, rendering temporary increases in wildlife populations of little consequence.

Another problem with diversion is that animals may become dependent on the food. Once a diversion program ends, the animals may take a long time to disperse or may return in subsequent years seeking food. Therefore, a considerable period may be required before wildlife numbers return to their original levels. For instance, Douglas squirrel populations that received supplemental food took two to three months to return to normal levels once the feeding ceased (Sullivan and Sullivan 1982b). In another study, red squirrel populations were still twice as high in plots where food had been provided than in untreated plots one year after a food diversion program ended (Sullivan 1990). However, after the diversionary food is removed and until wildlife populations return to normal levels, wildlife damage will be worse than before the diversion because more animals will be present and they will be hungrier.

COST EFFECTIVENESS

Major problems limiting the use of diversion are its expense and labor intensiveness. This cost can be justified only when high-valued crops or commodities are in need of protection. For instance, a 30-year-old Douglas fir tree has considerable value to a timber company, as do mature apple trees to an orchardist. The use of diversion to keep bears or voles from killing these trees would be cost effective. The key variable is that the item needing protection should be several times more valuable than the diversionary item being fed to wildlife.

Under most circumstances, planting a sunflower field to lure birds from other sunflower fields is misguided because we have not saved anything in this process, especially given the risk that the intended wildlife might not use the diversion field. In these cases, the more effective response is to let wildlife forage where they will and then compensate the farmers for the crops consumed by the wildlife. Also reducing the cost effectiveness of diversion is the consumption of food by nontarget species (e.g., grain placed out to divert waterfowl from farmers' fields may be consumed by blackbirds). Further, if diversion fields are planted, there is the danger the target species may consume the resulting crops either too early or too late to do any good as a diversion. For instance, if an alfalfa field is planted to divert deer from a plant nursery, there is the danger that all of the alfalfa will be eaten early in the fall, long before the deer start browsing in the nurseries during the winter.

In one diversion program, the federal government offered to purchase the grain in fields where ducks were foraging heavily (Fairaizl and Pfeifer 1987). Under this program, the field's grain yield was appraised by an independent crop adjustor, and the value determined by the price paid at the local grain elevator. Adjacent landowners were encouraged to scare birds from their fields towards the diversionary field. After the adjacent fields were harvested, the farmer was allowed to harvest the diversionary field and was paid the difference between the earlier appraised yield and the actual yield. The farmer also was compensated for any reduction in grain quality (Fairaizl and Pfeifer 1987).

One of the most promising uses of diversion, even for low-valued crops, is to protect the seeds in a newly planted field from being eaten. For instance, a bushel of genetically improved corn seeds may cost $50 while a regular bushel of corn sells for $2 to 4. Corn seeds acquire more value once planted, because each one may develop into an ear of corn containing hundreds of corn seeds. Sandhill cranes and American crows feed in newly planted corn fields, where they walk down the rows pulling up the seeds and eating them, leaving large gaps where no corn plants will grow. In this case, diversion is a viable option because a farmer can spread a few bushels of inexpensive corn grain on the ground in newly planted fields. Birds will consume the diversionary corn over the planted seeds because they can find grain lying on the ground much faster than they can locate and dig up buried seeds. To be successful, the diversion has to work only for the few weeks when the seeds are vulnerable to bird depredation; after the corn plants are a few centimeters high, birds will not bother them.

WHAT TYPE OF FOOD OR CROP SHOULD BE USED IN DIVERSION?

There really are only two important considerations regarding this question. First, as noted earlier, the food used at a feeder station or the crops grown for diversion should be less expensive than the resource needing protection. Second, we know from the optimal foraging theory that some animals attempt to maximize their intake of nutrients and energy, whereas others try to minimize the time or energy spent foraging. This implies that the diversionary food should have a higher nutritional value than the crop needing protection, and it needs to be easier to find. Crops with high nutritional value are hard to protect using diversion. Diverting deer away from an alfalfa field, except perhaps to another alfalfa field that is closer to them, is difficult because alfalfa is so nutritious. In contrast, young spruce, pine, and fir trees are not very palatable, so diverting deer from Christmas tree farms is relatively easy. Likewise, voles usually eat bark and girdle apple trees only when starving, making it possible to divert them from girdling trees by providing another food source. For example, Sullivan and Sullivan (1988) showed that voles preferred to eat pieces of wood saturated with soybean oil rather than apple bark and that vole damage to apple trees could be reduced by distributing these in orchards.

WHERE SHOULD A FEEDER STATION OR DIVERSION CROP BE LOCATED?

As we discussed earlier, diversionary food will attract wildlife, and it may be unwise to attract them to the area where the resource needing protection is located. The problem is that most animals seek a varied diet, and while they may fill most of their dietary needs from the diversionary food, they will still eat other things around them. For this reason, the vegetation around feeding stations usually is heavily browsed, sometimes to the point that there is no vegetation left (Conover 1991). Deer primarily forage in apple orchards to eat fallen apples and the ground cover but, while in the orchard, they also may reach up and nip the buds off apple trees. The total biomass of apple buds they consume in an orchard is low, but the damage to the trees is substantial; loss of these buds can alter the shape and growth rate of the apple tree. If feeder stations were placed in an apple orchard to deter deer from eating apple buds, the feeder stations would have the opposite effect as more deer would be lured into the orchard. Sullivan and Sullivan (1988) recognized this problem and recommended not placing diversionary food for voles by the trunks of apple trees because this might attract voles and increase the likelihood that trees would be girdled; instead, they thought diversionary food should be placed away from apple trees.

Another example of this dilemma involves the task of protecting the eggs of nesting ducks from mammalian predators. If diversionary food is distributed in the field where ducks are nesting, the effort may actually cause an increase in nest predation: the food may attract predators to the field and, while looking for the diversionary food, the predators may find duck nests. One way to visualize this is to consider the effect of a feeder containing an unlimited amount of food for skunks. This feeder causes skunks in its vicinity to shift their home ranges and foraging activities towards it. As a consequence, the skunks' activity in the distant parts of their home ranges is reduced. In doing so, the feeder creates a "halo zone," where skunk densities are reduced because that zone's former skunk residents have vacated

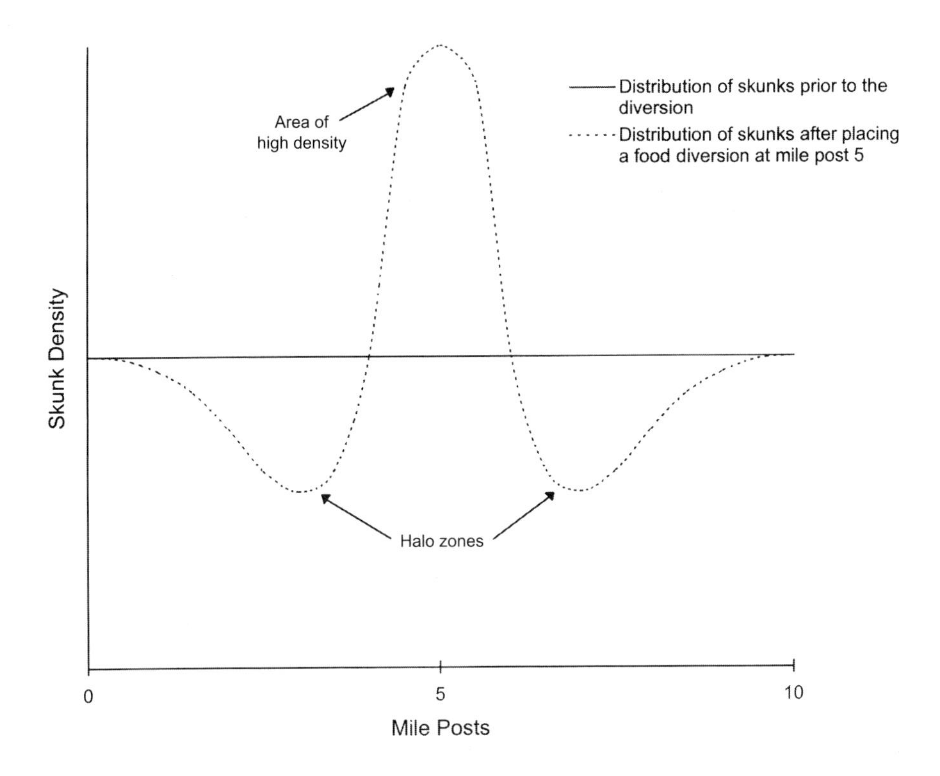

Figure 12.6 If a diversion causes skunks to move toward it and concentrate their foraging there, three zones of skunk density will be created: a zone of high skunk density close to the bait station, a halo zone further out where the skunk density is reduced, and a zone so far away that the diversion has no effect on skunk behavior.

it and moved closer to the feeder (Figure 12.6). Diversionary feeders or crops should be located at a distance such that the crop needing protection is within the halo zone created by the diversion.

If diversionary crops and feeder stations should be placed away from the crop that needs protection, then the question is how far. The problem is that if the diversion is placed too far away then some of the culprit animals will not find or use it because it is outside their home ranges. Additionally, as the distance between the diversionary food and the crop increases, a higher proportion of the animals feeding on the diversionary food would not normally feed on the protected crop because the latter is outside their home ranges. In that case, the consumed food provides no benefit. Obviously, diversionary feeders or fields should be located as close as possible to the crop needing protection, but at a distance where the feeders will not concentrate animals near the crop. Identifying this ideal distance is difficult, and it will vary depending upon the size of the species' home range and its mobility. A good rule of thumb is that a diversionary food should be separated from the crop by a distance equal to one half of the diameter of that species' home range.

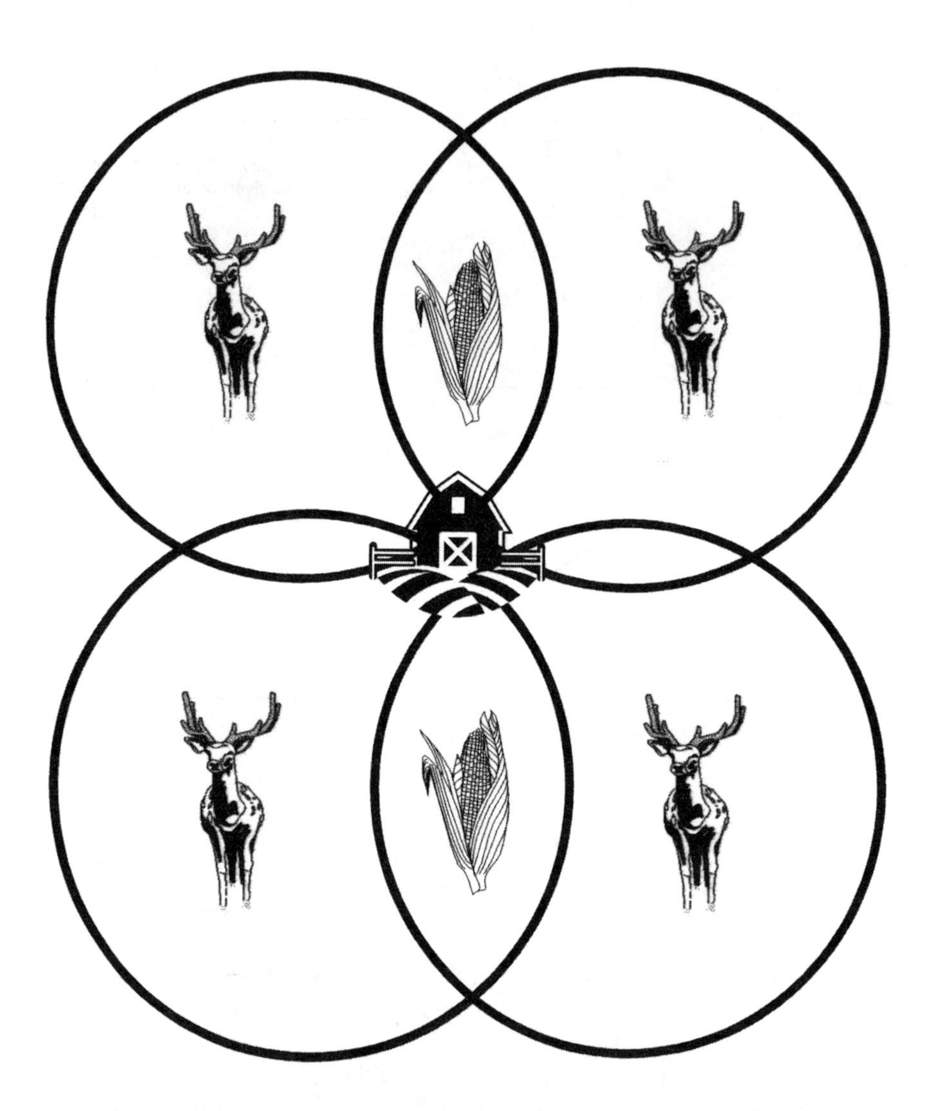

Figure 12.7 A hypothetical landscape showing a farm located within the overlapping home ranges of four deer. If bait stations need to be placed away from the farm so as not to attract more deer, a minimum of two bait stations (represented by the corn cobs) is required so all four deer have one within their home range.

If diversionary feeders or crops are going to be placed away from the crop needing protection, then more than one feeder will be required. Consider Figure 12.7, which depicts a landscape showing the circular and overlapping home ranges of four deer with a field being damaged in the center. If feeder stations are going to be placed away from the field, two would be required so at least one falls within the home range of each animal.

The problem of where to place the diversion is easy to solve for those animals which are central-place foragers and radiate out from a single location (often a roost

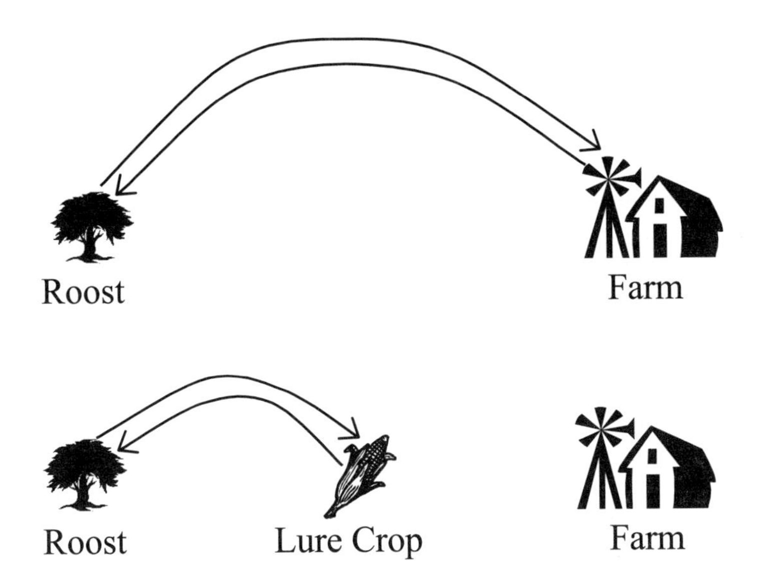

Figure 12.8 The ideal placement of a diversion to protect a crop from a species that is a central-place forager is between the roost or colony and the crop.

site, den, or colony) to forage. We know from research on the optimal foraging theory that animals attempt to minimize the time and energy required to forage. For central-place foragers, the diversion should be placed between the resource needing protection and the central place so the animals can reach the diversion sooner than they can reach the resource (Figure 12.8).

CAN WILDLIFE DAMAGE AND THE EFFECTIVENESS OF DIVERSION BE PREDICTED?

To use diversion successfully, damage to agricultural fields must be predictable because diversionary fields have to be planted and diversionary stations organized before damage begins. Unfortunately, wildlife damage to crops often varies substantially across years and fields, and often it cannot be predicted (Conover 1989; Conover and Kania 1995). Because of this, diversion will sometimes be used in areas and during years when there is no need to do so (Fairaizl and Pfeifer 1987).

The opposite problem also occurs. Wildlife may ignore the diversionary fields or bait stations and continue to cause damage. This problem is rather common. In one attempt to use diversion to protect grain fields from ducks, some diversion fields failed to attract ducks. As Fairaizl and Pfeifer (1987) stated, "Ducks appear to have their own criteria for feeding site selection and attempts to preselect alternative sites and attract ducks into those fields failed." In my own case, I once tried to divert white-tailed deer from nursery plants by placing out feeders containing hay and corn as soon as snow was on the ground. However, the experiment failed because no deer ever visited my feeders.

WHICH IS BETTER, DIVERSIONARY CROPS OR FEEDER STATIONS?

Several disadvantages are associated with diversionary crops. Growing them is labor intensive and requires specialized farming equipment. These problems can be overcome by having a local farmer grow the crop in exchange for the right to harvest a proportion of it (usually 50 to 66%). Moreover, there may be a great waste of food; for instance, food may be trampled into the mud by wildlife and lost. There also is the risk that the crop may not be used at all by the intended wildlife. Diversionary crops cannot be created quickly in response to unanticipated wildlife problems, and they cannot be shifted to a new location when needed. For all of these reasons, growing diversionary crops is a risky and inefficient way to feed wildlife (McBryde 1995).

Feeder stations, where food is placed out periodically, are much more efficient (a higher proportion of the food gets to the intended wildlife) and provide more flexibility because they can be easily moved, and the amount of food distributed can be changed daily. Under most circumstances, feeders are more cost effective than diversionary crops (McBryde 1995). Several practical problems, however, limit their effectiveness. One problem with feeder stations is that food must be distributed regularly. This turned out to be a serious problem in North Dakota where vehicle access to feeder stations was possible only in good weather (Fairaizl and Pfeifer 1987). Another problem with feeder stations is that, unless many are simultaneously employed or the diversionary food is distributed over a large area, animals will compete for the food and only the dominant animals will be able to feed. This was the apparent downfall of one attempt to provide diversionary food to deer (Schmitz 1990).

The consequences of terminating food at a feeder station will be more severe than at a diversionary field. In the latter, animals can observe the depletion of the food and decide when it is time to move and try to find an alternative foraging site. In contrast, when food is periodically distributed at a feeder station, animals cannot predict the arrival of food and thus cannot predict when there will be no more food. Hence, animals take longer to abandon a feeder station once it is closed. If this occurs during the harsh conditions of winter, animals may wait too long and starve before finding alternate foraging sites. For this reason, the amount of food distributed at a feeder should be slowly decreased over the course of a week or two.

CAN DIVERSION BE USED WITH OTHER TECHNIQUES?

Nonlethal techniques to reduce wildlife damage to crops work best when other food sources are available for the wildlife. Diversion can provide this alternate source of food, making other techniques to reduce wildlife damage more effective. Hazing or the use of fear-provoking stimuli in fields needing protection can also be used to move the wildlife culprits to the diversionary field.

Diversion can also be combined with lethal measures. Doing so can potentially overcome the major shortcoming of diversion: the increase in wildlife populations in response to the diversionary food. Philosophically, however, diversion and lethal control are not compatible. Diversion is used when animals causing the problem have high value. Usually there is no desire to kill them. More common than the simultaneous use of diversion and lethal measures is the use of the latter in those circumstances where diversion has failed owing to an increase in wildlife numbers. Stopping a diversion program can cause substantially higher levels of damage than when the program was initiated. Using lethal means to reduce quickly the number of problem animals may help alleviate this problem.

SUMMARY

Diversion is an appealing way to alleviate wildlife damage, but it is not commonly used, owing to its high cost and labor requirements. It works best when high-value crops need protection and when those crops are not very palatable. A major drawback with diversion is that wildlife numbers will rise over time, due to an increase in immigration, reproduction, and survival. This problem of increasing wildlife populations will become particularly acute if diversionary food is made available during the winter or whenever natural foods are scarce. Animals may become dependent upon the food. For these reasons, the amount of food that has to be distributed will increase over time. Stopping a diversion program may result in a higher level of damage than what occurred originally and cause high levels of wildlife mortality. For all of these reasons, diversion is most appropriate for problems of a short-term nature.

LITERATURE CITED

Boutin, S., Food supplementation experiments with terrestrial vertebrates: patterns, problems and the future. *Can. J. Zoo.*, 68, 203–220, 1990.

Campbell, D. L. and J. Evans, Establishing native forbs to reduce black-tailed deer browsing damage to Douglas fir. In *Proc. Vert. Pest Conf.*, 8, 145–151, 1978.

Conover, D. O. and M. R. Conover, Wildlife management in colonial Connecticut and New Haven during their first century: 1636–1736. *Trans. Northeast Section Wildl. Soc.*, 44, 1–7, 1987.

Conover, M. R., Relationships between characteristics of nurseries and deer browsing. *Wildl. Soc. Bull.*, 17, 414–418, 1989.

Conover, M. R., Herbivory by Canada geese: diet selection and effect on lawns. *Ecol. Appl.*, 1, 231–236, 1991.

Conover, M. R. and G. S. Kania, Annual variation in white-tailed deer damage in commercial nurseries. *Agric. Ecosystems Environ.*, 55, 213–217, 1995.

Crabtree, R. L. and M. L. Wolfe, Effects of alternate prey on skunk predation of waterfowl nests. *Wildl. Soc. Bull.*, 16, 163–169, 1988.

Cummings, J. L., J. L. Guarino, C. E. Knittle, and W. C. Royall, Jr., Decoy plantings for reducing blackbird damage to nearby commercial sunflower fields. *Crop Protect.*, 6, 56–60, 1987.

Fairaizl, S. D. and W. K. Pfeifer, The lure crop alternative. In *Great Plains Proc. Wildl. Damage Control Workshop*, 8, 163–168, 1987.

Flynt, R. D. and J. F. Glahn, Propagation of bamboo as blackbird lure roost habitat. In *Proc. East. Wildl. Damage Control Conf.*, 6, 113–119, 1995.

Giusti, G. A., Recognizing black bear damage to second growth redwoods. In *Proc. Vert. Pest Conf.*, 13, 188–189, 1988.

Glahn, J. F., R. D. Flynt, and E. P. Hill, Historical use of bamboo/cane as blackbird and starling roosting habitat: implications for roost management. *J. Field Ornithol.*, 65, 237–246, 1994.

Grabill, B. A., Reduced starling use of wood duck boxes. *Wildl. Soc. Bull.*, 5, 69–70, 1977.

Greenwood, R. J., D. G. Pietruszewski, and R. D. Crawford, Effects of food supplementation on depredation of duck nests in upland habitat. *Wildl. Soc. Bull.*, 26, 219–226, 1998.

Hamlin, K. L., S. J. Riley, D. Pyrah, A. R. Dood, and R. J. Mackie, Relationships among mule deer fawn mortality, coyotes, and alternate prey species during summer. *J. Wildl. Manage.*, 48, 489–499, 1984.

Hochachka, W. M. and D. A. Boag, Food shortage for breeding black-billed magpies (*Pica pica*): an experiment using supplemental food. *Can. J. Zoo.*, 65, 1270–1274. 1987.

Horn, E. E., Waterfowl damage to agricultural crops and its control. *North Am. Wildl. Conf.*, 14, 577–585, 1949.

Johnson, B. K., J. K. Straley, and G. Roby, Supplemental feeding of moose in western Wyoming for damage prevention. *Alces,* 21, 139–148, 1985.

Knittle, C. E. and R. D. Porter, Waterfowl damage and control methods in ripening grain: an overview. *U.S. Fish and Wildlife Service Technical Report 14*, 6–8, 1988.

Long, W. M., Habitat manipulation to prevent elk damage to private rangelands. In *Proc. Great Plains Wildl. Damage Control Workshop*, 9, 101–103, 1989.

McBryde, G. L., Economics of supplemental feeding and food plots for white-tailed deer. *Wildl. Soc. Bull.*, 23, 497–501, 1995.

Nelson, E. E., Black bears prefer urea-fertilized trees. *Western J. Appl. For.,* 4, 13–15, 1989.

Ozoga, J. J. and L. J. Verme, Physical and reproductive characteristics of a supplementally fed white-tailed deer herd. *J. Wildl. Manage.*, 46, 281–301, 1982.

Schmitz, O. J., Management implications of foraging theory: evaluating deer supplemental feeding. *J. Wildl. Manage.*, 54, 522–532, 1990.

Stoddart, L. C., R. E. Griffiths, and F. F. Knowlton, Coyote responses to changing jackrabbit abundance affect sheep predation. *J. Range Manage.*, 54, 15–20, 2001.

Sullivan, T. P., Response of red squirrel (*Tamiasciurus hudsonicus*) populations to supplemental food. *J. Mammal.,* 71, 579–590, 1990.

Sullivan, T. P. and W. Klenner, Influence of diversionary food on red squirrel populations and damage to crop trees in young lodgepole pine forests. *Ecol. Appl.*, 3, 708–718, 1993.

Sullivan, T. P. and D. S. Sullivan, The use of alternative foods to reduce lodgepole pine seed predation by small mammals. *J. App. Ecol.*, 19, 33–45, 1982a.

Sullivan, T. P. and D. S. Sullivan, Population dynamics and regulation of the Douglas squirrel (*Tamiasciurus douglasii*) with supplemental food. *Oecologia*, 53, 264–270, 1982b.

Sullivan, T. P. and D. S. Sullivan, Influence of alternative foods on vole populations and damage in apple orchards. *Wildl. Soc. Bull.*, 16, 170–175, 1988.

Van Vuren, D., Manipulating habitat quality to manage vertebrate pests. In *Proc. Vert. Pest Conf.*, 18, 383–390, 1998.

VerCauteren, K. C. and S. E. Hygnstrom, Effects of agricultural activities and hunting on home ranges of female white-tailed deer. *J. Wildl. Manage.*, 62, 280–285, 1998.

Vessey, S. H., Long-term population trends in white-footed mice and the impact of supplemental food and shelter. *Am. Zoo.*, 27, 879–890, 1987.

Ward, J. M. and P. L. Kennedy, Effects of supplemental food on size and survival of juvenile northern goshawks. *Auk,* 113, 200–208, 1996.

Wood, P. and M. L. Wolfe, Intercept feeding as a means of reducing deer–vehicle collisions. *Wildl. Soc. Bull.*, 16, 376–380, 1988.

CHAPTER 13

Exclusion

"Love your neighbor, but build sturdy fences."

Anonymous

"Good fences make good neighbors."

Robert Frost

The use of physical barriers is one of the best methods of resolving human–wildlife conflicts. The obvious assumption is that if animals cannot obtain access, they cannot cause damage. Exclusionary devices used in wildlife damage management include fences, tree guards, bird-proof netting, and overhead wires. Fences have been used since the dawn of agriculture to protect crops from wildlife damage and to separate captive animals from free-ranging animals, which might either compete with livestock for forage or prey upon them.

FACTORS INFLUENCING THE COST EFFECTIVENESS OF FENCING TO REDUCE WILDLIFE DAMAGE

Fences are often the most effective techniques to reduce damage by mammals and can be completely effective if they are properly constructed and maintained. In contrast, most other techniques used in wildlife damage management may reduce wildlife damage but not end it entirely. Cost is the major factor preventing the widespread use of fences, which can be so expensive that their use cannot be justified on economic grounds. The cost effectiveness of fencing is influenced by several factors.

Cost of Fence Construction

Fences vary tremendously in cost. A 2-m-high deer-proof fence would cost $4 per linear meter if made of chain-link mesh $2 per meter if made of woven wire and $1 per meter if made using electrified wires. As one might expect, cheaper fences are often less effective, so the challenge is to select the least expensive fence that can do the job.

Fences differ in their life expectancies. When a fence lasts for several years, its cost can be amortized over that time (Caslick and Decker 1979; Lokemoen 1984; Goodrich and Buskirk 1995). For instance, a fence made of woven wire (also called hog wire) is less expensive than a chain-link fence, but a woven-wire fence may have a life expectancy of 10 years while a chain-link fence may last 30 years.

Local topography affects the cost of fence construction. A corner post is needed each time a fence changes direction or elevation. Corner posts bear much of the weight of a fence and, therefore, must be sturdier and more securely set in the ground than other posts. This makes corner posts more expensive. Consequently, fences cost less when built in flat areas where they can run in straight lines and require fewer corner posts.

Area to Be Fenced

Construction costs for a fence are based on its length, but its cost effectiveness is related to the area of land enclosed within it. The important variable is the perimeter to area ratio; the length of a field's perimeter influences how much it costs to fence it, and its area determines the magnitude of the benefit. A fence of a set length encloses more area if built around a square or circular field than one which is oblong (Figure 13.1). Fences also become more cost effective as the area they enclose increases in size. For example, if the perimeter doubles around a square field, the area enclosed increases fourfold (Figure 13.2).

Crop Value

Crops, such as corn or soybeans, have such low value per hectare that enclosing them in a deer-proof fence may not be economical, regardless of the how severely the crop is damaged. Other crops are so valuable (e.g., nursery plants, vineyards, or orchards) that fencing is economically justified even when damage levels are low. For instance, a hectare of Japanese yews, which are used for landscaping, can be worth up to $30,000. If deer damage reduced the value of these plants by 10%, then fencing would be worthwhile, providing it cost less than $3000 per hectare annually and completely stopped deer damage. In contrast, a farmer could not afford to build a fence around a cornfield at that price, even if deer were going to destroy the entire crop. A cornfield yielding 250 bushels per hectare at $2.50 per bushel has a value of only $625 per hectare. As another example, putting a bird-proof net over a crop costs over $500 per hectare per year; this exceeds the value of a hectare of wheat. A wheat farmer, therefore, would be better off losing the entire crop rather than putting up a bird-proof net to save it. On the other hand, wine grapes are so valuable, and birds may eat so many, that the benefit to cost ratio for the use of netting at some vineyards may reach 7 to 1 (Fuller-Perrine and Tobin 1993).

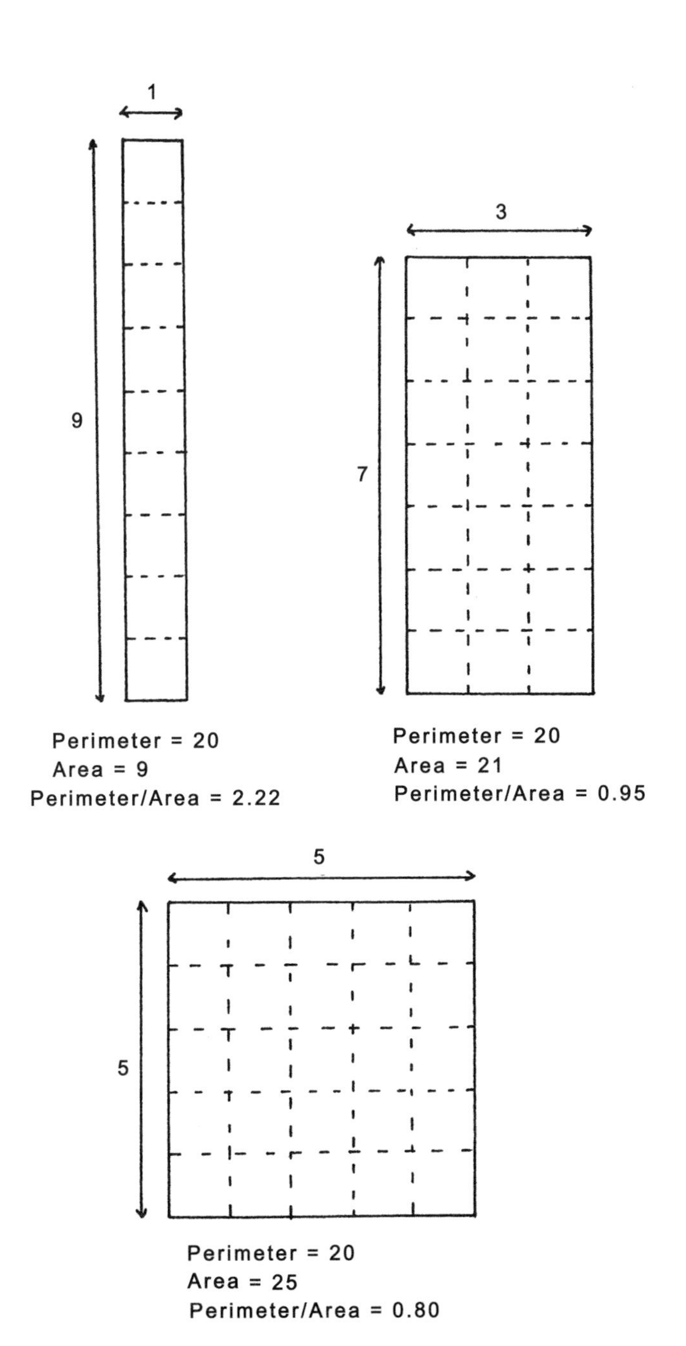

Figure 13.1 How the relationship between the length of a fence (perimeter) and the area enclosed within it changes as the shape becomes more square.

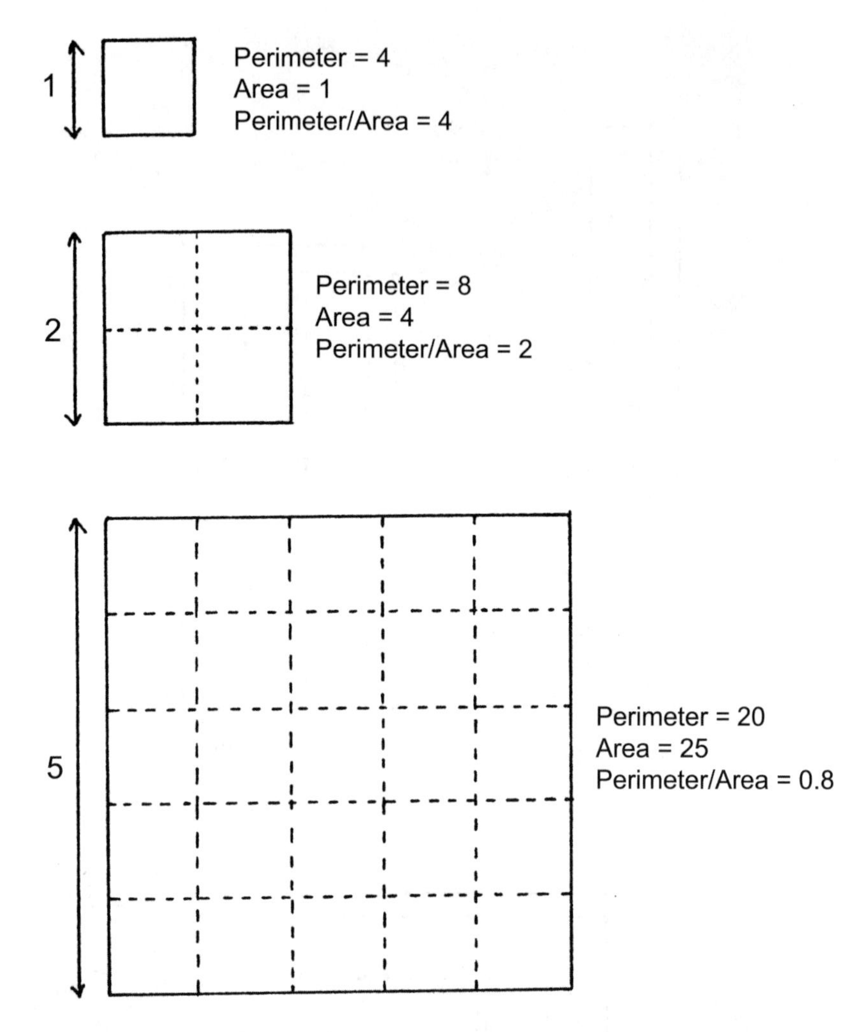

Figure 13.2 How the relationship between the length of a fence (perimeter) and the area enclosed within it changes as the area increases in size.

FENCES TO EXCLUDE DEER

In areas where deer are abundant, fencing may be the only effective method of reducing crop damage. The two major types of deer fences are woven-wire and electric fences.

Woven-Wire Fencing

The fencing material of chain-link and woven-wire fences consists of metal wires that are woven into a mesh grid. These fences function by physically preventing animals from passing through the mesh (Figure 13.3). Woven-wire and chain-link

Figure 13.3 Woven-wire deer-proof fence. (Photo courtesy of Don McIvor. With permission.)

fences last 10 to 30 years, require little maintenance, and are effective in all weather conditions (Caslick and Decker 1979). They are the most effective fences at excluding deer, but are expensive to build, costing $2 to $4 per meter (Caslick and Decker 1979; Hygnstrom and Craven 1988). These fences are used mainly for high-value crops or where effectiveness in excluding deer is more important than cost (e.g., along highways or at airports where deer pose a safety threat). Woven-wire and chain-link fences often are cost effective when their cost is amortized over their long life expectancy (Caslick and Decker 1979). To prevent deer from jumping over them, the recommended height for deer-proof fences is 2.5 to 3 m (Caslick and Decker 1979). Sometimes a single strand of wire is strung along the top of these fences to increase their height.

Sidebar 13.1 Australia's Monumental Wildlife Fences

The greatest attempt to reduce wildlife damage by building fences occurred in Australia during the second half of the nineteenth century, when European rabbits were spreading throughout the continent. Fences to stop them were built along the borders of South Australia, New South Wales, and Queensland (Figure 13.4). Subsequently, the governments of Queensland and Western Australia built several parallel fences to stop the range expansion of rabbits. Later, many of the rabbit fences were modified to exclude dingos. By 1908, there were over 10,000 km of dingo fencing in South Australia. The largest single effort was a dingo-proof fence which stretched for 5000 km through Queensland, New South Wales, and South Australia.

An interesting question is why such long wildlife fences were built in Australia, but not in other countries. The answer lies in the unique landscape of Australia and the nature of its wildlife problems. The topography of Australia is generally flat and the vegetation sparse, making it relatively easy to build long, straight fences. Further-

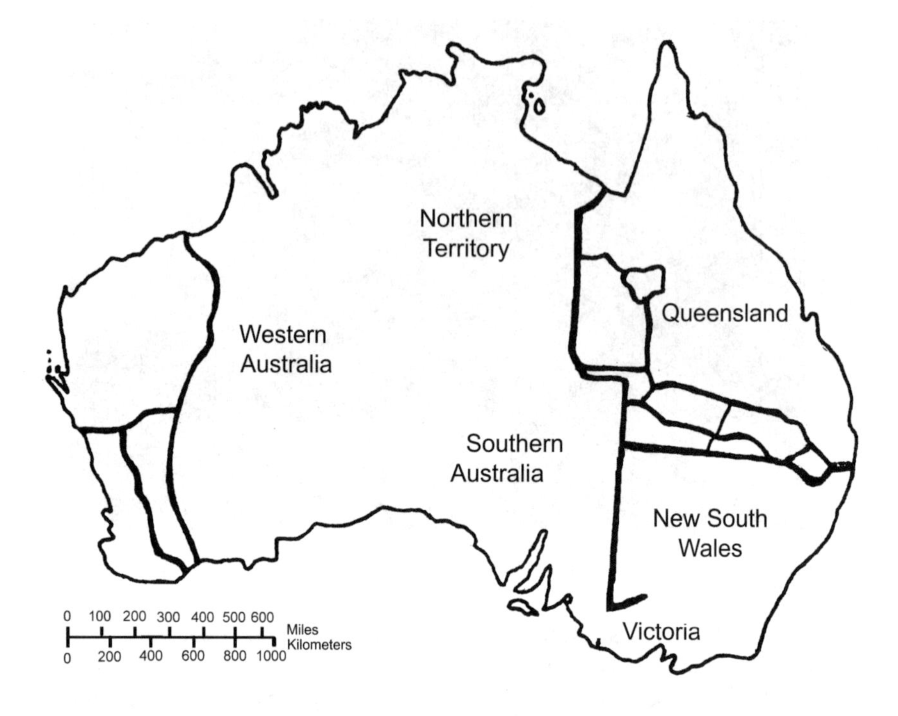

Figure 13.4 The wildlife fences (solid lines on map) of Australia in the early 1900s.

more, the vast interior of the continent is thinly settled, so that fence builders did not have to contend with a multitude of private landowners who would not want their property bisected by a fence. Moreover, animals causing problems in Australia, (e.g., rabbits, dingos, and kangaroos) could be excluded by fencing. Finally, Australians felt a sense of urgency, knowing that once rabbits reached an area, it would be impossible to remove them.

Unfortunately, the long fences in Australia were largely unsuccessful. It was too difficult to maintain such long fences and too easy for rabbits and dingos to find holes somewhere along those fences. Furthermore, rabbits were dispersing faster than some fences could be built and got around the ends of many fences before they were completed (McKnight 1969).

Electric Fencing

Electric fences are built from strands of metal wire spaced close enough together so that a deer cannot crawl through them without contacting a wire and being shocked. These fences are designed to make the task of crossing them painful enough that deer will avoid doing so.

To receive a shock from an electric fence, an animal must complete an electric circuit. Some fences are built with alternating positive and negative wires so that an animal will receive a shock when it simultaneously touches a positive and negative

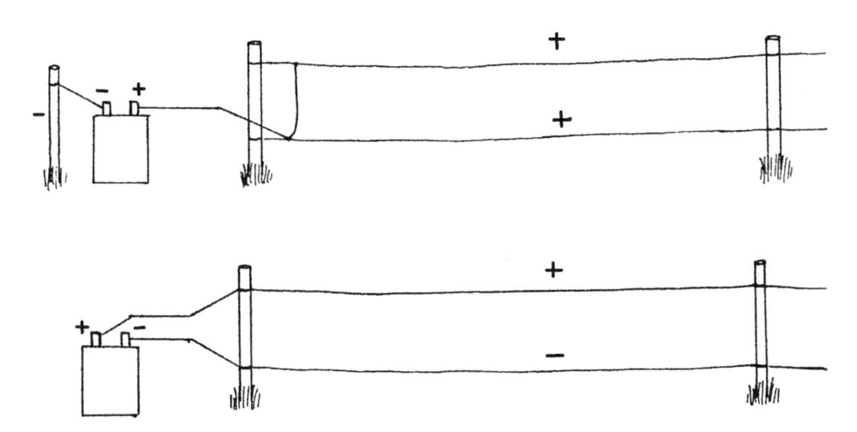

Figure 13.5 Two different designs for electric fences. In the top figure, all of the wires are positively charged, the negative lead is grounded, and the deer receives a shock whenever it touches any wire while standing on the ground. With the second design, the wires are either positively or negatively charged (bottom figure); a deer receives a shock when it simultaneously touches a negative and positive wire.

wire (Figure 13.5). More commonly, all of the wires are connected to the same pole on the charger and the other pole is connected to a metal post driven into the ground. With this design, the animal completes the circuit when it touches a wire while standing on the ground. The advantage of this design is that the animal only has to touch one wire to receive a shock. The disadvantage is that the wires must be insulated from the ground or the fence will short circuit. For this reason, plants must be kept from growing beneath an electric fence or contacting a wire. Hence, electric fences require constant weed control and must be checked every few days to ensure the fence is working properly and has not become grounded.

Electric fences are usually powered by a fence charger designed to send out a short pulse of power approximately once every second. The fence charger is usually powered by a 12-volt battery, which can be charged with a solar cell. Electric fences should only be powered by a fence charger specifically designed for that purpose. They should never be connected directly to an AC electric outlet because a person or animal could be electrocuted by contacting the fence.

Electric fences are not as effective at stopping deer as woven-wire fences because a deer can pass through an electric fence if willing to get shocked. They are most effective at protecting agricultural fields from deer browsing when food supplies outside the fenced area are abundant. During winter, when deer are hungrier and alternative food supplies are low, deer are likely to be more determined to get through a fence to the protected plants. Hence, an electric fence working well in the summer may not be effective during the winter (Porter 1983). There are several types of electric fences.

High-tensile electric fences — These electric fences are made from a stainless steel wire which can be pulled extremely tight without snapping. Fence posts can be separated by greater distances (up to 60 m) because there is less sag in the wires.

Powerful springs connect the wires to fence posts, allowing the wires to stretch without breaking if something runs into them.

High-tensile fences are most cost effective when built in a straight line and on flat ground. With this design, only the corner posts need to be sturdy enough to withstand the tremendous pull of the wires. The cost of a high-tensile deer fence with five wires is about $1 per meter (Palmer et al. 1985). High-tensile fences have been used to reduce deer damage to a variety of agricultural crops and are cost effective in many situations (Caslick and Decker 1979; Palmer et al. 1985; Hygnstrom and Craven 1988).

Double and slant-wire electric fences — Most fences are built vertically with the wires spaced directly above each other (Figure 13.6A). One problem with vertical deer fences is that they have to be at least 2 to 3 m tall to prevent deer from jumping over them. An alternative is a double fence (Figure 13.6B), which consists of two parallel fences 0.5 to 1.0 m apart from each other. The three-dimensional nature of this fence makes it more difficult for a deer to jump over it. Consequently, it need only be 1.5 m tall.

Another alternative is the slant-wire fence. This fence is built by placing the posts at approximately a 45° angle from the ground, with the most outward wire also the highest (Figure 13.6C). When a deer reaches the first wire, it normally will duck beneath the wire, only to have its path blocked by the more interior and lower wires. With the first wires now over the deer's back, it can no longer jump over the fence. However, the disadvantage of both the double and slant-wire fences is that they take up more land than a vertical fence. This leads to a loss in land which can be farmed. More importantly, double and slant-wire fences require a wider area that must be kept weed-free to prevent them from becoming grounded.

Temporary electric fences — Temporary electric fences can be assembled and disassembled quickly and cost only $0.35 per meter (Hygnstrom and Craven 1988; Owen et al. 1995). These fences are often made of polywire or polytape, which is a highly visible polyethylene ribbon interwoven with strands of stainless steel wire. Polytape has the benefit of being durable and easy to roll and unroll from a spool (Hygnstrom and Craven 1988). These temporary fences, however, are not very sturdy and require a great deal of maintenance (Porter 1983).

The drawback to temporary electric fences is that they have gaps along their length, which allow deer to crawl through or under the wires without getting shocked. Once deer learn these locations, these fences become ineffective. Because it takes time for deer to locate these gaps, these fences can be used for short-term problems (i.e., those lasting a few weeks). Temporary fences work best when the shock is so painful or frightening that considerable time is required before a deer is willing to attempt another fence crossing.

One variation of a temporary electric fence is called a peanut butter or baited fence (Figure 13.7). Peanut butter is smeared on the electric wire or on strips of aluminum foil attached to the wire (Porter 1983; Hygnstrom and Craven 1988). When a deer investigates the fence, it will touch the baited foil with its nose or tongue and get a more painful shock than if it brushed against the electrified fence with its fur. Such fences have been used successfully during the summer to reduce deer damage in apple orchards (Porter 1983) and cornfields (Hygnstrom and Craven 1988).

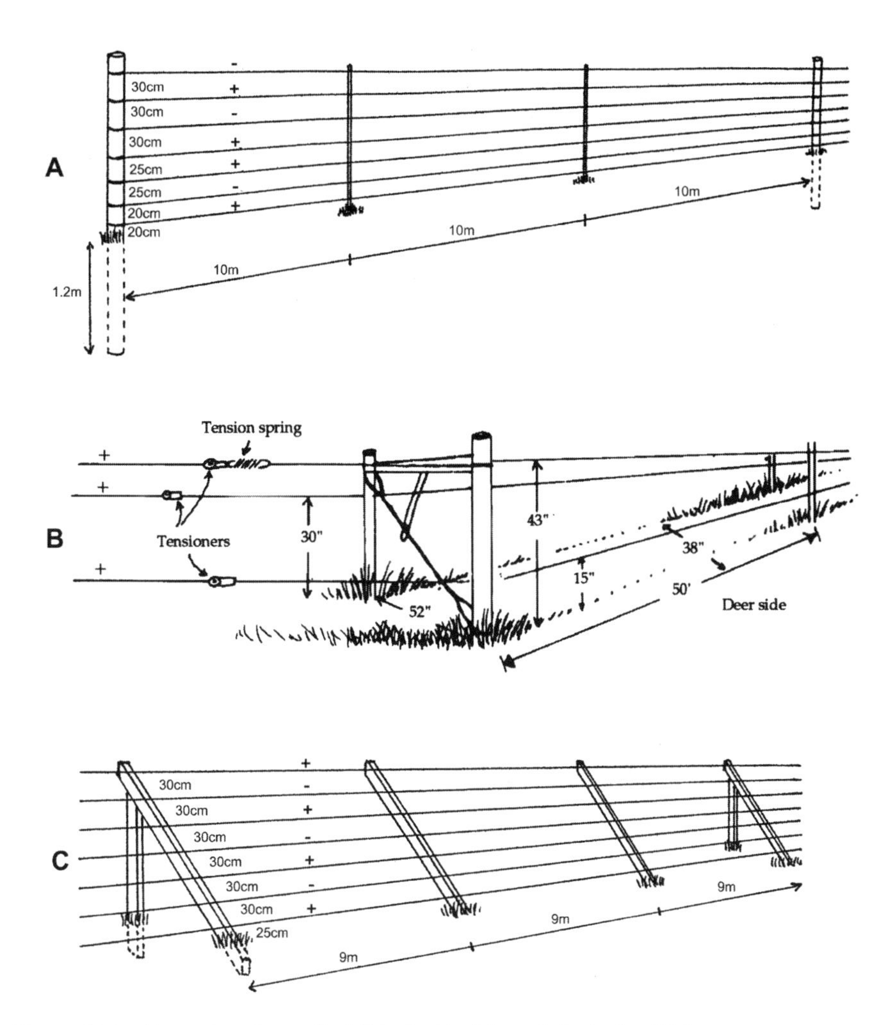

Figure 13.6 Examples of a vertical (A), double (B), and slant-wire (C) fence.

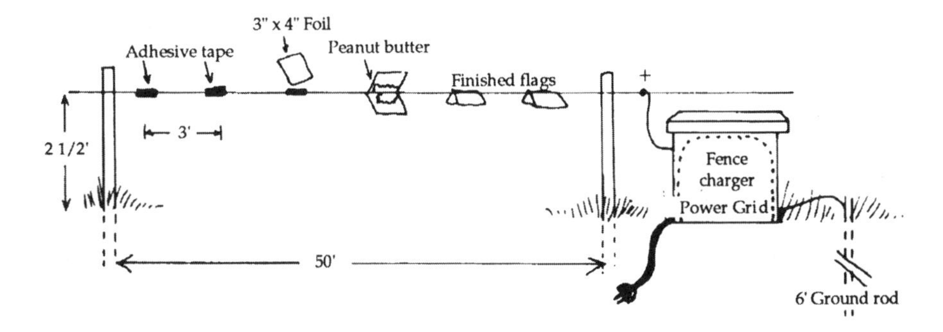

Figure 13.7 The peanut butter fence. (Figure from Hygnstrom et al. 1994.)

Sidebar 13.2 Using Captive Coyotes to Test Different Fence Designs

Dave deCalesta studied how captive coyotes managed to cross a number of different fences. He found that adult coyotes could pass through meshes as small as 15 × 23 cm. Coyotes had difficulty jumping over fences higher than 1.7 m. This led to the construction of a woven wire fence 1.7 m high with an overhang (Figure 13.8). No coyote jumped or climbed over this fence, but a few coyotes were able to crawl beneath it. Once a wire apron was added to the bottom of the fence (Figure 13.8), only one coyote crossed the fence, apparently by squeezing through the mesh (Wade 1982). When this fence design was later tested at sheep ranches, no sheep were killed by coyotes in pastures surrounded by this fence, while 38 sheep were killed in unfenced pastures (deCalesta 1978).

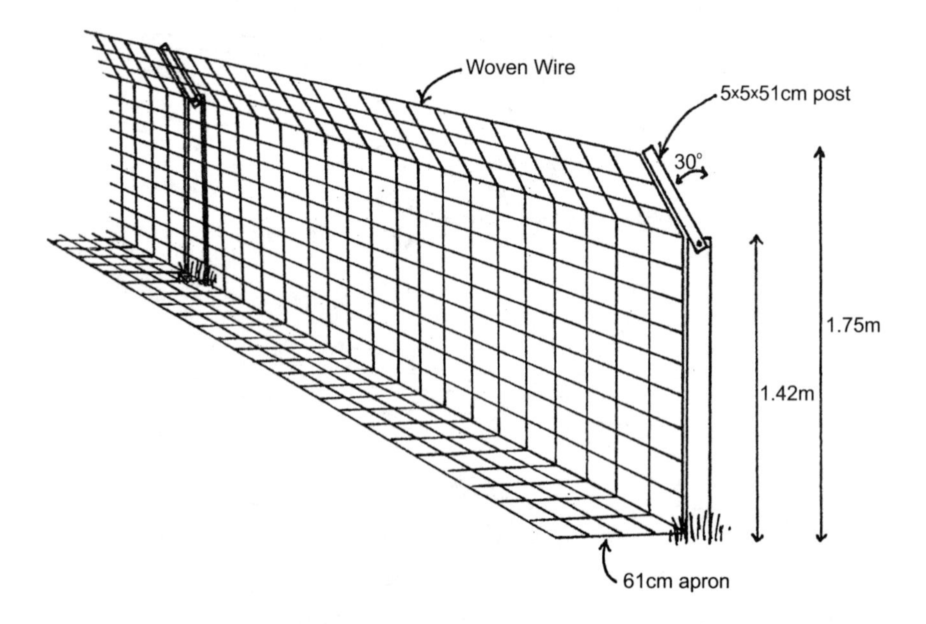

Figure 13.8 Coyote-proof fence.

USING FENCES TO REDUCE PREDATION ON LIVESTOCK

Fences have been used to reduce damage caused by mammalian predators including coyotes (Linhart et al. 1982; Nass and Theade 1988), arctic foxes (Estelle et al. 1996), black bears (Maehr 1982; Wade 1982), Asiatic black bears (Huygens and Hayashi 1999), polar bears (Davies and Rockwell 1986), badgers (Wilson 1993), skunks (LaGrange et al. 1995), and raccoons (LaGrange et al. 1995). Acorn and Dorrance (1994) interviewed 21 sheep ranchers who used electric fences to reduce coyote predation and found that 17 of them were pleased with their fences and would build another. Dorrance and Bourne (1980) examined coyote predation rates on five ranches using seven-wire electric fences and reported that the fences significantly reduced predation.

The ability of predators to crawl or burrow under fences poses a difficult challenge to fence builders. Because of this, the bottom wire must be placed close to the ground to prevent predators from going beneath it. This means that weeds do not have to grow very high or the wire does not have to sag much before contact is made and the fence shorts out. Hence, predator-proof electric fences require more maintenance than deer-proof fences.

Predator fences are often constructed of woven wire because it is difficult to keep electric fence wires close enough together to prevent small predators from passing through the wires without getting shocked. Despite the best efforts of their builders, most fences are not completely predator proof. Thus, some predator control within the fenced area may be needed to remove those predators that have learned how to penetrate the fence (Greenwood et al. 1990; LaGrange et al. 1995). Fences are useful even when they are not entirely effective in excluding predators, because access points predators use to go under or through a fence can be identified. Traps or snares can then be set at these points to kill those predators that have discovered how to breach the fence.

USING EXCLUSION TO REDUCE PREDATION ON NESTING BIRDS

High predation rates on ground-nesting birds and their eggs are a serious problem in many parts of North America. In extreme cases, predation on breeding birds has resulted in extirpation of local populations (Bailey 1993). More commonly, predation contributes to subtle long-term population declines, such as those experienced by dabbling duck populations in the Prairie Pothole region of North America (Cowardin et al. 1985; Greenwood et al. 1995; Beauchamp et al. 1996). One way to protect nesting birds is to keep mammalian predators away from them. There are numerous ways to accomplish this.

Fences to Protect Individual Nests

Wire-mesh fences built around individual nests have successfully protected those nests from predators (Table 13.1). Estelle et al. (1996) used wire-mesh fences to exclude arctic foxes from nests, improving the daily survival rate of pectoral sandpiper nests in Alaska. Nol and Brooks (1982) used mesh exclosures to exclude gulls from killdeer nests. Raccoons, however, were able to reach the eggs by inserting their paws through the holes in the fences, thus rendering the fences ineffective. In Florida, wire mesh and sheet metal exclosures were used to protect seaside sparrow nests from snakes, Norway rats, rice rats, and fish crows (Post and Greenlaw 1989). Deblinger et al. (1992) summarized the results of different studies to protect individual piping plover nests and concluded that fences were effective in reducing predation rates to below 10%. Successful fences were trian-

Table 13.1 Efficacy and Cost Effectiveness of Fences, Nesting Baskets, Artificial Islands, Fenced Peninsulas, or Moated Peninsulas to Improve Reproductive Success of Ground-Nesting Birds

Species	Location	Protected Unit	Effectiveness (treated vs. untreated)	Life Expectancy	Authors
			Single Nest Fences		
Pectoral sandpiper	Alaska	1 nest	Daily survival rate 0.98 vs. 0.72	1 season	Estelle et al. (1996)
Piping plover	Massachusetts	1 nest	Chicks fledged/pair 2.0 vs. 0.1	1 season	Melvin et al. (1992)
Piping plover	Massachusetts	1 nest	% nests hatched 92 vs. 25	1 season	Rimmer and Deblinger (1990)
Killdeer	Ontario	1 nest	% successful nests 71 vs. 33	1 season	Nol and Brooks (1982)
Seaside sparrow	Florida	1 nest	% successful nests 48 vs. 6	1 season	Post and Greenlaw (1989)
			Electrified Fences		
Terns and eiders	United Kingdom	1 colony	Kept foxes out	1 season	Patterson (1977)
Piping plover	North Dakota	7 ha	Nest success rate 0.41 vs. 0.27 Chicks fledged/pair 1.00 vs. 0.66	3 seasons	Mayer and Ryan (1991)
Dabbling ducks	North Dakota	45 ha	Nests/ha 0.83 vs. 0.38 Successful nests/ha 0.11 vs. 0.05	3 seasons	Arnold et al. (1988)
Dabbling ducks	North Dakota	9 ha	Nest success 0.65 vs. 0.45	20 years	Lokemoen et al. (1982)
Dabbling ducks	Minnesota	17 ha	Nest success 0.54 vs. 0.17	20 years	Lokemoen et al. (1982)
Dabbling ducks	North Dakota	40 ha	Nest success 0.36 vs. 0.07	Not provided	Greenwood et al. (1990)
Mallard	Iowa	19 ha	Nest success 0.39 vs. 0.14	Not provided	LaGrange et al. (1995)

			Nesting Baskets		
Mallard	Prairie Potholes	Not applicable	Production of 2.6 ducklings/basket/yr	20 years	Doty et al. (1975)
			Artificial Islands		
Gadwall	Saskatchewan	0.03 ha	Nest density 62/ha Nest success 0.65	Not provided	Hines and Mitchell (1983)
Mallard	North Dakota	0.0025 ha	Nest success 0.38–0.52 0.8 Ducklings/island	15 years	Higgins (1986)
			Peninsulas with Electric Fences		
Dabbling ducks	North Dakota	Not provided	Nest success 0.54 vs. 0.17 Ducklings/ha 18 vs. 1	20 years	Lokemoen and Woodward (1993)
			Peninsulas with Moats		
Dabbling ducks	North Dakota	Not provided	Nest success 0.75 vs. 0.14 Ducklings/ha 21.8 vs. 1.2	50 years	Lokemoen and Woodward (1993)

gular in shape, had walls higher than 120 cm, were built of 5 × 5-cm wire mesh, and had the bottom of the fence buried deeper than 10 cm.

Fences to Protect Habitat Patches

Fences can also be used to exclude predators from patches of nesting habitat (Figure 13.9). Beauchamp et al. (1996) analyzed 21 studies evaluating the use of electric fences to exclude mammalian predators from habitat patches where ducks nested (Table 13.1). They concluded that nest success in fenced areas was higher than in unfenced areas. In North Dakota and Minnesota, exclosures produced up to ten more ducklings per hectare than the surrounding areas (Lokemoen et al. 1982). Greenwood et al. (1990) reported nest success improved from 7% outside exclosures to 36% within them. Predator control inside the fences further increased nest success to 81%. However, some fences delayed the exit of the broods, and this increased duckling mortality. Pietz and Krapu (1994) and Howerter et al. (1996) subsequently demonstrated that the survival of ducklings could be improved by creating ground-level exits in the fences for duck broods.

Using Nesting Structures to Isolate Nesting Birds from Predators

Elevated nesting structures (e.g., elevated baskets, hay bales, floating platforms) can reduce mammalian predation on duck nests (Figure 13.10). In the Great Plains, over 80% of mallard nests built on structures hatched (Bishop and Barratt 1970; Doty et al. 1975). Raccoons were the only mammalian predator which could reach

Figure 13.9 Predator-proof fence placed around a habitat patch to provide a safe nesting site for waterfowl.

Figure 13.10 Elevated nesting platform occupied by a Canada goose. (Photos courtesy of North Dakota State University Extension Service and Jaime Jimenez. With permission.)

elevated nests, but they could be excluded using truncated metal cones or sheet metal on the support poles (Figure 13.11; Doty et al. 1975; Doty 1979). On a similar note, one interesting way to keep rat snakes from climbing pine trees used for nesting by endangered red-cockaded woodpeckers is to shave off the rough bark snakes use to climb the trees (Saenz et al. 1999).

Building Islands to Isolate Nesting Birds from Predators

Many birds typically nest on islands to gain some protection from mammalian predators. Hence, another way to separate nesting birds from mammalian predators is to construct islands in wetlands large enough to impede immigration of predators from the mainland (Lokemoen and Messmer 1993). This condition is met by large (greater than 5 ha), permanent wetlands with water depths of over 1 m.

Figure 13.11 Photo of a metal shield designed to keep mammalian predators from climbing up the support pole. (Photo courtesy of Gary San Julian.)

The simplest way to create an island is to make one out of an existing peninsula by cutting it off from the mainland with a moat or fence (Figure 13.12). Lokemoen and Woodward (1993) compared duck breeding on eight peninsulas isolated from mainland by electric fences, two isolated by water-filled moats, and ten attached to the mainland. Isolated peninsulas exhibited three times the nest success and produced nine times more ducklings per hectare than peninsulas which were still attached to the mainland (Table 13.1).

USING BARRIERS TO PROTECT INDIVIDUAL TREES FROM HERBIVORES

Tree Guards to Reduce Deer Browsing on Tree Shoots

It is often desirable to protect individual plants from herbivores (Marsh et al. 1990). During winter, deer often browse the apical meristem (the growing shoot) of trees. Because the meristem contains the bud for the plant's growth in the spring, its loss may have a major impact on the tree's growth. The browsed branch cannot grow until the bud has been replaced. Hence, a tree that has lost some of its buds can become misshapen when only some branches grow. This irregular growth is especially harmful to Christmas trees, nursery plants, and fruit trees. For instance, apple trees have to grow into a specific shape for maximum fruit production. Therefore, loss of a few buds has a profound impact on the tree's future productivity.

Buds can be protected from browsing by covering them with paper or plastic sleeves, known as tree guards. These sleeves are open at both ends so that one end

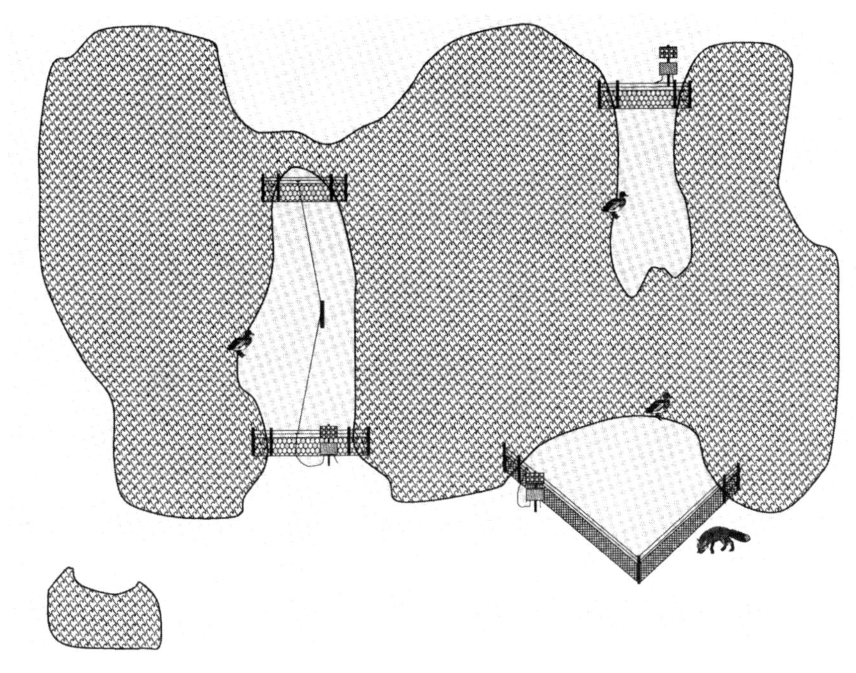

Figure 13.12 Diagram showing how fences can be used to isolate peninsulas from the mainland. (Figure courtesy of Berryman Institute. With permission.)

can be placed over the shoot and the bud can grow out the other end; they are often made of biodegradable material and will decompose in a few years.

Deer browsing on tree seedlings can be so severe and persistent that the trees are unable to grow high enough to escape the reach of foraging animals. In such cases, a 30-year-old tree may be less than a meter in height. Long tree guards (1 to 2 m in length) are used to alleviate this problem by preventing browsing until the tree is tall enough to grow out of the top of the sleeve (Figure 13.13). Tree guards have been used successfully to reduce damage to tree seedlings by deer (Anthony 1982), pocket gophers (Hooven 1971; Anthony et al. 1978), nutria (Conner and Toliver 1987), and pikas (Khan and Smythe 1980).

Wraps and Shields to Protect Tree Trunks from Being Girdled

Voles and other small mammals sometimes eat the bark of palatable trees, especially during winter when alternate food supplies are low. If all of the bark is removed from around the trunk (i.e., the tree is girdled), nutrients can no longer travel from the roots to the leaves and the tree will die. This is a serious problem in fruit orchards, where the loss of a single mature tree has a major effect on an orchard's productivity and may take ten years to replace. One way to protect trees from girdling is to wrap their trunks with weatherproof cardboard, plastic, or hardware cloth, so that small mammals cannot reach the bark (Marsh et al. 1990). Cylinders designed to encircle tree trunks also can be used for this purpose (Figure 13.14).

Figure 13.13 A newly planted orchard in which all of the young trees have been covered with plastic tree guards.

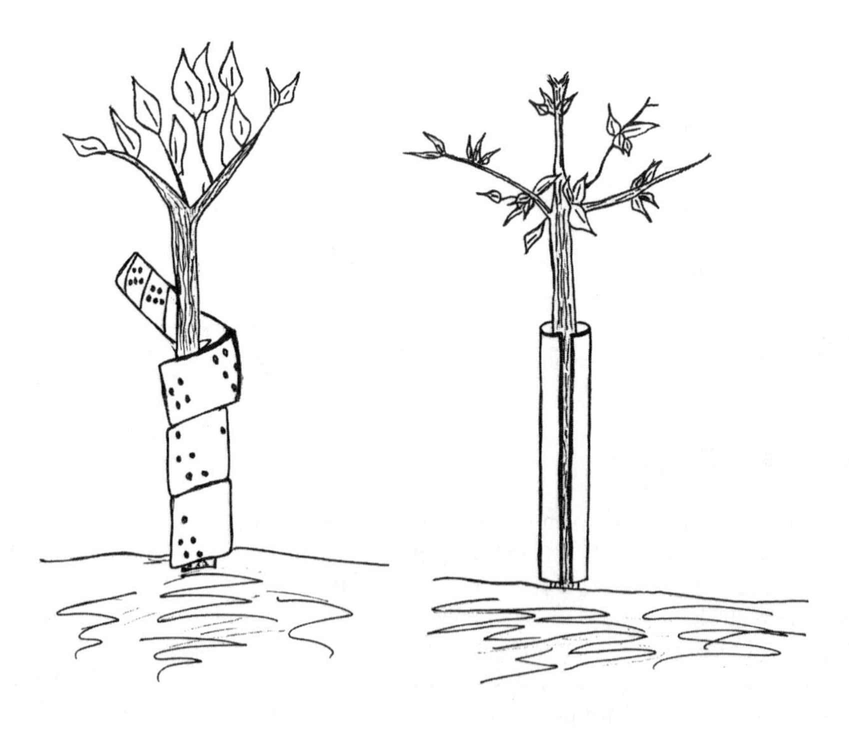

Figure 13.14 Wraps and shields to protect tree trunks from girdling by small mammals.

EXCLUSIONARY DEVICES TO PREVENT BEAVER FROM REBUILDING DAMS

Flooding caused by beaver dams can damage roads, homes, agricultural fields, and valuable timber. The short-term solution to this problem is to break open the dam and allow the impounded water to escape. However, beavers quickly repair breaks, thus requiring dams to be breeched repeatedly. A number of exclusionary methods exist which can discourage beavers from rebuilding dams, or at least keep them from rebuilding them so quickly. To prevent beavers from damming culverts, one can build a fence around the culvert on the upstream side (Figure 13.15) or put metal bars at both ends of the culvert to keep beavers from entering (Laramie 1963). Common limitations with these methods are that debris will collect along the fence or that beavers may use the fence to build a new dam. Removing these dams, however, is easier than removing those located inside culverts. Debris removal is facilitated if the metal bars and fence are constructed so they can be easily disassembled for cleaning. To accomplish this, metal bars placed at the mouths of culverts should be built so they can be individually removed. An easy-cleaning fence can be built from 1 × 2 m panels of woven wire and held to fence posts with metal clips that are easily unfastened. It is important that the fencing material be placed on the downstream side of the posts. If the fence panels were placed on the upstream side, the weight of the water and debris would push the fence panels against the posts with such force that they would be difficult to remove (Figure 13.15).

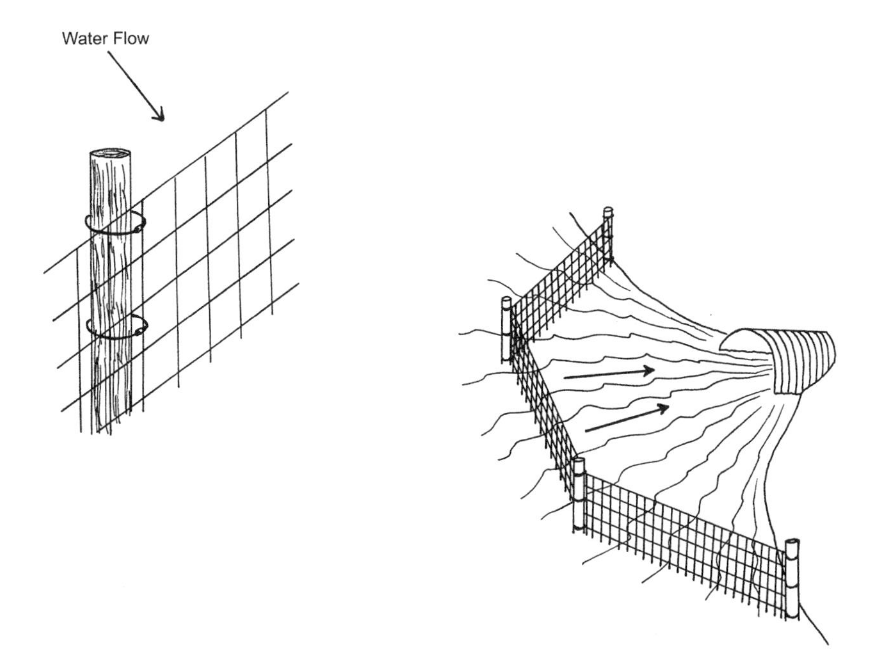

Figure 13.15 Keeping beaver from damming up a culvert by building a fence upstream of the culvert (the fence panels are attached to the downstream side of the posts with metal clips so the fence panel can be easily removed for cleaning).

Beaver locate breaches in their dams by listening for the sound of rushing water and feeling the flow of water around them. Thus, one way to discourage beaver from rebuilding a dam is to place a perforated pipe through the dam leading into the pond. This pipe should have a T-shaped end, with one end of the "T" sticking straight up out of the water and the other facing down. The T-shaped end will force water to flow up through the bottom of the pipe, making it harder for the beaver to dam it, especially if the bottom of the "T" is suspended several feet above the pond's bottom (Figure 13.16). A cleaning rod can be pushed down from the top end of the "T" to clean out any debris lodged inside the bottom end.

Another design, the Clemson Beaver Pond Leveler, also uses a perforated pipe, but it is encased in heavy-gauge woven-wire fencing which keeps the beaver from getting close to the perforated pipe (Wood and Woodward 1992). This device works best when it is submerged, as it is more difficult for beaver to locate and dam underwater breaches. This will require that a stand pipe be built into the Leveler downstream of the dam so that the Leveler will not drain all the water from the pond (Figure 13.16).

USING A TRAP-BARRIER SYSTEM TO REDUCE RAT DAMAGE IN RICE FIELDS

Rodent damage in rice fields causes severe economic losses in many parts of the world. One promising method to protect these crops is to use a trap-barrier

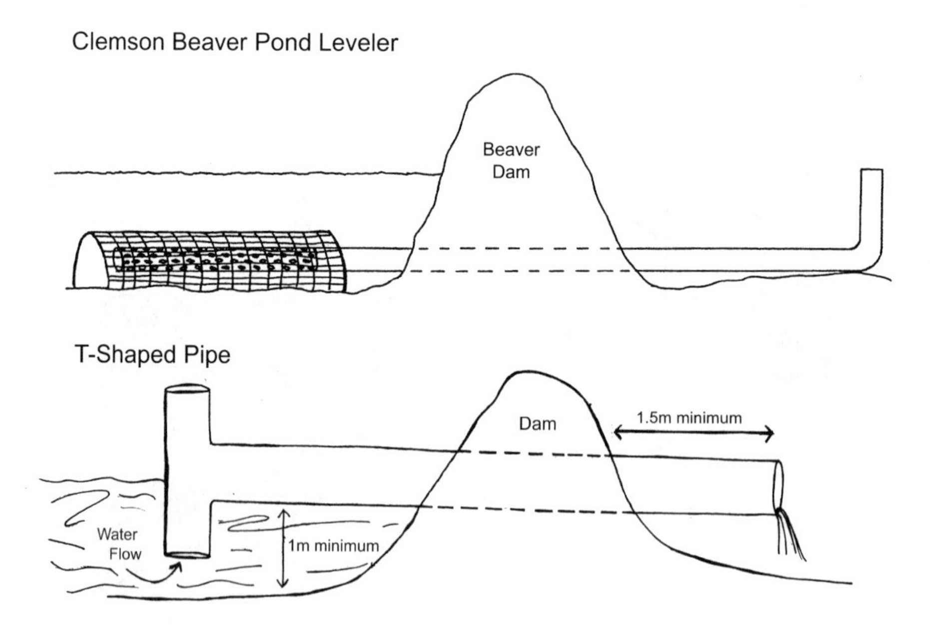

Figure 13.16 Using a Clemson Beaver Pond Leveler and a T-shaped pipe to prevent beaver from repairing a dam.

system in which a field is surrounded by a plastic fence. Holes are cut in the fence, through which rodents pass into the field. However, every few days, a multiple-capture cage trap is placed on the crop side of the fence so that any rats using the hole now enter the trap. This trap-barrier system is easy to build and is inexpensive, although labor is required to maintain the fence and check the traps. It is also effective: as many as 6000 rats have been caught in one night, and 44,000 were caught in one growing season using a trap barrier around a single field (Singleton et al. 1999). One major advantage of such a system is that the benefits also extend to adjacent fields located outside the fence. For this reason, the barrier-trap system works best as part of a large-scale rodent control program.

USING EXCLUSION TO SOLVE BIRD DAMAGE

One of the most effective ways of reducing bird damage is to exclude birds with netting (Stucky 1973; Foster 1979). Unfortunately, netting is one of the most expensive methods of bird control. For this reason, it is used only for valuable crops, such as vineyards, blueberry orchards, or aquaculture facilities.

Most netting deteriorates when exposed to the sun or weather. Thus, nets last longer if they are used sparingly and can be erected and disassembled annually. Much research has been conducted on the development of netting systems that can be easily erected and taken down with tractors (Fuller-Perrine and Tobin 1993; Taber and Martin 1998). Vineyards, for instance, can be protected by using nets 2 to 3 m wide and hundreds of meters long which can simply be draped over a row (Figure 13.17). In a simple backyard system, netting can be laid over the fruit trees. This system provides some protection, but birds can still land on the net and reach many berries through the netting. Another drawback is that people and equipment

Figure 13.17 Vineyard protected by bird-proof netting.

cannot work under the draped nets. For this reason, most bird nets are held aloft through the use of support poles that keep the netting above the plants.

A less expensive way to exclude birds is to erect a set of closely spaced wires overhead so that birds have to fly through the wires when landing or taking off. If the wires are close enough together, birds have difficulty flying through them without striking them with their wings. Overhead wires have been used to keep birds from reservoirs (Amling 1980), fish hatcheries (Lagler 1939; Barlow and Bock 1984), nesting colonies (Blokpoel and Tessier 1983; Morris et al. 1992), public places (Blokpoel and Tessier 1984), garbage dumps (Laidlaw et al. 1984), and agricultural fields (Pochop et al. 1990). Overhead wires are most effective against birds such as gulls and geese, which have long wingspans and are not agile fliers (Table 13.2).

The optimal spacing for parallel wires should be equal to the culprit bird's wingspan. For instance, Andelt and Burnham (1993) found that wires spaced 10 cm apart repelled pigeons, while widely spaced wires did not. The optimal spacing for grid wires will depend on how steep an angle the bird can climb when flying and on the height of the bird's wing stroke. These two variables determine if a bird flying at a right angle to the wires can pass under one wire and above the next one without hitting either (Figure 13.18). Close spacing between wires is required to deter birds like pheasants, quail, and ducks, which can gain altitude quickly.

Parallel lines often work as well as wire grids because birds taking off into the wind cannot always orient themselves to fly parallel to the wires (Table 13.2). This also suggests that if parallel lines are used, they should be placed perpendicular to the prevailing winds.

It is not known if the wires' height above the ground influences their effectiveness. At present, the height of the wires is usually determined by the most convenient height for the landowners. When birds first take flight, many use their feet to help them gain altitude by jumping. Because of this, their initial ascent (for the first 0.5 to 1 m) is steeper than their later flight when they can only use their wings to gain

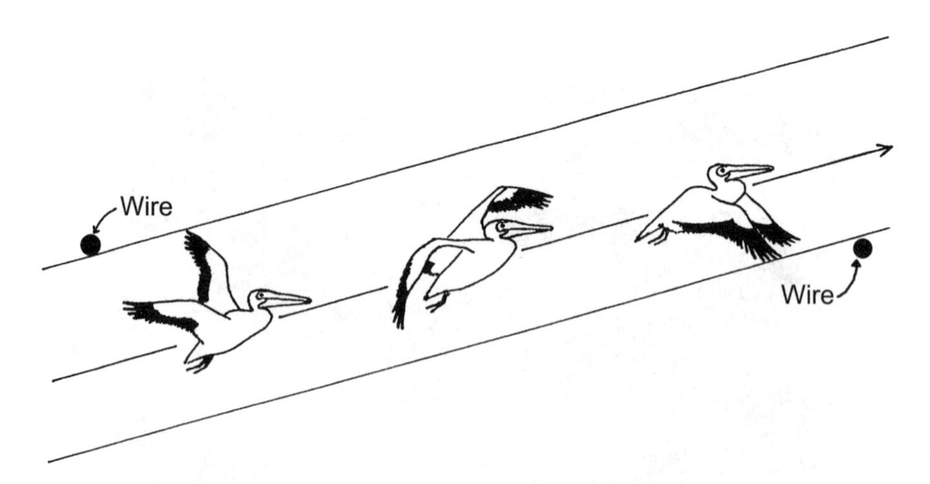

Figure 13.18 A bird risks hitting overhead wires with its wings when flying up through a grid or at an angle to parallel lines. That risk is increased if it cannot gain altitude quickly or has a large wing-beat, or if the wires are close together.

Table 13.2 Research on the Use of Overhead Lines to Repel Birds (Adapted from Pochop et al. 1990)

Species	Location	Line type	Line pattern	Gap between lines (m)	Effective?
Cormorants	Fish hatcheries	Nylon	Grid	10 × 10	Somewhat
Canada geese	Sewage lagoons	Metal wire	Parallel	6	Yes
	Lakes	Plastic	Grid	9 × 9	Yes
Brant	Agricultural fields	Nylon	Parallel	12 to 16	Yes
Duck species	Sewage lagoons	Metal	Grid	3 × 3, 6 × 6	Somewhat
Black vultures	Buildings	Monofilament	Parallel	1.8	Yes
Gull species	Nesting colonies	Monofilament	Parallel	0.6	Yes
	Public places	Stainless steel	Parallel	2.5	Yes
	Landfills	Wire	Grid	6 × 6	Yes
	Landfills	Wire	Parallel	6	Yes
Rock doves	Public places	Stainless steel	Grid	2.5 × 2.5	No
	Landfills	Metal	Parallel	3	No
Mourning doves	Citrus groves	Monofilament	Grid	3 × 3	No
Barn swallows	Houses	Monofilament	Parallel	0.3	Yes
American crows	Landfills	Monofilament	Grid	6 × 6	Yes
	Landfills	Metal	Parallel	3	No
Great-tailed grackles	Citrus groves	Monofilament	Grid	3 × 3	Somewhat
American robins	Grape plants	Monofilament	Parallel	0.3	No
European starlings	Grape plants	Monofilament	Parallel	0.3	No
House sparrows	Various plants	Monofilament	Diverging	00–0.6	Yes
	Bird feeders	Monofilament	?	0.3	Yes

Figure 13.19 Birds with strong leg muscles use them when taking flight and, because of this, their initial flight path is at a steeper angle than it is later.

altitude (Figure 13.19). This suggests that overhead wires would be more effective if placed more than 1 m off the ground, so birds cannot literally jump through them. They should not be so high, however, that birds taking off can change their flight direction before flying through the wires.

Not all of the variance among avian species in their response to overhead wires can be attributed to differences in their flying ability. American robins and European starlings are not deterred by overhead wires, while house sparrows seem particularly bothered by them. (Table 13.2). Because house sparrows can become nuisances when monopolizing bird feeders, one unique solution is to place wires above feeders where house sparrows are unwanted (Auguero et al. 1991; Kessler et al. 1994).

EXCLUDING WILDLIFE FROM BUILDINGS

Animals often become a nuisance and health hazard when they get inside buildings. Usually, the best way to correct this problem is to determine how the animals are entering the building and then seal up the opening. This often involves nothing more than capping a chimney, caulking a window, or boarding up a hole in the wall. It is best to do this when the animals are outside the building, rather than trapping them inside. It is also best not to exclude adults during the period when they may have young in the building because the young may then die there. When it is impossible to observe the comings and goings of animals and to know when they are all outside, a device can be installed, such as a one-way door or a check valve, that will allow animals to leave the building but not to return (Figure 13.20).

Birds can cause nuisance problems and health hazards due to their fecal material when roosting or nesting on ledges outside buildings, in aircraft hangers, under bridges, or in barns. To keep birds off ledges, the top of the ledge can be changed

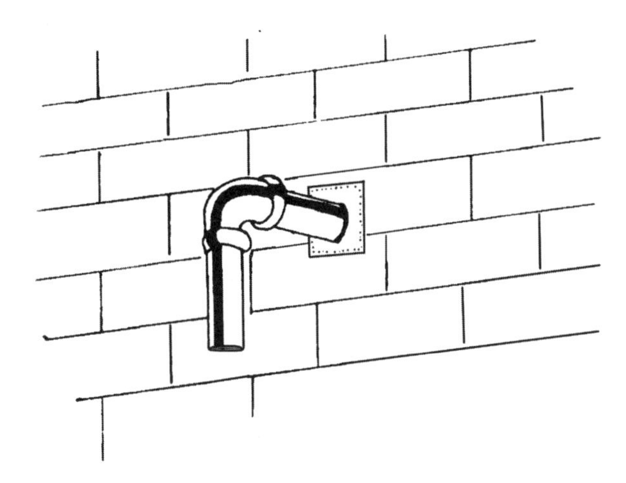

Figure 13.20 Example of a device which bats can use to exit from a house but does not let them return because bats cannot fly straight up.

from a flat surface to a surface pitched at an angle greater than 45° (Figure 13.21A). Another approach is to install porcupine wire on top of the ledges. Porcupine wire is well named: it is a tape consisting of hundreds of metal wires that stick up 10 to 20 cm at all angles (Figure 13.21B). When these wires are longer than the birds' legs, the birds cannot alight, sit, or walk on ledges treated with porcupine wire. Bird coils look like giant springs and function much like porcupine wire when installed

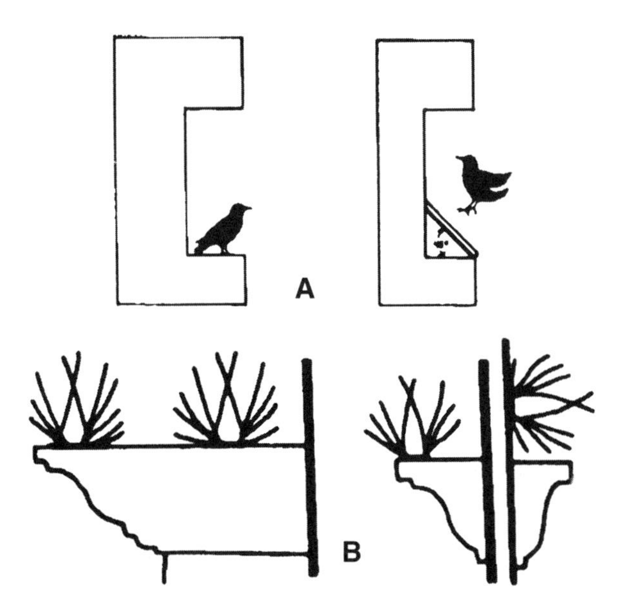

Figure 13.21 Keeping pigeons from roosting on a ledge by changing the upper surface to an angle of 45° (A) or by installing porcupine wire (B). (Figure courtesy of Hygnstrom et al. 1994. With permission.)

on ledges. However, some birds will build their nests on porcupine wire and bird coils. In another approach, Andelt and Burnham (1993) used closely spaced wires (about 10 cm apart) to prevent rock doves from roosting on ledges.

SUMMARY

Excluding wildlife through the use of fences or netting is one of the most effective methods to eliminate problems with wildlife. Unfortunately, building a fence to enclose a field or using a bird-proof net is very expensive and not economically justified unless the crop is of high value or the problem is particularly serious (deer–automobile collisions). Animals can be excluded by many techniques that differ in cost, life expectancy, and effectiveness. For instance, birds can be excluded with overhead wires or by complete netting systems. Deer can be excluded with a single-wire peanut butter fence, a slant-wire electric fence, or a chain-link fence. Exclusionary devices can also be used for many different problems, including the protection of nesting birds from predators, stopping mammals from climbing trees, preventing beaver from rebuilding their dams, or keeping wildlife out of buildings. Because of installation cost, fences, nets, and other exclusionary devices are most appropriate when effectiveness is more important than cost and when the wildlife problem is expected to persist over months or years.

LITERATURE CITED

Acorn, R. J. and M. J. Dorrance, An evaluation of anti-coyote electric fences. In *Proc. Vert. Pest Conf.*, 16, 45–50, 1994.

Amling, W., Exclusion of gulls from reservoirs in Orange County, California. In *Proc. Vert. Pest Conf.*, 9, 29–30, 1980.

Andelt, W. F. and K. P. Burnham, Effectiveness of nylon lines for deterring rock doves from landing on ledges. *Wildl. Soc. Bull.*, 21, 451–456. 1993.

Anthony, R. M., Protecting ponderosa pine from mule deer with plastic tubes. *Tree Planters' Notes,* 33, 22–26, 1982.

Anthony, R. M., V. G. Varnes, Jr., and J. Evans, "Vexar" plastic netting to reduce pocket gopher depredation of conifer seedlings. In *Proc. Vert. Pest Conf.*, 8, 138–144, 1978.

Arnold, P. M., R. J. Greenwood, B. G. McGuire, C. R. Luna, and R. F. Johnson, Evaluation of electric fence enclosures to improve waterfowl nest success in the Arrowwood wetland management district. P. 131–132 in *Proceedings of the Mallard Symposium.* Bismarck, ND, 1988.

Auguero, D. A., R. J. Johnson, and K. M. Eskridge, Monofilament lines repel house sparrows from feeding sites. *Wildl. Soc. Bull.*, 19, 416–422, 1991.

Bailey, E. P., Introduction of foxes to Alaskan Islands — history, effects on avifauna, and eradication. U.S. Fish and Wildlife Service, Resource Publication 193, 1993.

Barlow, C. G. and K. Bock, Predation of fish in farm dams by cormorants, *Phalacrocorax* spp. *Aust. Wildl. Res.*, 11, 559–566, 1984.

Beauchamp, W. D., T. D. Nudds, and R. G. Clark, Duck nest success declines with and without predator management. *J. Wildl. Manage.*, 60, 258–264, 1996.

Bishop, R. A. and R. Barratt, Use of artificial nest baskets by mallards. *J. Wildl. Manage.*, 34, 734–738, 1970.

Blokpoel, H. and G. D. Tessier, Monofilament lines exclude ring-billed gulls from traditional nesting areas. In *Proc. Bird Control Sem.*, 9, 15–19, 1983.

Blokpoel, H. and G. D. Tessier, Overhead wires and monofilament lines exclude ring-billed gulls from public places. *Wildl. Soc. Bull.*, 12, 55–58, 1984.

Caslick, J. W. and D. J. Decker, Economic feasibility of a deer-proof fence for apple orchards. *Wildl. Soc. Bull.*, 7, 173–175, 1979.

Conner, W. H. and J. R. Toliver, The problem of planting Louisiana swamplands when nutria (*Myocastor coypu*) are present. In *Proc. East. Wildl. Damage Control Conf.*, 3, 42–49, 1987.

Cowardin, L. M., D. S. Gilmer, and C. W. Shaiffer, Mallard recruitment in the agricultural environment of North Dakota. *Wildl. Monogr.*, 92, 1985.

Davies, J. C. and R. F. Rockwell, An electric fence to deter polar bears. *Wildl. Soc. Bull.*, 14, 406–409, 1986.

Deblinger, R. D., J. J. Vaske, and D. W. Rimmer, An evaluation of different predator exclosures used to protect Atlantic coast piping plover nests. *Wildl. Soc. Bull.*, 20, 274–279, 1992.

deCalesta, D. S. and M. G. Cropsey, Field test of a coyote-proof fence. *Wildl. Soc. Bull.*, 6, 256–259, 1978.

Dorrance, M. J. and J. Bourne, An evaluation of anti-coyote electric fencing. *J. Range Manage.*, 33, 385–387, 1980.

Doty, H. A., Duck nest structure evaluations in prairie wetlands. *J. Wildl. Manage.*, 43, 976–979, 1979.

Doty, H. A., F. B. Lee, and A. D. Kruse, Use of elevated nest baskets by ducks. *Wildl. Soc. Bull.*, 3, 68–73, 1975.

Estelle, V. B., T. J. Mabee, and A. H. Farmer, Effectiveness of predator exclosures for pectoral sandpiper nests in Alaska. *J. Field Ornithol.*, 67, 447–452, 1996.

Foster, T. S., Crop protection with Xironet. In *Proc. Bird Control Sem.*, 8, 254–255, 1979.

Fuller-Perrine, L. D., and M. E. Tobin, A method for applying and removing bird-exclusion netting in commercial vineyards. *Wildl. Soc. Bull.*, 21, 47–51, 1993.

Goodrich, J. M. and S. W. Buskirk, Control of abundant native vertebrates for conservation of endangered species. *Conserv. Biol.*, 9, 1357–1364, 1995.

Greenwood, R. J., P. M. Arnold, and M. G. McGuire, Protecting duck nests from mammalian predators with fences, traps, and a toxicant. *Wildl. Soc. Bull.*, 18, 75–82, 1990.

Greenwood, R. J., A. B. Sargeant, D. H. Johnson, L. M. Cowardin, and T. L. Shaffer, Factors associated with duck nest success in the Prairie Pothole region of Canada. *Wildl. Monogr.*, 59, 1–57, 1995.

Higgins, K. F., Further evaluation of duck nesting on small man-made islands in North Dakota. *Wildl. Soc. Bull.*, 14, 155–157, 1986.

Higgins, K. F., A reevaluation of mallard nesting on small man-made islands in North Dakota. P. 112–113 in *Proceedings of the Mallard Symposium*. Bismarck, ND, 1988.

Hines, J. E. and G. J. Mitchell, Gadwall nest-site selection and nesting success. *J. Wildl. Manage.*, 47, 1063–1071, 1983.

Hooven, E. F., Pocket gopher damage on ponderosa pine plantations in Southwestern Oregon. *J. Wildl. Manage.*, 35, 346–353, 1971.

Howerter, D. W., R. B. Emery, B. L. Joynt, and K. L. Guyn, Mortality of mallard ducklings exiting from electrified predator exclosures. *Wildl. Soc. Bull.*, 24, 667–672, 1996.

Huygens, O. C. and H. Hayashi, Using electric fences to reduce Asiatic black bear depredation in Nagano prefecture, central Japan. *Wildl. Soc. Bull.*, 27, 959–964, 1999.

Hygnstrom, S. E. and S. R. Craven, Electric fences and commercial repellents for reducing deer damage in cornfields. *Wildl. Soc. Bull.*, 16, 291–296, 1988.

Hygnstrom, S. E., R. M. Timm, and G. E. Larson, *Prevention and Control of Wildlife Damage*. University of Nebraska Cooperative Extension, Lincoln, 1994.

Kessler, K. K., R. J. Johnson, and K. M. Eskridge, Monofilament lines and a hoop device for bird management at backyard feeders. *Wildl. Soc. Bull.*, 22, 461–470, 1994.

Khan, A.A. and W. R. Smythe, The problem of pika control in Baluchistan, Pakistan. In *Proc. Vert. Pest Conf.*, 9, 130–134, 1980.

Lagler, K. F., The control of fish predators at hatcheries and rearing stations. *J. Wildl. Manage.*, 3, 169–179, 1939.

LaGrange, T. G., J. L. Hansen, R. D. Andrews, A. W. Hancock, and J. M. Kienzler, Electric fence predator exclosure to enhance duck nesting: a long-term case study in Iowa. *Wildl. Soc. Bull.*, 23, 256–260, 1995.

Laidlaw, G. W., H. Blokpoel, V. E. F. Soloman, and M. McLaren, Gull exclusion. In *Proc. Vert. Pest Conf.*, 11, 180–182, 1984.

Laramie, H. A., A device for control of problem beavers. *J. Wildl. Manage.*, 27, 471–476, 1963.

Linhart, S. B., J. D. Roberts, and G. J. Dash, Electric fencing reduces coyote predation on pastured sheep. *J. Range Manage.*, 35, 276–281, 1982.

Lokemoen, J. T., Examining economic efficiency of management practices that enhance waterfowl production. *North Am. Wildl. Nat. Res. Conf.*, 49, 584–607, 1984.

Lokemoen, J. T., H. A. Doty, D. E. Sharp, and J. E. Neaville, Electric fences to reduce mammalian predation on waterfowl nests. *Wildl. Soc. Bull.*, 10, 318–323, 1982.

Lokemoen, J. T. and T. A. Messmer, Locating, constructing, and managing islands for nesting waterfowl. *Jack Berryman Institute Publication 5*, Utah State University, Logan, 1993.

Lokemoen, J. T. and R. O. Woodward, An assessment of predator barriers and predator control to enhance duck nest success in peninsulas. *Wildl. Soc. Bull.*, 21, 275–282, 1993.

Maehr, D. S., Beekeeping enters the solar age. *Am. Bee J.,* 122, 280–281, 1982.

Marsh, R. E., A. E. Koehler, and T. P. Salmon, Exclusionary methods and materials to protect plants from pest mammals — a review. In *Proc. Vert. Pest Conf.*, 14, 174–180, 1990.

Mayer, P. M. and M. R. Ryan, Electric fences reduce mammalian predation on piping plover nests and chicks. *Wildl. Soc. Bull.*, 19, 59–63, 1991.

McKnight, T. L., Barrier fencing for vermin control in Australia. *Geogr. Rev.,* 59, 330–347, 1969.

Melvin, S. M., L. H. MacIvor, and C. R. Griffin, Predator exclosures: a technique to reduce predation at piping plover nests. *Wildl. Soc. Bull.*, 20, 143–148, 1992.

Morris, R. D., H. Blokpoel, and G. D. Tessier, Management efforts for the conservation of common tern (*Sterna hirundo*) colonies in the Great Lakes: two case histories. *Biol. Conserv.*, 60, 7–14, 1992.

Nass, R. D. and J. Theade, Electric fences for reducing sheep losses to predators. *J. Range Manage.*, 41, 251–252, 1988.

Nol, E. and R. J. Brooks, Effects of predator exclosures on nesting success of killdeer. *J. Field Ornithol.*, 53, 263–268, 1982.

Owen, J. T., J. B. Armstrong, H. L. Stribling, and M. K. Causey, An evaluation of Max Flex Fast Fence® for reducing deer damage to crops. In *Proc. Eastern Wildl. Damage Control Conf.,* 6, 98–101, 1995.

Palmer, W. L., J. M. Payne, R. G. Wingard, and J. L. George, A practical fence to reduce deer damage. *Wildl. Soc. Bull.*, 13, 325–327, 1985.

Patterson, I. J., The control of fox movement by electric fencing. *Biol. Conserv.*, 11, 267–278, 1977.

Pietz, P. J. and G. L. Krapu, Effects of predator exclosures design on duck brood movements. *Wildl. Soc. Bull.*, 22, 26–33, 1994.

Pochop, P. A., R. J. Johnson, D. A. Agueron, and K. M. Eskridge, The status of lines in bird damage control — a review. In *Proc. Vert. Pest Conf.*, 14, 317–324, 1990.

Porter, W. F., A baited electric fence for controlling deer damage to orchard seedlings. *Wildl. Soc. Bull.*, 11, 325–327, 1983.

Post, W. and J. S. Greenlaw, Metal barriers protect near-ground nests from predators. *J. Field Ornithol.*, 60, 102–103, 1989.

Rimmer, D. W. and R. D. Deblinger, Use of predator exclosures to protect piping plover nests. *J. Field Ornithol.*, 61, 217–223, 1990.

Saenz, D., C. S. Collins, and R. N. Conner, A bark-shaving technique to deter rat snakes from climbing red-cockaded woodpecker cavity trees. *Wildl. Soc. Bull.*, 27, 1069–1073, 1999.

Singleton, G. R., Sudarmaji, Jumanta, T. Q. Tan, and N. Q. Hung, Physical control of rats in developing countries. P. 178–198 in G. R. Singleton, L. Hinds, H. Leirs and Z. Zhang, Eds., *Ecologically-Based Rodent Management*. Australian Centre for International Agriculture Research Monograph 59, Canberra, Australia, 1999.

Stucky, J. T., Use of plastic netting. In *Proc. Bird Control Sem.*, 6, 195–197, 1973.

Taber, M. R. and L. R. Martin, The use of netting as a bird management tool in vineyards. In *Proc. Vert. Pest Conf.*, 18, 43–45, 1998.

Wade, D. A., The use of fences for predator damage control. In *Proc. Vert. Pest Conf.*, 10, 24–33, 1982.

Wilson, C. G., Badger damage to growing oats and an assessment of electric fencing as a means of its reduction. *J. Zoo.*, 231, 668–675, 1993.

Wood, G. W. and L. A. Woodward, The Clemson beaver pond leveler. In *Proc. Ann. Conf. Southeastern Assoc. Fish Wildl. Agencies*, 46, 179–182, 1992.

CHAPTER 14

Habitat Manipulation

A public meeting to discuss the optimal size of the local deer population became quite heated with farmers arguing for fewer deer and hunters wanting more. Finally one old farmer complained, "If there were a deer behind every single tree in the county, the hunters would still not be satisfied. They would simply argue for more trees."

Anonymous

Many human–wildlife conflicts can be avoided or reduced by making structural or habitat changes. This can be achieved by 1. changing the resource itself or the way it is managed, 2. modifying the habitat at the site where the resource is located (throughout this chapter, I refer to the object we are trying to protect as the resource), or 3. making changes to the surrounding landscape (Figure 14.1). Potential changes to the resource and its management include planting crop species or varieties that are less susceptible to wildlife damage, changing the time a crop is planted or harvested, altering animal husbandry practices to reduce predation, and designing buildings to be bird- or mammal-proof. Habitat changes include altering groundcover in forest clear-cuts or orchards and modifying surrounding habitat to make it less attractive to wildlife. A landscape ecology approach can be used to decide what crop to plant at a specific location or to alleviate a human–wildlife conflict by managing wildlife refugia or not allowing birds to roost in the vicinity.

REDUCING HUMAN–WILDLIFE CONFLICTS BY MODIFYING THE RESOURCE

Growing Unpalatable Plant Species to Reduce Wildlife Damage

As we discussed in Chapter 11, some plant species suffer greatly from herbivory while others are rarely browsed. Timber producers can minimize wildlife damage problems by planting clear-cuts with unpalatable tree species. In Great Britain, fallow

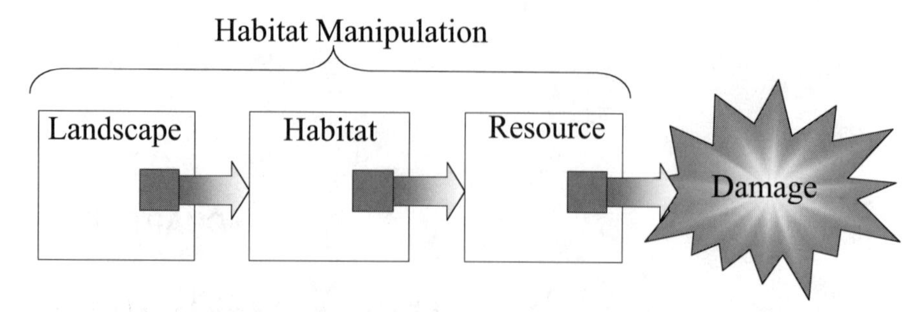

Figure 14.1 Human–wildlife conflicts can be reduced by changing the resource, modifying the resource's habitat, or making changes to the surrounding landscape.

deer commonly damage cherry and rowan trees but not poplar or sycamore trees (Moore et al. 1999). White-tailed deer exhibit a strong feeding preference for some conifers (e.g., Austrian pine and Douglas fir) over others (e.g., Scotch pine and white spruce). Hence, Christmas tree growers worried about deer browsing should consider planting spruce trees (Craven and Hygnstrom 1994).

Farmers can reduce the threat of wildlife damage by planting a crop that does not interest wildlife. For instance, birds prefer to eat wheat rather than barley. Therefore, growing the latter might be a better choice for people who farm next to a blackbird roost. Deer prefer to browse apple trees rather than pear trees (Craven and Hygnstrom 1994). Hence, an apple orchard with severe deer damage could instead be planted with pears to alleviate the problem. Homeowners experiencing deer damage to their landscape plants could reduce the problem by landscaping their yards with unpalatable plants. For instance, white-tailed deer readily browse Japanese holly but not American holly, which is unpalatable (Craven and Hygnstrom 1994).

In deciding which species are unpalatable, it is important to identify which wildlife species pose threats because food preferences vary among wildlife species. For example, soybeans are unpalatable to most birds. Consequently, growing soybeans is a wise choice for a farmer living near a large blackbird roost. However, deer readily eat soybean plants, so this would be a poor selection in areas where deer densities are high.

Growing Cultivars or Varieties Less Susceptible to Wildlife Damage

In many cases, farmers or timber growers are unable to grow unpalatable species because these species do not grow well on their farms or are not as valuable as other plants. When farmers cannot shift to an unpalatable crop, they may still be able to reduce wildlife damage by selecting a subspecies, variety, or cultivar of a palatable species which has chemical or morphological traits that make it less attractive to wildlife (Doggett 1957; Bullard 1988). Chemical traits include the presence of substances that give the crop an unpleasant taste or reduce its nutritional value (Chapter 11). For instance, deer find some cultivars of Japanese yews highly palatable but not others (Figure 14.2).

Figure 14.2 Deer browsing on two cultivars of Japanese yew (the cultivar on the right has fewer chemical defenses and has suffered more browsing).

Morphological traits reduce the feeding efficiency of wildlife by making the food source difficult to access or digest. Apple trees provide a great example of how morphological traits can affect the vulnerability of cultivars to wildlife damage because apple trees come in three sizes. There are dwarf trees, which are less than 3 m in height, standard-sized trees, which can reach up to 10 m, and semidwarfs, which are intermediate in height. In recent years, dwarf apple trees have become popular with growers because these trees yield well, reach maturity in just a few years, and are easy to harvest (no ladders required). The drawback with dwarf apple trees is that they never outgrow the reach of a hungry deer, whereas deer cannot reach most apples on a standard-sized tree. Hence, taller trees are a better option for orchards with a history of deer damage.

For several crops, there are varieties which rely upon a combination of chemical and morphological traits to gain protection from bird damage. These include grain sorghum, corn, and sunflowers.

Grain sorghum

Worldwide, sorghum is probably the crop most susceptible to bird damage. For instance, quelea can devastate sorghum fields in Africa and cause a complete crop failure in local areas (Bruggers 1980). However, there are three types of grain sorghum, and all differ in the presence or absence of tannins at different stages of maturity. Type I sorghums do not contain tannins, Type II sorghums only contain tannins in the immature stages, and Type III sorghums contain tannins in both the immature and mature stages (Bullard 1988). The presence of tannins in grain makes sorghum unattractive to birds but also reduces the palatability and nutritional quality of the grain for humans and livestock. Hence, Type III sorghums suffer less from bird damage but also sell for less because of their lower quality. Therefore, Type II sorghums are the best candidates for bird-resistant varieties, because they provide protection from bird damage during the immature stages but have similar nutritional qualities as Type I when ripened (Bullard 1988).

Some grain sorghum varieties also possess morphological characteristics which reduce their vulnerability to birds. Birds need to perch on the sorghum plants to reach the seed heads, so any plant characteristic that makes perching on the plants difficult or that makes the seeds harder to remove from the head will provide some protection against bird damage. Birds do not like to eat sorghum varieties that have long awns, thick husks, or hard seeds (Bullard and York 1985; Bullard 1988).

Corn

Corn hybrids with heavier husks that extend farther beyond the topmost kernels on the cobs were more bird-resistant than other varieties when fed to captive blackbirds (Dolbeer et al. 1982, 1984, 1986a). In field tests, bird damage varied among sweet corn varieties and was related to husk characteristics (Dolbeer et al. 1988), but this was not true for field corn (Dolbeer et al. 1984; Bollinger and Caslick 1985).

One problem with bird-resistant varieties of corn is that they may not produce as high a yield (in the absence of damage) as the varieties farmers normally grow. An experiment by Dolbeer et al. (1986b) in Ohio illustrated this problem. They selected fields that had suffered serious blackbird damage in previous years and compared bird damage and final yield for several corn hybrids: a bird-resistant variety, a high-yield variety susceptible to bird damage, and a hybrid of the farmer's choosing. The bird-resistant variety suffered less bird damage, but it still yielded less than the other varieties. Thus, its use would not have been warranted, even when farmers believed they were experiencing severe blackbird damage.

Sunflowers

Morphological characteristics that make it difficult for birds to reach the seeds or to open them increase bird resistance. In sunflowers, these characteristics include flat or concave heads, heads facing toward the ground, head-to-stem distances greater than 15 cm, long bracts surrounding the head, long chaffs, and tough hulls (Posey at al. 1982; Fox and Linz 1983; Parfitt 1984; Seiler and Rogers 1987; Mah et al. 1990; Mah and Nuechterlein 1991). Confectionery types of sunflowers, with heavy hulls and lower oil content, suffer less damage than oilseed types, which have thinner hulls and higher oil content (Bullard 1988). Sunflower seeds with purple hulls suffer less bird damage than others because they contain greater concentrations of anthocyanin, which is a taste repellent (Mason et al. 1989).

Two bird-resistant varieties of sunflower have been developed. In field tests, they suffered less bird damage than commercial varieties, especially when damage was caused by goldfinches rather than blackbirds (Dolbeer et al. 1986c). However, these bird-resistant varieties produced lower yields than commercial varieties.

Silvicultural Techniques to Reduce Wildlife Damage to Timber Production

One way to alleviate wildlife damage to tree seedlings is to speed up their growth rate so they will quickly outgrow the reach of deer and elk. This is accomplished by planting older, larger, and more vigorous seedlings (Hartwell 1973). Fertilizing the seedlings also increases their growth rate.

In some areas, black bears are a threat to timber production because of their propensity to strip the vascular tissues from 15- to 40-year-old trees during the spring, when such tissues have a high sugar content (Schmidt and Gourley 1992). This type of damage kills some trees and reduces the growth of others. Bears usually attack the most vigorous trees growing in the most productive stands, making the damage costly to timber producers (Nolte et al. 1998). Bear damage tends to be higher in thinned stands and after fertilization (Mason and Adams 1989; Nelson 1989; Schmidt and Gourley 1992), though results vary. In Oregon, a survey of young Douglas fir stands showed twice as much bear damage in thinned stands as in unthinned stands in the Cascade Mountains, but no difference between thinned and unthinned stands in the Coastal Mountain Range. At one site in Montana, bear damage to western larch was heaviest in stands with low densities of trees, and bears selected the largest trees to damage (Schmidt 1987, 1989). Elsewhere in Montana, Mason and Adams (1989) found bear damage to be five times higher in thinned stands than in unthinned ones.

Bears prefer a diet with high concentrations of sugars and low concentrations of terpenes. Nolte et al. (1998) showed that thinning increases the growth rate of trees and the concentration of sugar in the vascular tissues, while having little effect on terpene concentration (Kimball et al. 1998a). In contrast, pruning branches from the main trunks of trees to increase the value of the tree for lumber increases the concentration of terpenes in the tissue (Kimball et al. 1998b; Nolte et al. 1998); therefore pruned trees are less palatable. These findings help explain why bear damage occurs mainly in thinned stands and why unpruned trees are several times more likely to suffer damage than pruned trees (Kimball et al. 1998b; Nolte at al. 1998).

Changing Husbandry Practices to Reduce Predation on Livestock

Sheep farmers can change their operations to reduce the vulnerability of their livestock to predators (Sidebar 14.1). Keeping a herder with the sheep when they are on the range and bedding the sheep for the night close to the herder's camp reduces predation (Klebenow et al. 1978). Livestock confined to pastures suffer less predation than those that are on the open range. Enclosing pastures in predator-proof fences reduces predation even further (Linhart 1983; Knowlton et al. 1999). Shed lambing reduces losses to coyote predation, but this practice increases the cost of labor, feed, and shed construction (Wagner 1988). Sheep breeds also differ in size, aggressiveness, alertness, group cohesiveness, dispersion while grazing, and intensity of maternal protection. All of these characteristics influence the vulnerability of a particular breed to coyote predation (Gluesing et al. 1980; Knowlton et al. 1999).

Sidebar 14.1 Husbandry Techniques to Reduce Predation on Livestock

Robel et al. (1981) compared sheep losses at more than 100 sheep operations in Kansas. Sheep losses to predators were not equally distributed among producers; 22% of cooperators suffered over 80% of the losses, while 46% of cooperators did not lose a single sheep to predators during the 15-month study. Coyote predation was less at farms that removed or buried sheep carcasses, rather than just leaving them in the pastures or burning them. Sheep losses to predators were greater when lambing occurred during January to March rather than during October to December. Predation was less at farms where sheep were confined at night. Losses to coyotes were much lower in lighted corrals than in corrals without lights, but losses to dogs were somewhat higher. No sheep were lost from corrals in which at least one sheep was wearing a bell. The breed of sheep did not influence predation rates.

Most sheep and lamb losses to predators occurred in pastures (80% of losses to coyotes and 78% of losses to dogs). Predation rates were higher in larger pastures with tall grass and flat terrain. Predation by coyotes was higher in pastures with streams, but predation by dogs was lower. This may be due to differences in hunting behavior: coyotes tend to hunt along stream courses, but dogs do not (Robel et al. 1981). The authors concluded that the most effective method of reducing predation on sheep was to confine them in lighted corrals at night.

Many people have assumed that poor husbandry practices by cattle ranchers predispose some farms to wolf depredation. Fritts et al. (1992) suggested that ranchers who calve or pasture their livestock in brushy areas or fail to properly dispose of livestock carcasses are more prone to lose cattle to wolves. However, Mech et al. (2000) tested these hypotheses and found no differences between Minnesota cattle farms with chronic wolf problems and nearby cattle farms with no problems.

Agronomical Techniques to Reduce Agricultural Losses Due to Wildlife Damage

Several agronomical practices can change the vulnerability of crops to damage by small mammals. For example, no-till farming allows higher densities of voles and mice to survive in fields and increases the damage those animals cause to newly planted crops (Hygnstrom et al. 1996, 2000). In contrast, deep plowing kills voles and pocket gophers and destroys burrow systems located within 30 cm of the surface (Whisson and Giusti 1998). Mowing vineyards and orchards or using flood irrigation can also be used to reduce vole and pocket gopher densities (Sullivan and Hogue 1987; Whisson and Giusti 1998). Ground squirrel densities also are lower in agricultural fields which are frequently plowed or cultivated (Fitzgerald and Marsh 1986).

Controlling insects and weeds in agricultural fields can reduce bird damage to crops. Mesurol® (methiocarb) is an insecticide with bird-repellent properties and has been shown to reduce bird damage to sweet corn (Stickley and Ingram 1976). However, Woronecki et al. (1981) hypothesized that the reduction in bird damage might actually have resulted from a reduction in insect populations in the fields treated with Mesurol, rather than from the bird-repellent properties of the compound. To test this, they compared bird damage to sweet corn among three types of fields: fields treated with Mesurol, fields treated with Sevin® (an insecticide without bird-repellent properties), and untreated fields. Both insecticide treatments were equally effective in reducing insect populations, bird numbers, and bird damage to sweet corn. These findings support the hypothesis that fields with fewer insects were less attractive to birds. Similarly, sweet corn treated with another insecticide, Chlorpyrifos®, had fewer European corn borers and less blackbird damage than untreated cornfields (Straub 1989).

Red-billed quelea damage to ripening wheat in Tanzania is more extensive in weedy portions of fields than in nonweedy areas (Luder 1985). These birds have difficulty standing on wheat plants and find it easier to forage in fields where weeds are entangled with the wheat, providing stable perches (Luder 1985). In sunflower fields, severity of blackbird damage was related to the field's weediness (Otis and Kilburn 1988), but this was not true of cornfields (Gartshore et al. 1982; Bollinger and Caslick 1985).

Reducing Wildlife Damage by Changing Planting and Harvesting Schedules

Many birds are migratory; thus large numbers can be present in an area during some seasons and absent during others. One way to reduce crop damage by migratory species is to adjust planting schedules so that crops reach the stage in which they are vulnerable to bird damage during the period when birds are absent (Bruggers et al. 1992). For example, in Louisiana rice fields, 98% of the rice sprouts were eaten by birds in fields planted in March, but less than 15% were lost in fields planted in mid-April (Wilson et al. 1989). Several factors may be responsible for this seasonal pattern. Many blackbirds leave Louisiana between late February and April to migrate north, so fewer birds are around when late-season fields are planted in rice. Furthermore, there is an increase in the availability of alternate food sources as spring progresses, and blackbirds undergo a dietary change that reduces their damage to sprouting rice as spring progresses (Wilson et al. 1989).

Wildlife damage can sometimes be reduced by moving up the harvest date. In the Great Plains of Canada, waterfowl damage to grain fields mainly occurs in those years when wet weather keeps farmers from harvesting their crops in a timely manner. In Oklahoma, blue jay populations peak after October 1, and most blue jay damage to pecans occurs after that date. One potential solution to this problem is to harvest pecans before then, and dry the nuts in large ovens (Batcheller et al. 1984). This practice can only be justified economically if the cost of drying the pecans is less than the value of the pecans lost to blue jays.

In Florida, bird damage to early season cultivars of highbush blueberries can exceed 50% in some years (Nelms et al. 1990). Yet, late-season cultivars rarely suffer from bird damage. The reason is that large numbers of cedar waxwings winter in Florida but have left the state before late-season blueberries ripen. However, early ripening cultivars are popular with Florida growers because they ripen when no other blueberries are available; hence, they bring a premium price (Nelms et al. 1990).

Reducing the Vulnerability of Buildings to Bird Problems

Birds can cause nuisance problems or health hazards when they roost, loaf, or nest on the outside of or inside buildings because of their droppings, noise, and diseases they may carry. Many of these problems can be avoided if buildings are built without exposed ledges, rafters, I-beams, or architectural decorations that offer birds places to roost or nest. Instead, rafters should be enclosed so that birds cannot access them. Other building features (Figure 14.3) that provide access into the building or good nesting sites for birds include the following: 1. eaves; 2. vents for kitchen exhaust fans, clothes dryers, and bathrooms; 3. construction flaws, such as poor installation of siding, that leave gaps through which birds enter the wall; 4. spaces between the gutter and roof; and 5. flat or slightly pitched roofs (Geis 1976; Slater 1992).

Woodpeckers drill holes in buildings when searching either for nest sites or insect prey. Such damage is more likely to occur when the outside walls are made of wood or have a wooden veneer, or when the roof contains wooden shingles. Thus, many woodpecker problems could be avoided if other building materials were used.

Existing buildings can be made less attractive to birds by modifying ledges so their upper surface is slanted at an angle greater than 45° to prevent nesting and roosting. If ornamental features are already present, they can be screened with netting to keep birds away. Eaves, vents, and other openings can be screened with netting or hardware cloth. To prevent access by house sparrows or bats, all openings larger than 2 cm should be closed or screened. In warehouses or other high-traffic areas,

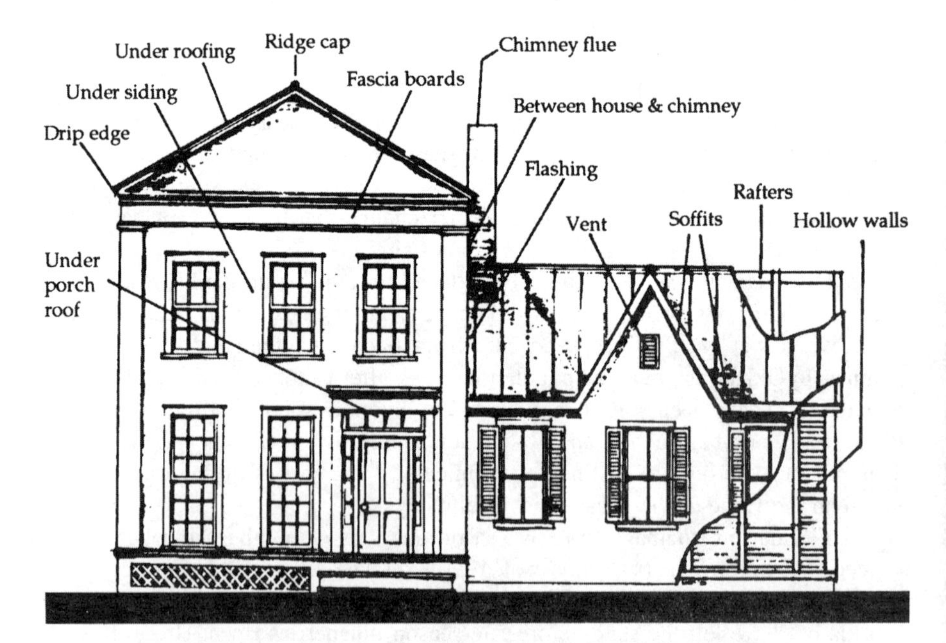

Figure 14.3 Common entry points into a home and nesting sites for birds. (From Hygnstrom et al. 1994.)

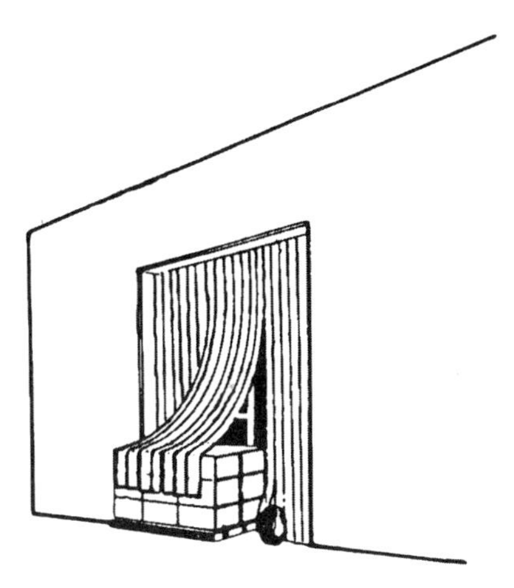

Figure 14.4 Bird damage in warehouses or other high-traffic areas can be reduced by hanging plastic or rubber strips in doorways without being too much of an obstacle to people. (From Hygnstrom et al. 1994.)

plastic or rubber strips hung in doorways are effective in preventing bird access while presenting no obstacle to humans (Figure 14.4). Overhead wires can also be used to keep birds away (Chapter 13).

The type of vegetation surrounding a building can also affect the extent of bird damage to it. For example, ivy-covered walls look appealing to people, but they also appeal to birds seeking roosting sites during winter (the wall and the ivy provide ideal thermal cover for roosting birds).

Making Buildings More Rodent-Proof

The best way to avoid rodent infestations is to design buildings to be rodent-proof. Building foundations should be deep enough to prevent rodents from burrowing beneath them. A 30-cm-long horizontal footing extension placed about 60 cm below ground level or a buried wire-mesh fence (Figure 14.5) can be used to deflect burrowing rodents away from the wall (Timm 1987; Baker et al. 1994). Rats can climb along the outsides of vertical pipes up to 7 cm in diameter. Such pipes can be equipped with metal discs or cones that act as rodent guards. To prevent rodents from climbing walls, the lower portions of walls should be covered with materials that have a hard, smooth surface. This surface needs to be higher than surrounding plants, fences or other structures that rodents can use to climb.

Common entry points of rodents into buildings include vents, drains, pipes, spaces at the base of doors, and openings in walls for electric, water, and gas lines. By gnawing on the edges to enlarge them, rats and mice can gain entry into buildings through holes as small as 1.3 cm and 0.6 cm, respectively. Thus, all holes larger than 0.6 cm should be sealed. Large holes should be covered with 0.6 cm woven hardware cloth before being plugged. Sheet metal patches can also be used (Baker et al. 1994). Vents should be covered with 0.6×0.6 cm galvanized hardware cloth.

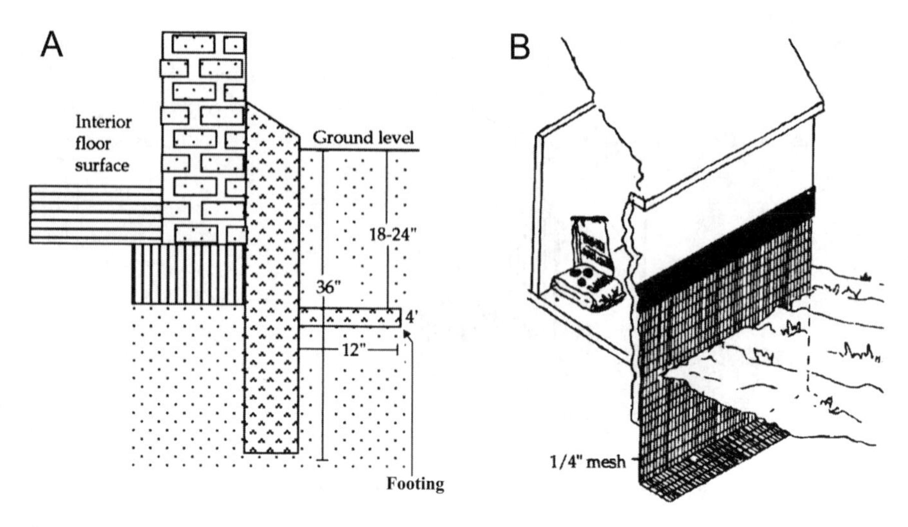

Figure 14.5 A concrete footing extending out from the wall (A) or a buried wire screen (B) can be used to keep rodents from burrowing into the basement. (From Hygnstrom et al. 1994.)

Rodents are particularly attracted to food handling and storage areas. Such areas should be kept as clean as possible and should be designed to allow cleaning beneath and behind equipment such as refrigerators, counters, and dishwashers. Food should be stored in rodent-proof containers. Warehouses should have self-closing doors; loading docks should be equipped with barriers preventing rodent access. Areas around garbage dumpsters should be kept clean, and the dumpster lids should fit tightly and be kept close. Dumpster drains should be equipped with rodent screening (Baker et al. 1994).

Weeds, woodpiles, and construction debris around buildings attract rodents by providing them with shelter and food (seeds and insects) and should be removed. Maintaining a 1-m-wide, weed-free area around buildings by close mowing or installing heavy gravel can discourage rodent burrowing.

REDUCING HUMAN–WILDLIFE CONFLICTS BY MODIFYING THE HABITAT AROUND THE RESOURCE

Habitat Modification to Minimize Wildlife Damage to Timber Production

After a timber stand is harvested, piles of dead trees and branches usually litter the site and can be either burned or left in place. One advantage of leaving the slash piles is that they protect tree seedlings from deer and elk browsing by making it more difficult for wildlife to reach the seedlings. Bergquist and Örlander (1998a) tested this in Sweden by removing slash from half of several clear-cuts while leaving it in the other half. They found that slash removal had no significant effect on browsing damage at the clear-cut level. However, seedlings growing in slash piles suffered less damage and were more vigorous. Thus, the net effect of leaving slash piles was beneficial (Bergquist and Örlander 1998b). However, mountain beaver prefer areas with woody debris and slash piles. Because damage by this species can prevent forest clear-cuts from regenerating, it may be better to remove slash piles from clear-cuts in the Pacific Northwest where mountain beaver occur, even though this makes the seedlings more vulnerable to ungulates (Van Vuren 1998).

When forest clear-cuts are replanted, the vulnerability of tree seedlings to wildlife damage is influenced by the site's groundcover. The existing vegetation can reduce the growth of tree seedlings by competing with them for sunlight, water, or nutrients so that the seedlings take longer to outgrow the reach of wildlife (Gourley at al. 1990). However, ground vegetation does not always have a detrimental effect on seedlings. In some cases, vegetation may offer visual or physical protection to the seedlings or provide an alternate food source for deer, thereby reducing browsing pressure on the seedlings (Campbell and Evans 1978).

Pocket gopher densities and their damage to tree seedlings can be reduced through vegetation management to reduce the gophers' food supply. Pocket gophers prefer succulent, fleshy roots (such as the taproots of herbaceous plants) over the fibrous roots of grasses. In Oregon, treatment with atrazine herbicide killed forbs and some grasses in the groundcover and substantially reduced Mazama pocket gopher densities and their damage to ponderosa pine seedlings (Engeman et al. 1995). Elsewhere in Oregon, an application of 2,4-D herbicide decreased forb cover and northern pocket gopher damage to lodgepole pine seedlings (Engeman et al. 1997). In both cases, herbicide was first applied two years before the tree seedlings were planted to provide time for the treatment to suppress pocket gopher populations.

Habitat Modification to Minimize Wildlife Damage to Agricultural Production

Voles cause problems in vineyards and fruit orchards by girdling the vines and trees during the winter (Servello et al. 1984), but during most of the year, voles forage on the orchard's groundcover. By creating an unpalatable groundcover, it should be possible to reduce vole densities and the damage voles cause in vineyards and apple orchards (Servello et al. 1984; Whisson and Giusti 1998). Mowing can reduce vole populations by half (Edge et al. 1995). Whisson and Giusti (1998) noted that newly planted vineyards are very susceptible to pocket gopher damage and suggested no groundcover be planted until the vines are a couple of years old and large enough to tolerate some vole damage. Woodchuck densities in orchards also can be reduced by planting the groundcover with an unpalatable plant species and by decreasing the abundance of forbs (Swihart 1990).

Most birds prefer to forage close to trees to which they can fly if threatened. For example, bird damage in sunflower fields was related to the presence or absence of adjacent trees (Otis and Kilburn 1988). Removal of isolated trees in the middle or along the edge of fields can reduce the vulnerability of the field to birds by denying them a secure loafing site.

On grain farms in Victoria, Australia, controlling weeds along fence rows and the borders of fields by mowing them or spraying them with a herbicide during the spring reduced mouse densities in late summer (Brown et al. 1998). Another management practice to reduce mouse densities is to use livestock to graze fields immediately after harvest (Brown et al. 1998).

One problem with changing the habitat to reduce wildlife damage is that the resource we are trying to protect may be adversely affected by the change. For example, wood rats cause damage in Malaysian oil palm and cocoa plantations. These rodents spend most of their time hiding in the piles of palm fronds left between the tree rows (Buckle et al. 1997). Removing frond piles by burning or shredding should reduce rat populations in plantations. However, these frond piles also reduce soil erosion, water loss through evaporation, and weed problems (Buckle et al. 1997). Hence, eliminating them may not be wise.

Manipulating Habitat to Increase an Animal's Fear of a Site

One interesting approach to reducing damage by small mammals is to attract free-ranging predators to the site. For instance, farmers with vole and bird problems erect perches in their fields to attract hawks and owls (Howard et al. 1985). In Malaysia, farmers reduced rat damage in their palm plantations and rice fields by placing nest boxes for barn owls (H. M. Noor, unpublished report). In contrast, barn owls were ineffective in reducing pocket gopher damage to orchards and vineyards in California (Moore et al. 1998; Salmon 1998).

Wildlife avoid areas where they are vulnerable to predators. Hence, one way to reduce the density of a particular wildlife species is to change the habitat so that the wildlife culprits are more vulnerable. For example, voles are vulnerable to raptors when vegetation is sparse or absent (Baker and Brooks 1982). One way to reduce vole densities in orchards would be to remove cover by mowing or burning. In contrast, ground squirrels and prairie dogs rely on their vision to detect predators, and some ground squirrels avoid areas where thick or tall vegetation obstructs their view (Fitzgerald and Marsh 1986; Cable and Timm 1988). Several scientists have tested whether visual barriers can be used to keep prairie dogs from expanding their colonies, but they obtained mixed results (Hygnstrom 1996; Sidebar 14.2).

Sidebar 14.2 Using Visual Barriers to Control Objectionable Prairie Dog Colonies

Prairie dogs live in colonies in open habitat and depend on unobstructed visibility to be able to detect predators or trespassers at a distance and to keep in visual contact with other members of their family. Sometimes, a colony may become so large that landowners or neighbors seek to curtail its expansion. Franklin and Garrett (1989) hypothesized that prairie dogs would not colonize areas where their view was obstructed and that it may be possible to control a prairie dog colony's expansion by erecting visual barriers. They tested this using two colonies of prairie dogs in Wind Cave National Park, SD: a small, recently established colony and a large, old one. Along the east edge of the small colony, they built a series of three parallel visual barriers made of burlap cloth attached to steel stakes, perpendicular to the direction of expansion of the colony. A similar area along the west edge of the colony was used as control. At the large colony, four sets of three parallel visual barriers made of 1 to 2 m high ponderosa pines were erected along the edge of the colony. Each set of barriers was associated with adjacent and distant control plots, also located at the edge of the colony.

Visual barriers were successful in reducing prairie dog activity and colony expansion in the areas where they were erected. In the small colony, prairie dog activity in the area with the barriers was reduced to one third of its original level after two months, whereas activity in the control area almost doubled during the same period of time. In the large colony, prairie dog activity in the plots with the barriers and in the plots located just outside the barriers was significantly less than in distant control plots (Franklin and Garrett 1989). However, these barriers required considerable maintenance, especially when elk and bison rubbed against them during the rut. No livestock were present in the area, but livestock damage to the barriers could be a problem on rangelands. Because most colony expansion occurs in the three months after juvenile prairie dogs emerge in May, barriers only have to be maintained during this period (Franklin and Garrett 1989).

Canada geese forage in open fields or lawns where they have an unobstructed view of their surroundings (Figure 14.6). In such areas, they are safe from mammalian predators because if a predator tries to approach, the geese will detect it soon enough to fly away. Thus, one way to discourage urban Canada geese from using a site where they are unwanted is to landscape the site with trees, bushes, hedges, boulders, or anything that geese would have difficulty seeing around (Figure 14.7). Geese also prefer to forage on lawns which are close to a pond or lake to which they can flee if they see a mammalian predator. Such sites can also be made less inviting to geese by draining the pond or blocking access to it with fences or overhead wires (Chapter 13).

Figure 14.6 A typical site where Canada geese like to forage because nothing impedes their view of any approaching danger.

Figure 14.7 A pond landscaped to be unattractive to Canada geese (trees, boulders, and benches all obstruct a goose's view of its surroundings).

REDUCING HUMAN–WILDLIFE CONFLICTS AT THE LANDSCAPE LEVEL

The amount of wildlife damage which occurs at any site depends, in part, on what the landscape and land-use patterns are in the broad area surrounding the site. In some cases, we can manipulate the landscape to resolve human–wildlife conflicts. One reason why airports experience bird problems is due to zoning regulations. Cities often place airports and garbage dumps close to each other and locate both of them in flood plains or wetlands where there are few homes. (Most people do not want either a garbage dump or an airport in their neighborhood.) Unfortunately, both wetlands and garbage dumps attract birds and will increase bird densities at nearby airports and the danger of bird–aircraft collisions.

Predation on livestock is influenced by the landscape surrounding ranches and farms. Robel et al. (1981) reported that sheep farms located near towns had fewer problems with coyotes than rural farms but experienced more problems with dogs. Keeping sheep flocks in areas where there is human activity can also help reduce predation by coyotes (Knowlton et al. 1999).

In the northeastern U.S., deer damage in plant nurseries is correlated with the size of adjacent woodlots (Conover 1989). For this reason, palatable plant species are best grown in fields located away from woodlots, while fields near large woodlots should be used to grow unpalatable plant species. Van Vuren (1998) suggested that agricultural damage by white-tailed deer could be reduced by managing the vegetation in adjacent woodlots so these woodlots do not provide good cover for deer. This could be accomplished, for instance, by allowing the trees to grow tall and the forest canopy to close, so the understory vegetation is shaded out (Figure 14.8).

Figure 14.8 Deer are uncomfortable in woodlots where there is no understory vegetation or other vegetation that they can use for cover. (Photo courtesy of Mark McClure.)

In Great Britain, squirrel damage to timber stands is related to the presence of mast-bearing trees in adjacent areas (Kenward et al. 1996), probably because these trees allow higher densities of squirrels to survive by providing a winter food source. Squirrel damage could presumably be reduced by eliminating mast-bearing trees from the area. As another example, forest clear-cuts attract pocket gophers, which then kill the young trees. Leaving buffer strips 150 to 300 m wide between areas occupied by gophers and clear-cuts can prevent or delay pocket gopher invasions of the areas to be planted. The buffers can be logged once the newly planted trees grow past the stage where they are most vulnerable to gopher damage (Marsh and Steele 1992).

Avoiding Damage by Clustering Vulnerable Resources Together

In New England, it is not uncommon for blackbirds to reduce corn yields by 25%, whereas bird damage rarely exceeds 2% in the Midwest. The main reason for this difference is the "economics of scale." In New England, forests dominate the landscape and cornfields tend to be widely scattered (Figure 14.9). Hence, all of the blackbirds within 10 km of a New England cornfield may be foraging in that one field because it may be the only one in the area. In the Midwest corn belt, there can be thousands of hectares of cornfields within a similarly sized area. Therefore, blackbird damage will be spread out over hundreds of fields in the corn belt, but this will not be true in New England. The economics of scale principle is based on the fact that, as production increases, the proportion of it that is damaged by wildlife declines. Thus, one way to reduce wildlife damage is to concentrate vulnerable crops in a single area.

In addition to clustering vulnerable fields in one area, wildlife damage often can be reduced by clustering them by time. Most crops are vulnerable to wildlife damage

Figure 14.9 In New England, forests dominate the landscape and agricultural fields tend to be small and isolated, making them more vulnerable to wildlife damage than areas in the Midwest where most of the land is in agricultural production.

only during a particular growth stage. For instance, cornfields are vulnerable to birds when first planted (birds eat the planted seeds) and again when the ears are maturing. One way to minimize damage during these two periods is to plant all of the fields in the same area at the same time. One problem with this approach, however, is that farmers normally prefer to space out their fields in time because it is difficult to harvest them all at once. By spacing them out, farmers can diversify their risks; that is, they can reduce the danger that a single weather event (such as a hail storm) will destroy all their fields.

Avoiding Damage by Increasing Field Size

The size of a field has a great impact on the severity of wildlife damage. Usually, the absolute amount of damage caused by wildlife increases slightly as the size of a field increases, while the proportion of the crop that is damaged decreases as field size increases, because the perimeter of the field increases more slowly than its area. Consider a blueberry field where the ripe berries attract all of the mockingbirds, American robins, and northern orioles nesting within 200 m of it. Let us assume that 100 birds foraging in the field remove 100 kg of berries during the growing season. If the blueberry field yields 400 kg per ha, then the birds would remove 25% of the blueberries from a 1 ha field. In contrast, a 100 ha field would attract 450 birds, which would remove 450 kg of berries, representing only 1% of the field's total yield (Figure 14.10).

The same principle applies to predation on livestock and farm animals. One way to think about this is that, if a raccoon breaks into a chicken house one night, the number of chickens it kills will be the same regardless of whether the chicken coop

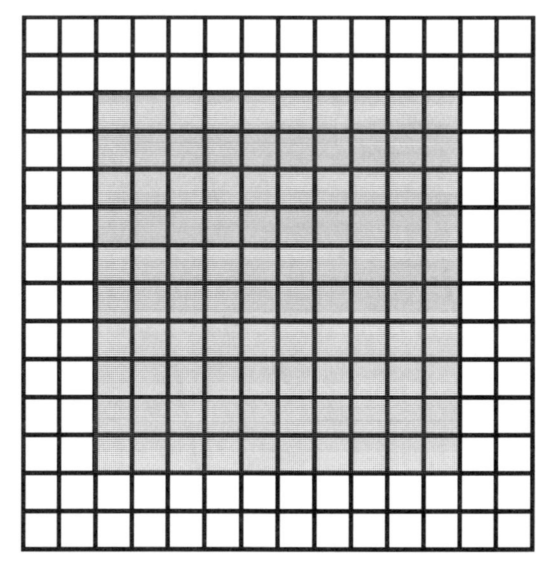

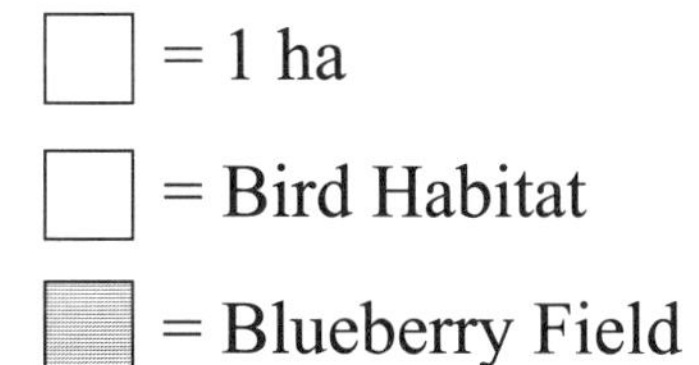

Figure 14.10 If songbirds within 200 m of a blueberry field forage in it, then the total amount of blueberries consumed by birds increases as the field increases in size, but the proportion of the total crop they consume declines.

contains 100, 1000, or 1 million chickens. Several authors have noted this economics of scale principle. For instance, Robel et al. (1981) found that large sheep flocks suffered a smaller percentage of losses to predators than small flocks, even though the number of sheep killed was higher in larger flocks.

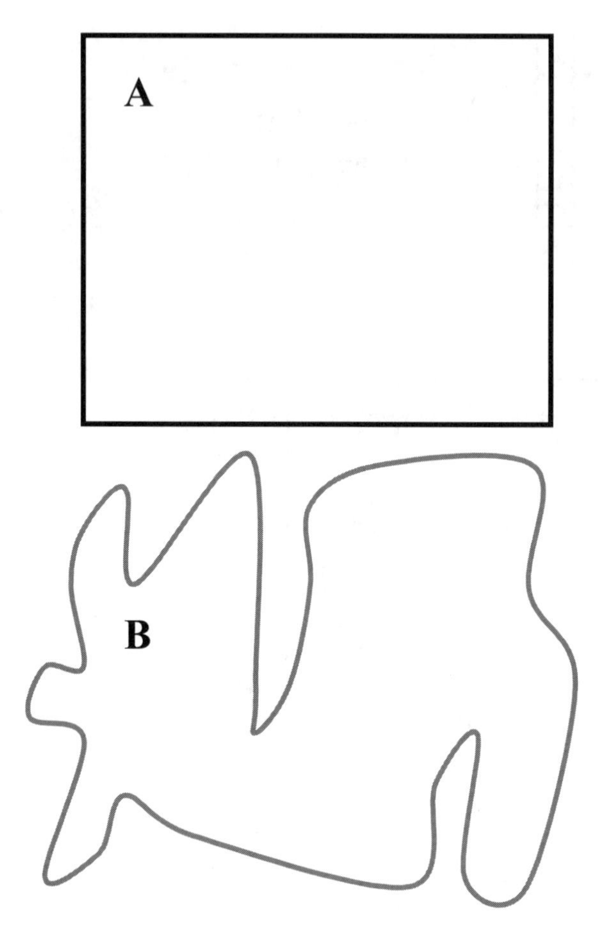

Figure 14.11 Fields with straight edges (A) are less vulnerable to deer damage than fields with a convoluted edge (B) because a higher proportion of the convoluted field is located close to the field's edge.

Another factor influencing the level of wildlife damage is the shape of the agricultural field. Most animals, such as deer and elk, do not like to venture too far from secure cover when foraging. For this reason, most wildlife damage occurs along the edges of a field, rather than in its center. Fields with a straight edge will be less vulnerable to wildlife damage than those that have a convoluted edge (Figure 14.11).

Reducing Damage by Managing Distant Bird Roosts

Sometimes, human–wildlife conflicts are best addressed at distant sites that serve as the source for the wildlife culprits. One example of this principle is reducing bird damage by preventing birds from roosting close to the resource. Blackbird damage to corn and sunflowers tends to be higher in the vicinity of blackbird roosts (Dolbeer 1980; Linz et al. 1996). For instance, Bollinger and Caslick (1985) showed that blackbird damage in 68 cornfields in New York decreased exponentially with distance from a large blackbird roost located in Montezuma National Wildlife Refuge. None of the fields located more than 6 km from the roost received more than 1% damage, whereas fields located 2 to 4 km from the roost received up to 29% damage. In Ohio, Dolbeer (1981) observed losses of 5 to 16% in cornfields located less than 7 km from blackbird roosts, whereas fields located more than 16 km from roosts received less than 2% damage. In the southeastern U.S., the level of blackbird damage at livestock operations was related to their proximity to large blackbird roosts (Glahn and Otis 1986).

For birds that are central-place foragers, it is not surprising there is a close relationship between the severity of bird damage and proximity to a roost or nesting colony. Optimal foraging theory predicts that birds will not waste either time or energy flying to a distant food source if there is a closer one (Chapter 12). Scientists have taken this observation a step further and found that they can reduce bird damage to sunflowers by forcing birds to abandon a nearby roost for one further away (Linz et al. 1992, 1996). The same approach has been used to reduce the threat of bird–aircraft collisions by discouraging gulls from nesting near airports (Dolbeer et al. 1993).

Birds can be forced from roosts using distress or alarm calls, propane exploders, and/or harassment (Chapter 10). This can also be accomplished by thinning out the vegetation at the roost site. When roosts are located in marshes, this can be achieved by killing the cattails with a herbicide (Chapter 16). Roosts in woodlots can be opened up by thinning out the trees (Sidebar 14.3).

Sidebar 14.3 Management of Objectionable Bird Roosts

Lyon and Caccamise (1981) studied the characteristics of communal blackbird and starling roosts in central New Jersey. Most of the roosts were located in hardwood stands, and the authors compared woodlots used for roosts with other woodlots which were avoided by roosting blackbirds. Roost sites were located in discrete patches of vegetation rather than large forests. They were typically dense stands of trees in early successional stages (woodlots with 21- to 26-year-old trees). In contrast, woodlots not used for roosts were older (mean age was 41 years). The species of tree did not seem to be as important as having a closed canopy.

Possible ways to discourage blackbirds from roosting at a particular site include clear-cutting the tree stand, waiting for the stand to get old enough that it becomes unsuitable for roosting, or opening up the canopy through heavy pruning or stand thinning (Lyon and Caccamise 1981). In Ohio, a thinning of 30 to 50% was sufficient to disperse most roosts, but the practice may have to be continued every few years to prevent the canopy from closing again.

Crows roosting in towns can be a nuisance because of their droppings and noise; they also can pose a health hazard. Gorenzel and Salmon (1995) studied the characteristics of American crow roosts in the town of Woodland, CA. Crows preferred to roost in deciduous trees located in residential areas during summer but used more evergreen trees in commercial areas during winter, probably because temperatures tended to be higher in the canopies of such trees. Most roost trees were located over asphalt or concrete. The authors recommended that tree species often used as roosts by crows, such as ashes, sycamores, mulberries, elms, and conifers, not be planted in parking areas. They also suggested that pruning branches from roost trees to open up their canopies may make these trees less attractive as roosts.

One drawback to the approach of trying to resolve human–wildlife conflicts by forcing birds to abandon nearby roosts is that once they abandon one site we have little ability to get them to move to the site of our choosing. Hence, we may chase birds from one roost, only to have them roost the next night at a site even closer to the resource we are trying to protect. Of course, we can keep forcing the birds to leave their newly selected roost site until we are finally satisfied with the result. This approach works best when suitable roost sites are few and far between. Flynt and Glahn (1995) proposed that, when there is no roost site suitable to both birds and people, we can select an optimal location for a roost from society's standpoint and then create a bird roost there by planting a dense stand of vegetation to attract blackbirds (see Sidebar 12.2).

Reducing Damage by Managing Distant Refugia

Some wildlife populations fluctuate widely depending upon weather or habitat conditions. During poor conditions, these species may only be able to survive in small refugia that provide the food or security they need. When conditions improve, wildlife venture out of these refugia and repopulate the landscape. When these species cause human–wildlife conflicts, one approach is to suppress the wildlife populations when they are at their lowest level and the animals are concentrated in the refugia. For example, when sugarcane is planted, the land is first prepared by burning the residual vegetation, followed by repeated cultivation. During this period, the canefield rat must leave the field and take refuge in noncrop areas (e.g., swamps, woodlots). Because of this, the extent of rat damage to sugarcane is related to the distribution and location of suitable refuges (Whisson 1996). By removing these refugia, densities of canefield rats can be suppressed in the surrounding sugarcane fields.

This approach was used to manage rabbit populations in semiarid regions of Australia, where habitat conditions fluctuate with rainfall (Parker et al. 1976). In wet years, rabbits spread out over large areas, but during droughts, only a few warrens are active, and they are near drainage channels and swamps. Destruction of these warrens can make it impossible for rabbits to survive during droughts and permanently lower populations over wide areas (Sidebar 14.4).

Sidebar 14.4 Reducing Rabbit Population by Destroying Their Refugia

During a drought in 1965, an effort to control rabbits through the systematic destruction of warrens was started on a large property (about 21,000 ha) in western New South Wales (Parker et al. 1976). Earlier attempts to bulldoze warrens and then refill the excavations were unsuccessful because rabbits from adjacent areas recolonized the area and easily dug new warrens in the loose soil. Efforts to bulldoze warrens and leave the soil in mounds next to the excavations were equally unsuccessful: rabbits simply created new warrens in the mounds. However, when warrens were destroyed using a ripper and a crawler tractor, the practice was more successful. Hence, this technique was used for a campaign to eradicate rabbits. Surveys were started in 1965 to document changes in rabbit populations. By the end of the eradication campaign in 1974, 1140 warrens had been ripped in an area of 15,000 ha and only 13 warrens with a total of 16 entrances remained. This corresponded to a density of one warren and one active entrance per 1000 ha of suitable habitat. In contrast, untreated neighboring properties had a mean density of 31 warrens and 367 active entrances per 1000 ha (Parker et al. 1976).

SUMMARY

Sometimes, human–wildlife conflicts can be avoided by making structural changes to the resource or the surrounding area. This can be achieved by 1. changing the resource itself or the way it is managed, 2. modifying the habitat at the site where the resource is located, or 3. making changes to the surrounding landscape. Structural changes to the resource include the use of species or varieties that are less susceptible to wildlife damage, changes in the timing of planting or harvesting date for a crop, and altering animal husbandry practices to reduce predation. Wildlife damage can be reduced by making habitat changes near the resource. Some of these changes can include altering the groundcover to reduce wildlife densities or changing the surrounding habitat so that wildlife will be more vulnerable to predation and ill at ease when in the area. A landscape approach to reducing human–wildlife conflicts might involve clustering vulnerable resources together in time and space, growing crops in large fields with straight edges, not allowing birds to roost nearby, and removing wildlife refugia so that the wildlife subpopulation will collapse during droughts or periods when habitat conditions have deteriorated.

LITERATURE CITED

Baker, J. A. and R. J. Brooks, Impact of raptor predation on a declining vole population. *J. Mammal.*, 63, 297–300, 1982.

Baker, R. O., G. R. Bodman, and R. M. Timm, Rodent-proof construction and exclusion methods. P. B137–B150 in S. E. Hygnstrom, R. M. Timm, and G. E. Larson, Eds., *Prevention and Control of Wildlife Damage*. University of Nebraska Cooperative Extension, Lincoln, 1994.

Batcheller, G. R., J. A. Bissonette, and M. W. Smith, Towards reducing pecan losses to blue jays in Oklahoma. *Wildl. Soc. Bull.*, 12, 51–55, 1984.

Bergquist, J. and G. Örlander, Browsing damage by roe deer on Norway spruce seedlings planted on clearcuts of different ages: 1. Effect of slash removal, vegetation development, and roe deer density. *For. Ecol. Manage.*, 105, 283–293, 1998a.

Bergquist, J. and G. Örlander, Browsing damage by roe deer on Norway spruce seedlings planted on clearcuts of different ages: 2. Effect of seedling vigor. *For. Ecol. Manage.*, 105, 295–302, 1998b.

Bollinger, E. K. and J. W. Caslick, Factors influencing blackbird damage to field corn. *J. Wildl. Manage.*, 49, 1109–1115, 1985.

Brown, P. R., G. R. Singleton, D. A. Jones, and S. C. Dunn, The management of house mice in agricultural landscapes using farm management practices: an Australian perspective. In *Proc. Vert. Pest Conf.*, 18, 156–159, 1998.

Brugger, K. E., R. F. Labisky, and D. E. Daneke, Blackbird roost dynamics at Millers Lake, Louisiana: implications for damage control in rice. *J. Wildl. Manage.*, 56, 393–398, 1992.

Bruggers, R. L., The situation of grain-eating birds in Somalia. In *Proc. Vert. Pest Conf.*, 9, 5–16, 1980.

Buckle, A. P., T. H. Chia, M. G. P. Fenn, and M. Visvalingam, Ranging behaviour and habitat utilisation of the Malayan wood rat (*Rattus tiomanicus*) in an oil palm plantation in Johore, Malaysia. *Crop Protect.*, 16, 467–473. 1997.

Bullard, R. W., Characteristics of bird-resistance in agricultural crops. In *Proc. Vert. Pest Conf.*, 13, 305–309, 1988.

Bullard, R. W. and J. O. York, Breeding for bird resistance in sorghum and maize. P. 193–222 in *Progress in Plant Breeding*. Butterworths, London, 1985.

Cable, K. A. and R. M. Timm, Efficacy of deferred grazing in reducing prairie dog reinfestation rates. In *Proc. Great Plains Wildl. Damage Control Workshop*, 8, 46–49, 1988.

Campbell D. L. and J. Evans, Establishing native forbs to reduce black-tailed deer browsing damage to Douglas-fir. In *Proc. Vert. Pest Conf.*, 8, 145–151, 1978.

Conover, M. R., Relationships between characteristics of nurseries and deer browsing. *Wildl. Soc. Bull.*, 17, 414–418, 1989.

Conover, M. R. and G. S. Kania, Characteristics of feeding sites used by urban–suburban flocks of Canada geese in Connecticut. *Wildl. Soc. Bull.*, 19, 36–38, 1991.

Craven, S. R. and S. E. Hygnstrom, Deer. P. D25–D40 in S. E. Hygnstrom, R. M. Timm, and G. E. Larson, Eds., *Prevention and Control of Wildlife Damage*. University of Nebraska Cooperative Extension, Lincoln, 1994.

Doggett, H., Bird-resistance in sorghum and the quelea problem. *Field Crop Abstr.,* 10, 153–156, 1957.

Dolbeer, R. A., Blackbirds and corn in Ohio. U.S. Fish and Wildlife Service Resource Publication 136, Washington, D.C., 1980.

Dolbeer, R. A., Cost–benefit determination of blackbird damage control for cornfields. *Wildl. Soc. Bull.*, 9, 44–51, 1981.

Dolbeer, R. A., J. L. Belant, and J. L. Bucknall, Shooting gulls reduces strikes at John F. Kennedy International Airport. *Wildl. Soc. Bull.*, 21, 442–450, 1993.

Dolbeer, R. A., P. P. Woronecki, and J. R. Mason, Aviary and field evaluations of sweet corn resistance to damage by blackbirds. *J. Am. Soc. Hortic. Sci.,* 113, 460–464, 1988.

Dolbeer, R. A., P. P. Woronecki, and R. A. Stehn, Effect of husk and ear characteristics on resistance of maize to blackbird (*Agelaius phoeniceus*) damage in Ohio, USA. *Protect. Ecol.,* 4, 127–139, 1982.

Dolbeer, R. A., P. P. Woronecki, and R. A. Stehn, Blackbird (*Agelaius phoeniceus*) damage to maize: crop phenology and hybrid resistance. *Protect. Ecol.,* 7, 43–63, 1984.

Dolbeer, R. A., P. P. Woronecki, and R. A. Stehn, Resistance of sweet corn to damage by blackbirds and starlings. *J. Am. Soc. Hortic. Sci.,* 111, 306–311, 1986a.

Dolbeer, R. A., P. P. Woronecki, and R. A. Stehn, Blackbird-resistant hybrid corn reduces damage but does not increase yield. *Wildl. Soc. Bull.*, 14, 298–301, 1986b.

Dolbeer, R. A., P. P. Woronecki, R. A. Stehn, G. J. Fox, J. J. Hansel, and G. M. Linz, Field trials of sunflower resistant to bird depredation. *N.D. Farm Res.,* 43(6), 21–24, 28, 1986c.

Edge, W. D., J. O. Wolff, and R. L. Carey, Density-dependent responses of gray-tailed voles to mowing. *J. Wildl. Manage.*, 59, 245–251, 1995.

Engeman, R. M., V. G. Barnes, Jr., R. M. Anthony, and H. W. Krupa, Vegetation management for reducing mortality of ponderosa pine seedlings from *Thomomys* spp. *Crop Protect.,* 14, 505–508, 1995.

Engeman, R. M., V. G. Barnes, Jr., R. M. Anthony, and H. W. Krupa, Effect of vegetation management for reducing damage to lodgepole pine seedlings from northern pocket gophers. *Crop Protect.,* 16, 407–410, 1997.

Fitzgerald, W. S. and R. E. Marsh, Potential of vegetation management for ground squirrel control. In *Proc. Vert. Pest Conf.*, 12, 102–107, 1986.

Flynt, R. D. and J. F. Glahn, Propagation of bamboo as blackbird lure roost habitat. In *Proc. East. Wildl. Damage Control Conf.*, 6, 113–119, 1995.

Fox, G. J. and G. Linz, Evaluation of red-winged blackbird resistant sunflower germplasm. In *Proc. Bird Control Sem.*, 9, 181–189, 1983.

Franklin, W. L., and M. G. Garrett, Nonlethal control of prairie dog colony expansion with visual barriers. *Wildl. Soc. Bull.*, 17, 426–430, 1989.

Fritts, S. H., W. J. Paul, L. D. Mech, and D. P. Scott, Trends and management of wolf–livestock conflicts in Minnesota. U.S. Fish and Wildlife Service, Resource Publication 181, 1992.

Gartshore, R. G., R. J. Brooks, F. F. Gilbert, and J. D. Somers, Census techniques to estimate blackbirds in weedy and nonweedy field corn. *J. Wildl. Manage.*, 46, 429–437, 1982.

Geis, A. D., Effects of building design and quality on nuisance bird problems. In *Proc. Vert. Pest Conf.*, 7, 51–53, 1976.

Glahn, J. F. and D. L. Otis, Factors influencing blackbird and European starling damage at livestock feeding operations. *J. Wildl. Manage.*, 50, 15–19, 1986.

Gluesing, E. A., D. F. Balph, and F. F. Knowlton, Behavioral patterns of domestic sheep and their relationship to coyote predation. *Appl. Anim. Ethol.*, 6, 315–330, 1980.

Gorenzel, W. P. and T. P. Salmon, Characteristics of American crow urban roosts in California. *J. Wildl. Manage.*, 59, 638–645, 1995.

Gourley, M., M. Vomocil, and M. Newton, Forest weeding reduces the effect of deer browsing on Douglas fir. *For. Ecol. Manage.*, 36, 177–185, 1990.

Hartwell, H. D., A comparison of large and small Douglas-fir nursery stock outplanted in potential wildlife damage areas. Washington Department of Natural Resources Note 6, Olympia, WA, 1973.

Howard, W. E., R. E. Marsh, and C. W. Corbett, Raptor perches: their influence on crop protection. *Acta Zoo. Fennica,* 173, 191–192, 1985.

Hygnstrom, S. E., Plastic visual barriers were ineffective at reducing recolonization rates of prairie dogs. In *Proc. Great Plains Wildl. Damage Control Workshop*, 12, 74–76, 1996.

Hygnstrom, S. E., R. M. Timm, and G. E. Larson, *Prevention and Control of Wildlife Damage*. University of Nebraska Cooperative Extension, Lincoln, 1994.

Hygnstrom, S. E., K. C. VerCauteren, and J. D. Ekstein, Impacts of field-dwelling rodents on emerging field corn. In *Proc. Vert. Pest Conf.*, 17, 148–150, 1996.

Hygnstrom, S. E., K. C. VerCauteren, R. A. Hines, and C. W. Mansfield, Efficacy of in-furrow zinc phosphide pellets for controlling rodent damage in no-till corn. *Int. Biodeterioration Biodegradation*, 45, 215–222, 2000.

Kenward, R. E., J. C. F. Dutton, T. Parish, F. I. B. Doyle, S. S. Walls, and P. A. Robertson, Damage by grey squirrels. I. Bark-stripping correlates and treatment. *Q. J. For.*, 90, 135–142, 1996.

Kimball, B. A., E. C. Turnblom, D. L. Nolte, D. L. Griffin, and R. M. Engeman, Effects of thinning and nitrogen fertilization on sugars and terpenes in Douglas-fir vascular tissues: implications for black bear foraging. *For. Sci.*, 44, 599–602, 1998a.

Kimball, B. A., D. L. Nolte, D. L. Griffin, S. M. Dutton, and S. Ferguson, Impacts of live canopy pruning on the chemical constituents of Douglas-fir vascular tissues: implications for black bear tree selection. *For. Ecol. Manage.*, 109, 51–56, 1998b.

Klebenow, D. A., K. McAdoo, and J. D. Kauffeld, Predation on range sheep as related to predator control and sheep management. In *Proc. Int. Rangelands Congr.*, 1, 270–272, 1978.

Knowlton, F. F., E. M. Gese, and M. M. Jaeger, Coyote depredation control: an interface between biology and management. *J. Range Manage.*, 52, 398–412, 1999.

Linhart, S. D., Managing coyote damage problems with nonlethal techniques: recent advances in research. In *Proc. East. Wildl. Damage Control Conf.*, 1, 105–118, 1983.

Linz, G. M., D. L. Bergman, and W. J. Bleier, Progress on managing cattail marshes with Rodeo® herbicide to disperse roosting blackbirds. In *Proc. Vert. Pest Conf.*, 15, 56–61, 1992.

Linz, G. M., D. L. Bergman, H. J. Homan, and W. J. Bleier, Effects of herbicide-induced habitat alterations on blackbird damage to sunflowers. *Crop Protect.*, 15, 625–629, 1996.

Luder, R., Weeds influence red-billed quelea damage to ripening wheat in Tanzania. *J. Wildl. Manage.*, 49, 646–647, 1985.

Lyon, L. A. and D. F. Caccamise, Habitat selection by roosting blackbirds and starlings: management implications. *J. Wildl. Manage.*, 45, 435–443, 1981.

Mah, J., G. M. Linz, and J. J. Hanzel, Relative effectiveness of individual sunflower traits for reducing red-winged blackbird depredation. *Crop Protect.*, 9, 359–362, 1990.

Mah, J. and G. L. Nuechterlein, Feeding behavior of red-winged blackbirds on bird-resistant sunflowers. *Wildl. Soc. Bull.*, 19, 39–46, 1991.

Marsh, R. E. and R. W. Steele, Pocket gophers. P. 205–230 in H. C. Black, Ed., *Silvicultural Approaches to Animal Damage Management in Pacific Northwest Forests*. USDA, Forest Service, Pacific Northwest Research Station, General Technical Report PNW–GTR-287, Portland, OR, 1992.

Mason, A. C. and D. L. Adams, Black bear damage to thinned timber stands in Northwest Montana. *Western J. Appl. For.*, 4, 10–13, 1989.

Mason, J. R., R. W. Bullard, R. A. Dolbeer, and P. P. Woronecki. Red-winged blackbird (*Agelaius phoeniceus* L.) feeding response to oil and anthocyanin levels in sunflower meal. *Crop Protect.*, 8, 455–460, 1989.

Mech, L. D., E. K. Harper, T. J. Meier, and W. J. Paul, Assessing factors that may predispose Minnesota farms to wolf depredations on cattle. *Wildl. Soc. Bull.*, 28, 623–629, 2000.

Moore, N. P., J. D. Hart, and S. D. Langton, Factors influencing browsing by fallow deer *Dama dama* in young broad-leaved plantations. *Biol. Conserv.*, 87, 255–260, 1999.

Moore, T., D. van Vuren, and C. Ingels, Are barn owls a biological control for gophers? Evaluating effectiveness in vineyards and orchards. In *Proc. Vert. Pest Conf.,* 18, 394–396, 1998.

Mountford, E. P. and G. F. Peterken, Effect of stand structure, composition and treatment on bark-stripping of beech by grey squirrels. *Forestry*, 72, 379–386, 1999.

Nelms, C. O., M. L. Avery, and D. G. Decker, Assessment of bird damage to early-ripening blueberries in Florida. In *Proc. Vert. Pest Conf.*, 14, 302–306, 1990.

Nelson, E. E., Black bears prefer urea-fertilized trees. *Western J. Appl. For.*, 4, 13–15, 1989.

Nolte, D. L., B. A. Kimball, and G. J. Ziegltrum, The impact of timber management on the phytochemicals associated with black bear damage. In *Proc. Vert. Pest Conf.*, 18, 111–117, 1998.

Otis, D. L. and C. M. Kilburn, Influence of environmental factors on blackbird damage to sunflowers. U.S. Fish and Wildlife Service Technical Report 16, Washington D.C., 1988.

Parfitt, D. E., Relationship of morphological plant characteristics of sunflower to bird feeding. *Can. J. Plant Sci.*, 64, 37–42, 1984.

Parker, B. S., K. Myers, and R. L. Caskey, An attempt at rabbit control by warren ripping in semi-arid western New South Wales. *J. App. Ecol.*, 13, 353–367, 1976.

Posey, A. F., J. G. Burleigh, and R. Katayama, Sunflower head droop and curl may affect bird depredation. *Arkansas Farm Res.,* 31, 10, 1982.

Robel, R. J., A. D. Dayton, F. R. Henderson, R. L. Meduna, and C. W. Spaeth, Relationships between husbandry methods and sheep losses to canine predators. *J. Wildl. Manage.*, 45, 894–911, 1981.

Salmon, R. E., Barn owl nest boxes offer no solution to pocket gopher damage. In *Proc. Vert. Pest Conf.*, 18, 414–415, 1998.

Schmidt, W. C., Bear damage — a function of stand density in young larch forests. P. 145 in *Proceedings of the Symposium on Animal Damage Management in Pacific Northwest Forests*. Washington State University, Pullman, 1987.

Schmidt, W. C., Management of young *Larix occidentalis* forests in the northern Rocky Mountains of the United States. P. 246–253 in H. C. H. Thomasius, Ed., *Treatment of Young Forest Stands: Proceedings of the Symposium.* IUFRO Working Party, S 1.05–03, Dresden, Germany, 1989.

Schmidt, W. C., and M. Gourley, Black bear. P. 309–331 in H. C. Black, Ed. *Silvicultural Approaches to Animal Damage Management in Pacific Northwest Forests*. USDA, Forest Service, Pacific Northwest Research Station, General Technical Report PNW–GTR-287, Portland, OR, 1992.

Seiler, G. J. and C. E. Rogers, Influence of sunflower morphological characteristics on achene depredation by birds. *Agric. Ecosystems Environ.*, 20, 59–70, 1987.

Servello, F. A., R. L. Kirkpatrick, K. E. Webb, Jr., and A. R. Tipton, Pine vole diet quality in relation to apple tree root damage. *J. Wildl. Manage.*, 48, 450–455, 1984.

Slater, A. J., Management of birds associated with buildings at the University of California, Berkeley. In *Proc. Vert. Pest Conf.*, 15, 79–82, 1992.

Stickley, A. R., Jr. and C. R. Ingram, Methiocarb as a bird repellent for mature sweet corn. In *Proc. Bird Control Sem.*, 7, 228–238, 1976.

Straub, R. W., Red-winged blackbird damage to sweet corn in relation to infestations of European corn borer (Lepidoptera: Pyralidae). *J. Econ. Entomol.*, 82, 1406–1410, 1989.

Sullivan, T. P. and E. J. Hogue, Influence of orchard floor management on vole and pocket gopher populations and damage in apple orchards. *J. Am. Soc. Hortic. Sci.,* 112, 972–977, 1987.

Swihart, R. K., Common components of orchard ground cover selected as food by captive woodchucks. *J. Wildl. Manage.*, 54, 412–417, 1990.

Timm, R. M., Commensal rodents in insulated livestock buildings. In C. G. Richards and T. Y. Ku, Eds., *Control of Mammal Pests.* Taylor and Francis, New York, 1987.

Van Vuren, D., Manipulating habitat quality to manage vertebrate pests. In *Proc. Vert. Pest Conf.*, 18, 383–390, 1998.

Wagner, F. H., *Predator Control and the Sheep Industry.* Regina Books, Claremont, CA, 1988.

Whisson, D. E., The effect of two agricultural techniques on populations of the canefield rat (*Rattus sordidus*) in sugarcane crops of North Queensland. *Wildl. Res.*, 23, 589–604, 1996.

Whisson, D. E. and G. A. Giusti, Vertebrate pests. P. 126–131 in C. A. Ingels, R. L. Bugg, G. T. McGourty, and L. P. Christensen, Eds., *Cover Cropping in Vineyards: A Grower's Handbook.* University of California, Division of Agriculture and Natural Resources Publication 3338. Oakland, CA, 1998.

Wilson E. A., E. A. LeBoeuf, K. M. Weaver, and D. J. LeBlanc, Delayed seeding for reducing blackbird damage to sprouting rice in southwestern Louisiana. *Wildl. Soc. Bull.*, 17, 165–171, 1989.

Woronecki, P. P., R. A. Dolbeer, and R. A. Stehn, Response of blackbirds to Mesurol and Sevin applications on sweet corn. *J. Wildl. Manage.*, 45, 693–701, 1981.

CHAPTER 15

Human Dimensions

"Wildlife management is, at its core, the management of people."

Anonymous

"Knowledge speaks but wisdom listens."

Jimi Hendrix

"When resource managers work within the limits of ecological parameters, they must solve problems; but when they work within the limits and forces of sociological parameters, they must resolve issues."

Minnis and Peyton (1995)

For wildlife damage to occur, three elements must come together. There must be a resource being damaged, an animal causing the damage, and an injured person. If no one has been injured or has suffered a loss, then there has been no damage. Usually, the injured person is the owner of the damaged resource, but other people can be injured as well. When urban Canada geese roost on a public beach, the injured people include swimmers, who have to step around goose manure, and local citizens, who, through their taxes, have to pay more to keep the beach clean. If the state decides to address the problem by killing the geese, the injured people would also include those individuals who like to feed the geese at the beach, neighbors who like to see them, hunters who have been looking forward to the fall's goose season, and those individuals who have empathy for the geese and worry about their pain and suffering. Hence, individuals who are affected by wildlife damage issues can form a complicated web of people. The area of wildlife management that deals with people and society is called human dimensions; it is the focus of this chapter.

SOCIETAL RESPONSES TO WILDLIFE DAMAGE

There are many different ways that society can and does respond to complaints about wildlife damage (Figure 15.1). In some states, wildlife species such as coyotes, striped skunks, or European starlings are classified as "vermin" and are not protected by law. For conflicts involving these animals, society may provide the injured persons a free hand to solve the wildlife problem any way they want. However, government may restrict the use of some techniques for solving a human–wildlife conflict because the methods are either considered too cruel or inhumane, or because they pose too great a threat to human safety, nontarget animals, or the environment. Society may also prevent the injured person from killing the offending animal in some cases, especially when the animal is rare or valuable. Sometimes, government may use its own employees to help reduce wildlife damage; for instance, U.S.D.A. Wildlife Services fulfills this function. At other times, government may compensate landowners for their losses caused by wildlife. How society responds to wildlife damage depends, in part, on people's attitudes about wildlife.

DIFFERENCES IN ATTITUDES TOWARD WILDLIFE

Kellert (1979, 1980a, b, c) examined how people differed in their attitudes toward wildlife. These studies identified several attitude types that can be used to categorize attitudes toward wildlife. The most widely held attitudes of Americans about wildlife are listed next.

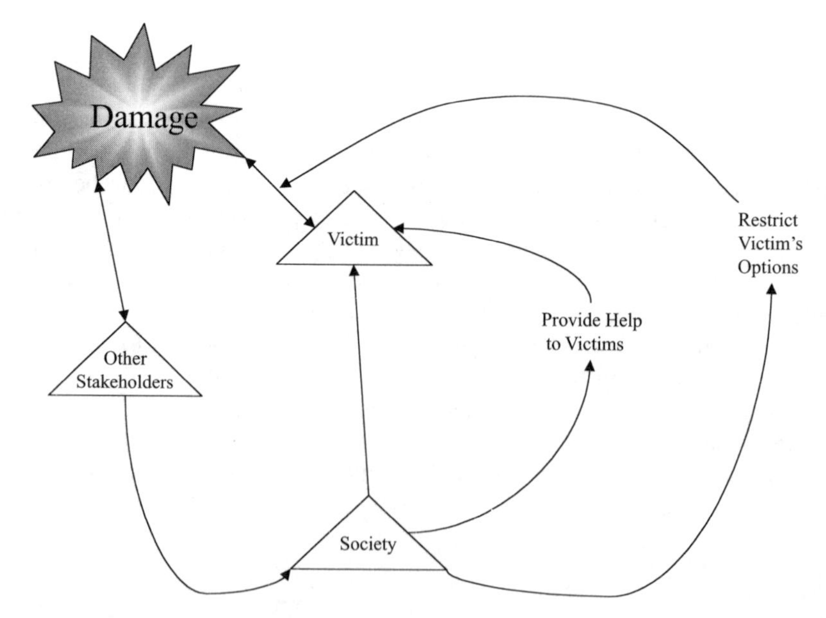

Figure 15.1 Conceptual model of how human dimensions fit into the science of resolving human–wildlife conflicts.

Negativistic and Neutralistic Attitudes

People with these attitudes dislike animals and either actively avoid them due to fear (negativistic attitude) or passively avoid them due to indifference (neutralistic attitude). These attitudes are held by 37% of the U.S. population (percentages add up to more than 100% because some people hold more than one attitude).

Humanistic Attitude

People with a humanistic attitude have strong emotional ties to individual animals, especially pets. They are interested in the welfare of the animals they love. Many enjoy wildlife tourism and trips to zoos. This group comprises 35% of the U.S. population (Figure 15.2).

Figure 15.2 People with a humanistic attitude have strong emotional ties to individual animals, especially pets. (Photo courtesy of Tony DeNicola.)

Moralistic Attitude

People holding this attitude are concerned about human morality and ethics. They are opposed to human exploitation of or cruelty toward animals. Many of them oppose hunting, trapping, or fishing. This attitude is held by 20% of the U.S. population and is much more common in women than in men.

Utilitarian Attitude

The utilitarian viewpoint is interested primarily in the practical and economical uses of wildlife. These people believe that animals should serve some human purpose. They view animals based on how they contribute to personal gain. They have a high opinion of game species and a low opinion of animals that cause wildlife damage. This group includes 20% of the U.S. population and is prevalent among Caucasian males.

Aesthetic Attitude

The aesthetic type's primary interest is art. They enjoy wildlife art and photography and, for this reason, particularly like large mammals and brightly colored birds. This attitude is held by 15% of U.S. residents.

Naturalistic Attitude

These people enjoy outdoor recreation and getting "close to nature." Hunters, trappers, fishermen, bird watchers, hikers, and campers often have a naturalistic attitude. About 10% of the U.S. population holds this attitude.

STAKEHOLDER PERCEPTIONS OF WILDLIFE DAMAGE

People who suffer from or are concerned about wildlife damage are called "stakeholders." Other segments of society also are stakeholders because they are concerned about the methods used to alleviate the damage. The views of the various stakeholders influence how the government responds to wildlife problems. The major stakeholders involved with the management of wildlife damage are listed below.

Farmers, Ranchers, and Private Landowners

There are approximately 2 million farmers and ranchers in the U.S. (U.S. Bureau of the Census 1992). Although they comprise less than 2% of the U.S. population, agricultural producers are an important group because they control 45% of the land (Berg 1986). Their support is essential if wildlife conservation in the U.S. is going to occur on private land (Leopold 1933). Unfortunately, relationships between wildlife agencies and private landowners are often strained. The difficulty stems from the paradox that wildlife is publicly owned but is dependent upon a habitat base that is largely owned by individuals, who may have land use objectives other than producing wildlife (Berryman 1981; Matthews 1986).

Agricultural producers are a unique occupational group; their attitudes about wildlife differ from those of people employed in other jobs (Kellert 1980c). They view wildlife in utilitarian terms and focus on how wildlife affects them economically. Farmers are more supportive of hunting to eliminate depredating animals than the general public and more likely to favor programs that compensate landowners for wildlife damage (McIvor and Conover 1994a).

Agricultural producers who are economically dependent on their land or who derive most of their income from farming are less tolerant of wildlife damage than part-time or "hobby" farmers (Tanner and Dimmick 1983; Campa et al. 1997). Farmers producing high-value crops that are vulnerable to wildlife damage (e.g., apples, nursery plants) are less tolerant of wildlife than other farmers (Decker and Brown 1982).

Yet, despite suffering from wildlife damage, agricultural producers appreciate wildlife (Brown et al. 1978). Half of the nation's agricultural producers spend some of their time and money to enhance wildlife habitat on their property (Conover 1998). The American Farm Bureau Federation and National Farmers' Union are two groups which represent U.S. agricultural producers. There also are several active farmer groups organized around specific commodities (e.g., American Soybean Association, National Cattlemen' s Association, National Woolgrowers' Association).

Hunters and Fur Trappers

Hunters and trappers want large and healthy populations of game and fur-bearing species. Hence, their ideas of what constitutes the optimal size of a wildlife population are often higher than those of agricultural producers (Diefenbach et al. 1997). Their views, however, are similar to those of farmers and ranchers with regard to other environmental issues and in seeking to preserve the right to hunt and trap. Many hold a utilitarian or naturalistic attitude towards wildlife.

Only a small proportion of the U.S. population hunts (16 million in 1991) and even fewer engage in fur trapping (U.S. Department of Interior 1993). However, many hunters and trappers are passionate about their sports, receive great enjoyment from them, and spend considerable time and money engaging in them (Figure 15.3). During 1991, people spent $12 billion annually in hunting-related activities in the U.S. (U.S. Department of Interior 1993). Funds obtained from hunters through hunting licenses, federal duck stamps, and special taxes on hunting supplies provide most of the funding for state and federal wildlife agencies. Boone and Crockett Club, Ducks Unlimited, Mule Deer Foundation, National Rocky Mountain Elk Foundation, National Trappers Association, National Wild Turkey Foundation, Pheasants Forever, and Quail Unlimited are some of the national organizations in the U.S. whose memberships contain a large number of hunters and trappers and that represent their interests.

Figure 15.3 Many hunters receive great enjoyment from this recreational activity. (Courtesy of the U.S. Fish and Wildlife Service's Bear River Bird Refuge.)

Wildlife Enthusiasts

Wildlife enthusiasts are people who place a high value on wildlife or derive great pleasure from it. They often have a naturalistic attitude toward wildlife. Bird watchers and outdoor recreationists are members of this group, as are many hunters and fur trappers. Also included are people who sit at home and enjoy watching nature programs on television. When they hear about wildlife damage, the sympathy of wildlife enthusiasts is often with the wildlife rather than with the person suffering the damage. They want healthy wildlife populations and the preservation of wildlife habitat; they are concerned about the impact of increasing human populations on nature. Defenders of Wildlife, National Audubon Society, National Wildlife Federation, and World Wildlife Fund are some of the organizations in the U.S. that represent these stakeholders.

Animal Welfare Activists

Many people are concerned with the humane treatment of animals (Schmidt 1989, 1991a). They have empathy for individual animals and do not want to see them hurt. Animal welfare activists are sensitive to human activities that result in animal suffering. For instance, many activists oppose the use of leg-hold traps, preferring either that trapping be banned or that only live traps be used. Many oppose hunting or the use of lethal control to resolve wildlife problems. Instead, they support the use of nonlethal techniques, especially if those techniques do not cause pain to the animals. Animal welfare activists often have a humanistic attitude toward animals and are more knowledgeable about animals than the general public. Richards and Krannich (1991) reported that the "typical profile of an activist is a middle-aged woman, who is well educated, well-to-do, and a pet owner. She considers herself to be a liberal and an environmentalist." During interviews, 77% of the national leaders of animal activists groups reported that it was unacceptable for someone to hunt, fish, or trap for sport (Hooper 1992). In contrast, 82% believed that it was acceptable for native people to hunt, fish, or trap for subsistence and 100% thought it acceptable if it was a matter of human survival (e.g., someone lost in the wilderness). The Humane Society of the United States (1.3 million members) and American Society for the Prevention of Cruelty to Animals (400,000 members) are the largest animal welfare groups in the U.S. (Ehrenfeld 1991).

Animal Rights Activists

Some people believe that animals should have the same moral rights as humans. They believe that "specism" — the attitude that humans have a right to exploit other animals — is just as immoral as racism or sexism (Schmidt 1991a). They feel that the suffering of animals is morally the same as the suffering of humans — that the rights of an animal used in medical research are the same as those of a sick child. They are opposed to hunting, trapping, or the lethal control of animals because they believe that it is morally and ethically improper to subject animals to such treatment (Schmidt 1991b). Many animal rights activists feel that animals should have legal status, which means that animals possess legal rights and could sue in a court of law if their rights are violated. These people have a moralistic attitude toward animals.

A major difference between conservation biologists and wildlife managers on one hand, and those people concerned with animal rights and animal welfare on the other, is that the latter are concerned primarily with the sanctity of the individual animal while the former are concerned with the integrity of ecosystems and the populations incorporated within them (Ehrenfeld 1991). Animal rights activists are opposed to what they perceive as an alliance between hunters and wildlife management agencies. They believe that wildlife agencies try to create a surplus of animals which can be harvested by hunters, while hunters, in turn, provide political and economical support to the wildlife agencies (Decker and Brown 1987). People for the Ethical Treatment of Animals (PETA) and In Defense of Animals are two of the more active animal rights groups in the U.S.

Metropolitan Residents

In the U.S., most people (160 million) live and work in metropolitan areas. Most metropolitan residents enjoy watching wildlife in their neighborhoods, while visiting parks, or while engaging in outdoor recreation (Figure 15.4), and 57% of them spend time and money each year on efforts to encourage wildlife, such as erecting bird feeders (Conover 1997).

While metropolitan residents enjoy having wildlife in their neighborhoods, they are also concerned with the problems these animals cause (Decker and Gavin 1987;

Figure 15.4 There is a celebration in Hamden, CT, when the eggs of the mute swan hatch (note the sign on the fence announcing the blessed event).

Stout et al. 1993; Green et al. 1997), and most experience some type of wildlife problem each year (see Chapter 3). Common wildlife problems for metropolitan residents include raccoons eating food placed out for pets, birds damaging gardens, moles burrowing in yards, or mice living indoors (Figure 15.5). These are mostly nuisance problems that reduce the quality of life of metropolitan residents. More serious wildlife problems that worry urban residents include deer–automobile collisions and serious zoonotic diseases (Decker and Gavin 1987).

Most metropolitan residents prefer nonlethal techniques when dealing with wildlife problems due to ethical considerations but disagree over the use of lethal

Figure 15.5 One of the more common, but unpleasant, ways in which metropolitan residents interact with wildlife occurs when they find mice in their kitchen. (Photo courtesy of C. R. Madsen.)

means (Stout et al. 1997). Some people staunchly believe that the life of a deer is more valuable than the risk it poses to human health and safety; others believe the opposite. When wildlife damage is perceived to be severe and chronic, the attitudes of some metropolitan residents toward the depredating species can become irreversibly negative (Loker et al. 1999). When this happens, some people become convinced that only extreme measures of limiting the wildlife population are satisfactory. In such situations, wildlife management agencies can find themselves in an impossible situation between these people and their neighbors, who enjoy wildlife or have emotional ties to the animals and will oppose the use of the same measures (Sidebar 15.1).

Sidebar 15.1 Attitudes of Urban Residents towards Prairie Dogs

Public opinion of urban residents in Fort Collins, CO varied widely when it came to the management of prairie dog colonies within the city, with some people feeding them and others wanting them eradicated. Caught in the middle were city officials who lacked objective information about their constituents' feelings towards prairie dogs or the extent of prairie dog–human conflicts. To provide answers, Zinn and Andelt (1999) surveyed the public. They found that 58% of the people living within one block of a prairie dog colony had problems with prairie dogs. People living near prairie dogs were more likely than other people to view prairie dogs as pests, more likely to favor the use of toxicants to control them, and less likely to consider them as a valuable part of nature. People living near prairie dogs lost tolerance for the animals over time. In response, the city of Fort Collins adopted a preservation and control policy in which reserves were established for prairie dogs because of their aesthetic benefits. In residential areas where prairie dogs were causing specific conflicts, control measures were implemented. Zinn and Andelt (1999) recommended a public education program so that all residents would understand both the benefits and disadvantages of prairie dogs.

Figure 15.6 Prairie dogs. (Figure from Hygnstrom et al. 1994.)

They also suggested that an educational program be provided to individuals considering purchasing a home near a reserve so that they would have a more realistic expectation about living with prairie dogs (Figure 15.6).

Rural Residents

The 92 million people in the U.S. who live outside a metropolitan area are a diverse group (U.S. Bureau of the Census 1992). They include farmers and ranchers along with people in farm-related professions, residents of small cities and towns, urban commuters, retired people, and urban refugees. There is an increasing exchange of people and ideas between rural and urban areas, so that the views and attitudes of the residents of rural America are becoming more similar to those of urban residents. Still, differences exist between rural and metropolitan residents. Rural residents are more knowledgeable than urban residents about wildlife (Kellert 1980b), have more first-hand experience with wildlife, are more likely to have farmers and ranchers as friends, and have more empathy for their problems. A higher proportion of rural residents fish, hunt, or engage in wildlife-related recreational activities than urban residents (Table 15.1). Rural residents have a more utilitarian attitude towards animals than metropolitan residents and are more interested in the practical or material value of animals (Kellert 1980a).

Table 15.1 Percent of Rural and Urban Residents Who Participated in Various Wildlife-Related Activities in 1991 (U.S. Department of Interior 1993)

	Rural	Urban
Fish	26	16
Hunt	14	5
Observe wildlife	39	25
Photograph wildlife	13	7
Feed free-ranging birds	44	29

IMPACT OF WILDLIFE DAMAGE ON A PERSON'S ATTITUDES TOWARD WILDLIFE

Wildlife damage can alter a person's perceptions about wildlife, especially when damage exceeds his/her tolerance. For example, farmers who had experienced deer damage were more likely to believe that deer populations were increasing and to want a reduction in the deer population than other farmers (Decker and Brown 1982; Decker et al. 1983). In a national survey of agricultural producers, 53% of respondents reported that the amount of wildlife damage they experienced exceeded their level of tolerance (Conover 1998). The same survey found that 40% of all agricultural producers reported that wildlife damage on their farm or ranch was so severe that they would oppose the creation of a wildlife sanctuary near them; 26% said wildlife damage reduced their willingness to provide wildlife habitat on their property. In many parts of the world, people who live or farm near a park are the ones who suffer from damage by the park's wildlife. Hence, they often are opposed to the park and its wildlife (Sidebar 15.2). This local opposition can cause problems for the park until human–wildlife conflicts are resolved.

Sidebar 15.2 Resolving Conflicts between Nature Preserves and Local Residents

In underdeveloped countries, the creation of a nature preserve often is opposed by local residents, who are most impacted by it. These people fear restrictions on their historic use of resources within the preserve, crop damage from herbivores venturing out of the preserve, and loss of livestock and human lives due to an increase in local predator populations. In fact, crop damage is often cited as the main reason why neighbors dislike nature reserves (Parry and Campbell 1992; Heinen 1993; Newmark et al. 1993). Yet, without local support, nature preserves cannot function. For this reason, the opinions of local residents are being incorporated into the management plans of many nature reserves.

Elephants venturing out of Cameroon's Mạza National Park caused considerable damage to local villagers, most of whom were subsistence farmers. In 1993, 400 African elephants destroyed about 10,000 ha of farmland and killed four people (Tchamba 1996). Another cost imposed on local farmers by the elephants was wasted time; farmers had to spend the whole day and night guarding their crops. Much of this guarding was done by school children who could not attend school while on guard duty. In a survey, 54% of the local villagers around the park believed the elephants provided no value to them at all, and 75% reported that the elephants posed a threat to their personal safety and caused crop damage. Most local villagers (73%) thought that more elephants needed to be killed. Tchamba (1996) pointed out that these conflicts between local villagers and elephants pose a serious threat to the elephant's continued survival.

Most of the local villagers outside the Maputo Elephant Reserve in Mozambique had a higher opinion of that elephant reserve. Most (88%) reported that they liked the Maputo Reserve and used it; 71% harvested plants from within it and 21% hunted there (De Boer and Baquete 1998). The neighbors' ability to continue using resources within the reserve helped improve their attitude toward it. The local residents reported that the biggest problem with the reserve was crop damage by marauding wildlife, and the opinion of local residents was correlated with their crop damage experiences. People who had suffered crop damage by elephants, hippos, or bushpigs were the most opposed to the reserve. De Boer and Baquete (1998) believed that participation of the local communities was critical to the development of management strategies for the Maputo Elephant Reserve and that crop damage by reserve animals must be addressed.

Even among metropolitan residents, individuals who have previously experienced a wildlife problem are more likely to be concerned about local wildlife populations and more willing to have their government take steps to control wildlife populations. For instance, metropolitan residents who had experienced problems caused by urban Canada geese, beaver, deer, or prairie dogs were more likely to favor a reduction in these populations and to support the use of lethal methods to achieve it than those who do not experience wildlife problems (Decker and Gavin 1987; Stout et al. 1993; Loker et al. 1999; Zinn and Andelt 1999). Metropolitan residents who saw themselves at risk of having a deer–vehicle collision or contracting Lyme disease were more likely to prefer a decrease in local deer populations (Stout et al. 1993).

Tolerance of human–wildlife conflicts depends upon what is at risk. That is, people are least tolerant of threats to human health and safety, more tolerant of threats to their personal or community economic well-being, and most tolerant of nuisance problems. Loker et al. (1999), however, found that threats to health and safety were no more influential in shaping the opinions of suburban residents about wildlife than were the other types of wildlife problems. Perhaps the reason is that few people see themselves as future victims of an uncommon calamity, such as a deer–automobile collision or a life-threatening zoonotic disease. In contrast, people who constantly experience the same economic or nuisance problem are more likely to see themselves as future victims of wildlife damage.

THE CONCEPT OF CULTURAL CARRYING CAPACITY

In recent decades, the emphasis of wildlife management has shifted from trying to maintain wildlife populations at their biological carrying capacity to trying to maintain them at their cultural carrying capacity (CCC). The latter is defined as the maximum wildlife population which society will accept within an area (Decker and Purdy 1988), or the number of animals that can compatibly coexist with the local human population (Minnis and Peyton 1995). In contrast, the biological carrying capacity is the maximum wildlife population which can be sustained in an area.

The CCC is difficult to determine because it is based on the views of many diverse groups. Each stakeholder group has its own view about what is the maximum wildlife population that is acceptable to them in an area, which Decker and Purdy (1988) defined as the wildlife acceptable capacity. If the wildlife density exceeds this level, members of a stakeholder group may perceive the wildlife density to be unsatisfactory because there are too many animals causing too many problems. When the wildlife density starts to deviate from the optimal range, people may prefer fewer animals but are willing to accept the current situation and take no action to change the wildlife population. However, as the wildlife density drifts further away from the optimal level,

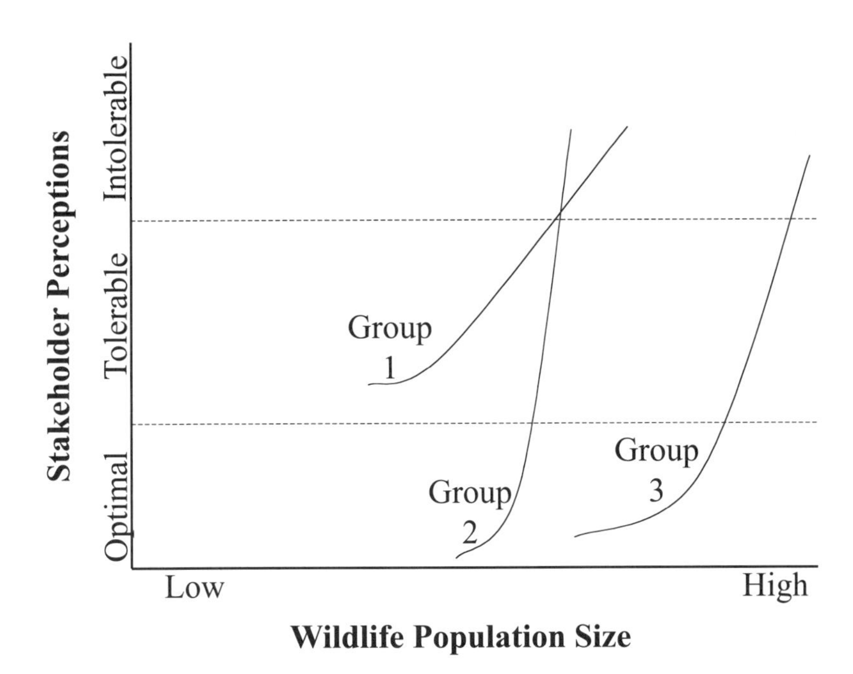

Figure 15.7 The optimal and tolerable upper levels of wildlife densities for three different groups of stakeholders.

stakeholders become more concerned and more intolerant of the wildlife and start taking action. For instance, stakeholders can 1. take steps to insulate themselves by trying to alleviate problems caused by the wildlife density, 2. try to reduce the wildlife population to an acceptable level, 3. encourage others to take action on their behalf, or 4. give up and abandon the stakeholder group (e.g., stop farming) (Minnis and Peyton 1995). One way to set the cultural carrying capacity is to determine the stakeholder group with the lowest wildlife acceptable capacity and manage for that level (Decker and Purdy 1988). This approach minimizes human–wildlife conflicts because the wildlife density does not exceed anyone's tolerance (Figure 15.7).

Each stakeholder group also has an optimal lower level, which they do not want a wildlife density to drop beneath, along with a lowest tolerable level, which is the level that causes them to take action whenever a wildlife density drops below it. Hence, each stakeholder group has a range of wildlife densities that they consider optimal and tolerable (Minnis and Peyton 1995). Another way to define the cultural carrying capacity is the wildlife density that is within the tolerable range of every stakeholder group (Figure 15.8).

WHY IS THE MANAGEMENT OF HUMAN–WILDLIFE CONFLICTS SO CONTROVERSIAL?

Given the diversity of stakeholders, it is not surprising that people often disagree about whether wildlife populations are too large or too small, or what actions should be

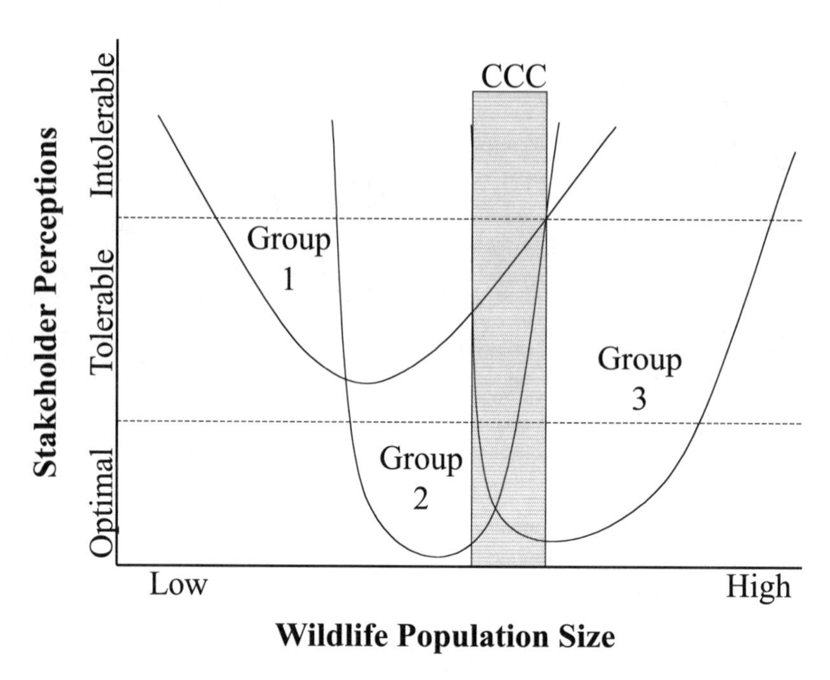

Figure 15.8 The optimal and tolerable range of wildlife densities for three different groups of stakeholders (the cultural carrying capacity, or CCC, is the range of wildlife densities that all three groups consider to be tolerable or optimal).

taken to alleviate human–wildlife conflicts. Sometimes, the ranges of tolerance for all stakeholder groups do not overlap (Figure 15.9). When this happens, there is a lack of consensus within society about how the wildlife population should be managed. That is, for any wildlife density, some people will complain that there are too many animals and, concomitantly, other people will complain that there are too few. This creates a difficult situation for the government agency charged with managing the wildlife resource, because any action it implements will displease some people. In these situations, one can argue that the ideal solution is the one that displeases everyone equally.

This lack of consensus among stakeholders usually occurs because the benefits and liabilities of wildlife are not distributed evenly among all segments of society. Consider the question of whether wolves should be reintroduced into a national park. Park visitors will support the reintroduction of wolves because they look forward to the opportunity to see a wolf. Most park concessionaires will be in favor of the wolves because more tourists means they will sell more items. Likewise, most park managers will support the introduction because more tourists will generate more money from admission fees. Most of these people benefit from having wolves but suffer few, if any, of the liabilities of having wolves in the area. However, consider the wolf reintroduction from the standpoint of local ranchers. Some ranchers will suffer economic losses when wolves leave the park and kill their animals. They will receive few, if any, of the benefits of having wolves in the area but will bear most of the problems. Hence, it is not surprising that many ranchers will oppose any effort to reintroduce wolves to an area. In this case,

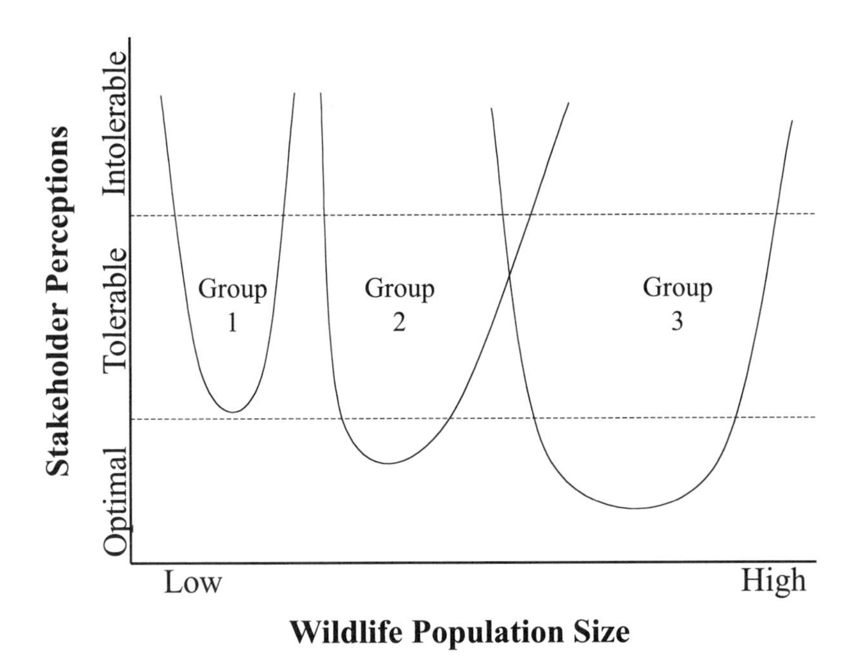

Figure 15.9 The optimal and tolerable range of wildlife densities for three different groups of stakeholders (in this case there is no cultural carrying capacity (CCC) because there is no wildlife density that all three stakeholder groups consider tolerable).

everyone is acting in his own self-interest; it is just that one person's self-interest is different from another's.

One way to reduce these controversies is to align things so that the benefits and liabilities of wildlife fall more evenly upon all segments of society. For example, farmers seek a reduction in deer numbers because the deer destroy their crops, whereas hunters want more deer because deer are not doing them any harm. The interests of farmers and hunters can be better aligned if farmers charge deer hunters for the hunting rights to their farms (Sidebar 15.3). Under this scenario, both hunters and farmers might benefit from a large and healthy deer herd and no longer disagree about how deer should be managed. The difficulty for wildlife agencies and governments is knowing how best to respond when stakeholders want opposite things.

Sidebar 15.3 Aligning Landowner Interests with Public Interests through Cooperative Hunting Units

For wildlife to exist on private land, it must have inherent values for the landowner that enable it to be competitive with alternative uses of its habitat that the landowner might be considering (Berryman 1981). Unfortunately, private landowners have had little economic incentive to invest in wildlife (Wigley and Melchiors 1987). However, increasing demands for high-quality hunting areas and a decreasing supply of wildlife habitat (Doig 1986) have stimulated interest in hunting by permit among private landowners in the U.S. (Wallace et al. 1989).

In response, Utah created the Cooperative Wildlife Management Unit (CWMU) program to: 1. provide income for landowners, 2. create satisfying hunting opportunities, 3. increase wildlife habitat, 4. provide adequate trespass protection for landowners who open their lands for hunting, 5. increase access to private lands for hunting big game, and 6. increase tolerance of wildlife damage.

Hunting in CWMUs is by permit only. Some of these permits are distributed free to the public through a public drawing conducted by the Utah Division of Wildlife Resources; others are given to participating landowners, who can then sell them. Permits obtained through the public drawing guarantee successful applicants free access and hunting opportunities comparable to those of hunters who buy their permits through CWMU landowners.

Hunters enjoy hunting in CWMUs because there are fewer hunters, greater chances of harvesting an animal, and opportunities for a better quality hunt than most regular season hunts. Hence, CWMUs satisfy hunter experiences and enhance hunter access to private lands (Messmer and Dixon 1997). CWMUs benefit landowners by providing increased income and a mechanism to manage hunter access and control trespass. CWMUs also provide an economic incentive to landowners to improve wildlife habitat. One survey reported that more than 5000 ha of rangeland habitat in Utah had been improved at a cost of $51,400 by CWMUs, and over 150 water sources had been created to benefit wildlife.

CWMUs increased landowner tolerance of wildlife damage because landowners were able to recoup some economic losses by selling hunting permits (Messmer and Schroeder 1996). For instance, one CWMU pays local farmers $200 for every acre of irrigated alfalfa they grow. The payment is designed to compensate them for any big-game damage to their alfalfa and has gone a long way toward increasing farmer tolerance of wildlife damage.

MAKING POLICY DECISIONS REGARDING WILDLIFE DAMAGE MANAGEMENT

It is the responsibility of elected legislators to decide how to manage the wildlife resource so that the greatest good is obtained for the greatest number of people. They may, in turn, delegate the decisions to wildlife agencies. Hence, the success of wildlife policies will largely rest on the ability of our elected officials and wildlife agencies to recognize, embrace, and incorporate differing stakeholder values, attitudes, and beliefs in the decision-making process.

Government officials need public input to make sound management decisions. Seeking it requires time and resources, but will result in more public ownership in the outcome and enhanced credibility for wildlife agencies (Hewitt and Messmer 1997; Messmer et al. 1997). Input from stakeholders can be sought by holding public forums in which citizens are allowed to air their views. Governments may also create citizen boards to oversee wildlife policy (Sidebar 15.4).

Sidebar 15.4 Using Citizen Task Forces to Resolve Controversial Wildlife Problems

In the metropolitan area surrounding Rochester, NY, management of the local deer population had become heated. In fact, local citizens had formed three very vocal and active groups. The Deer Action Committee was composed of people who were concerned about deer problems and wanted deer populations to be reduced. The Alliance For Wildlife Protection opposed the use of lethal means to control deer numbers, preferring an emphasis on defensive driving education to reduce the frequency of deer–automobile collisions and the experimental use of contraceptives to reduce deer reproduction. Another group called Save Our Deer also opposed the killing of deer.

To resolve the dispute, the New York Department of Environmental Conservation, Cornell Cooperative Extension, and the Human Dimensions Research Group formed a Citizens Task Force (CTF) to give local citizens an opportunity to determine the optimal deer density for the area. CTF functioned by allowing people from diverse groups (farmers, sportsmen, foresters, conservationists, animal welfare and animal rights activists, and members of the three citizen groups) to meet face to face to discuss the costs and benefits of increasing, decreasing, or making no change to the deer population and to recommend management plans on how to achieve this goal. After a series of meetings where the members received input from wildlife biologists, health officials, and various stakeholders, the CTF decided on a population goal of 8 to 10 deer per km^2. They decided not to recommend a specific percent decrease from current deer densities because they were unsure of the current density. When an aerial survey revealed that deer densities in parts of the area were four times higher than their population goal, the CTF recommended that 1. the number of deer removed equal the number of deer killed in local deer–automobile collisions and 2. that the number of deer removals be doubled in the following year if there was no decrease in the number of deer–automobile collisions. The CTF recommended that hunters be used to reduce deer densities where feasible and sharpshooters be used in other areas.

Local governments took steps to implement the CTF recommendations, but controversy over how to manage the local deer population continued. Save Our Deer, Alliance For Wildlife Protection, Fund For Animals, and the Humane Society began a public relations campaign to discredit the CTF's recommendations and to seek a court injunction to stop them from being implemented. Still, the CTF provided many benefits. Wildlife agencies were able to understand better the concerns and attitudes of local residents about deer and build credibility in the local area by employing a bottom-up approach to deciding deer management issues. By involving community and stakeholder leaders in these decisions, the local community developed a sense of ownership in the deer management program (Curtis et al. 1993).

Many wildlife agencies use a consensus-building approach to address disagreements over controversial wildlife policy (Bingham 1997). This approach is based on a series of face-to-face meetings in which leaders of different stakeholder groups seek to achieve resolution of their differences. Usually these processes are led by a mediator, who serves as a neutral third party in a negotiation process and helps the group establish a framework for discussions (Figure 15.10). This consensus-building approach has been used successfully to resolve conflicts regarding endangered species, wildlife depredation, and grazing. Elements of a successful consensus-building process include 1. identification of clear objectives, 2. agreement among participants on how group decisions will be made, 3. inclusion of team-building activities, such as dining together or participating in activities unrelated to the issues, and 4. achievement of success with smaller issues prior to addressing larger ones (Guynn 1997).

Figure 15.10 The consensus-building approach to resolving disagreements over controversial wildlife policy is based on a series of face-to-face meetings with leaders of different stakeholder groups. (Photo courtesy of Joe Caudell.)

FORMING PARTNERSHIPS BETWEEN WILDLIFE AGENCIES AND PEOPLE SUFFERING FROM WILDLIFE DAMAGE

One task of wildlife managers is to persuade landowners to adopt management practices that are beneficial to wildlife. Unfortunately, wildlife damage makes it more difficult to convince landowners that it is in their best interests to take steps to improve wildlife habitat. As Tanner and Dimmick (1983) stated, "The incentives for a farmer to [manage for] wildlife are ethereal and few: aesthetic values, sporting opportunities, perhaps an important source of food. The disincentives, however, are glaring and many: damage to crops and/or livestock, nuisance animals, negative interactions with fellow citizens (e.g., hunters, animal-rights groups), and the myriad of social and legal entanglements that may arise from these problems."

The formation of a working partnership between landowners and wildlife managers requires time, trust, and a shared sense of mission. These do not come easily. To achieve them, wildlife managers should have good interpersonal skills, tolerance for diverse views, and a willingness to listen. They also need to address the problem of wildlife damage in an open, proactive manner. Too often, wildlife managers are unconcerned about wildlife damage or gloss over the problem (Kirby et al. 1981). This attitude is counterproductive and conveys the impression to farmers that the wildlife manager is naive or uncaring (Tanner and Dimmick 1983).

RESOLVING HUMAN–WILDLIFE CONFLICTS THROUGH THE HUMAN DIMENSION

Human–wildlife conflicts can be resolved by reducing wildlife populations (Chapters 7–8), translocating animals (Chapter 9), changing wildlife behavior (Chapters 10–12), excluding wildlife from specific sites (Chapter 13), or modifying the habitat or the object being damaged (Chapter 14). Likewise, wildlife problems can be resolved by focusing on the people who are the victims.

Alleviating Human–Wildlife Conflicts by Changing Human Behavior

In some cases, wildlife problems can be ameliorated by changing human behavior. For instance, over half the people bitten by poisonous snakes in the U.S. were deliberately handling the snakes before they were bitten. Such bites could have been avoided if the snakes had simply been left alone. Some predators may first develop a taste for sheep by feeding on dead carcasses and then become sheep killers when the supply of carcasses is depleted. Under these conditions, some predation problems could be avoided if carcasses were disposed of rather than simply left to decompose (Robel et al. 1981). As another example, skunks may be drawn to a house to eat food left out for the family's pet and then create a nuisance problem. In this case, the problem can be resolved by simply not leaving food out. Likewise, it is dangerous when bears or other animals seek food from park visitors. Certainly, the simplest way to resolve this problem is to stop visitors from feeding the bears. Many parks have prohibited the feeding of animals due to concerns both for the health of the animals and for human safety. Some parks have established public education programs to inform the public why they should not feed animals. Unfortunately, some tourists believe they have a right to feed animals and do so even though they know it is against the rules. Hence, changing human behavior is not always as easy as it seems.

Often, the person whose behavior is precipitating the problem is not the one who is experiencing the damage. For example, when urban homeowners are confronted with pigeon problems, it often turns out that one of their neighbors is attracting the birds to the neighborhood by feeding them. In these situations, the neighbors should be encouraged to discuss the problem and come up with a mutually satisfying solution.

Alleviating Human–Wildlife Conflicts by Increasing the Injured Person's Appreciation for Wildlife

When private property is damaged by wildlife, the loss falls mainly on the owner of the property. If a cat is killed by coyotes, it is the owners of the cat who grieve. If blackbirds destroy a sunflower field, it is the farmer who suffers the financial loss. If bats roost in an attic, it is the homeowner who is inconvenienced.

One way to solve wildlife damage problems is to change the perceptions of people experiencing the damage so that they are more willing to tolerate it. This can be accomplished by enhancing their appreciation for both the tangible and nontangible benefits of wildlife. For instance, most suburban residents like having deer in their neighborhoods, although they recognize that the price they pay for the presence of deer includes damage to gardens and landscaping plants and some risk of Lyme disease or deer–automobile collisions (Decker and Gavin 1987; Stout et al. 1993). Some urban residents living with problems caused by prairie dogs also realized that prairie dogs were beneficial because these animals attracted eagles and hawks to the neighborhood (Zinn and Andelt 1999). Some farmers tolerate bird damage to their ripening grain fields because they realize that the birds are also eating harmful insects (Gabrey et al. 1993).

Farmers who hunt are more tolerant of deer damage to their crops than those who do not. These individuals still suffer economic losses from deer herbivory but, because the deer provide them enjoyment during the hunting season, they are more willing to tolerate the damage. In fact, they may actually manage their farms to maximize the local deer population by planting special crops for deer. Many farmers who complain about starling or blackbird damage to their grain crops are tolerant of turkey damage because they enjoy hunting turkeys or allowing their friends to do so (Craven 1989; Gabrey et al. 1993). Many landowners suffering from damage by game species tolerate the damage because they can generate money by leasing out the hunting rights to their land. In some rural areas, access fees can be quite high and can exceed the income landowners receive from timber production, farming, or grazing (Burger and Teer 1981). Other farmers and landowners who do not hunt may derive so much pleasure from seeing wildlife on their land that they do not mind the economic losses they suffer from wildlife damage.

Everyone likes to be recognized, and many people suffering from wildlife damage seek nothing more than empathy and understanding. Hence, another way to increase the value of wildlife for landowners is to ensure that these people feel appreciated for their willingness to tolerate damage. Letters of thanks, phone calls, personal visits, and offers of help all convey the notion that other people appreciate the efforts of farmers and landowners to help wildlife. As one innovative example, color calendars were sent to farmers who participated in a wildlife enhancement project as a token of thanks. Landowners who received these calendars had a deeper appreciation of wildlife and were more willing to take steps to help wildlife than other landowners (Messmer et al. 1996).

Increasing Tolerance for Human–Wildlife Conflicts through Education

When perceptions of risk differ from reality, public education can be used to provide accurate information to stakeholders and reduce their unfounded fears and, thus, increase public tolerance for wildlife damage (Stout et al. 1993; Loker et al. 1999). Education can also increase a person's appreciation of wildlife.

Rural landowners are most likely to take actions that are beneficial to wildlife if information is delivered through personal contacts rather than through distant communication (Kelly 1981; Warner 1983; Svoboda 1984; Feder and Slade 1985; Miller and Bromley 1989; Pease 1992). This creates a dilemma for wildlife management agencies because, with their small staffs and smaller budgets, they have little time and money available to establish personal contacts. Consequently, managers who wish to communicate and motivate rural stakeholders should consider combining nontraditional and traditional means of communication. Communication efforts should be developed with the target audience in mind and then evaluated to determine if the intended message motivates the recipients.

Increasing Tolerance for Human–Wildlife Conflicts through Compensation

One source of conflict involving wildlife management in the U.S. arises from the fact that wildlife are owned by society and managed by the state and federal governments, but inflict damage on private property. Some landowners resent the fact that somebody else's animals are damaging their property. They note that, if their livestock escaped and damaged another person's property, they would be liable for any damage. They question why the government is not responsible for wildlife damage if they own and manage the wildlife. Several landowners have sued the government over wildlife damage and the courts have ruled that, although the government owns the wildlife resource, it is not responsible for wildlife damage because the individual animals are not under its direct control (Musgrave and Stein 1993).

Despite immunity from lawsuits for wildlife damage, 19 U.S. states and 7 Canadian provinces financially compensated landowners for certain types of wildlife damage during 1994 (Wagner et al. 1997). During the same year, 34 states and 7 provinces provided landowners with damage-abatement material (such as propane cannons to scare away birds or fencing material). However, none of these states or provinces paid for all forms of wildlife damage, raising the interesting question of why some forms of wildlife damage are compensated for but not others. The answer seems to be that compensation is more likely to be available for damage caused by a game species or a large predator. More governments (14 states and provinces) compensated for damage caused by deer than any other species, followed by bear (12), elk (10), moose (7), waterfowl (6), pronghorn antelope (6), wolf (5), and cougar (4). For all of these species, governments restrict the ability of landowners to kill depredating animals because the animals are so valuable. In many cases, government agencies also have taken actions that have increased the populations of these species. Under these circumstances, some governments acknowledge that their actions may be responsible for an increase in wildlife damage and compensate landowners for it. In contrast, no state currently compensates for damage caused by voles because the government does not take actions to increase vole numbers, nor does it restrict the ability of landowners to kill voles (although there are restrictions on the use of toxicants due to environmental concerns).

Although compensation programs are often perceived to be an appealing alternative to lethal control, not everyone likes them (Olsen 1991). One problem is that they can be expensive. Idaho's compensation program cost $500,000 in 1988, and Wisconsin's costs about 1 million dollars annually (Rimbey et al. 1991; Wagner et al. 1997). Wildlife agencies are generally opposed to these programs because they do not want to become trapped in a payment system for an indefinite period of time. They also worry that compensation payments do not provide incentives for landowners to solve their own problems (Wagner et al. 1997). To avoid this latter problem, some states help landowners acquire resources needed for damage prevention rather than compensating farmers with money. For instance, some states supply the materials for a deer-proof fence and require the landowner to provide the labor to build it.

In Idaho and Utah, only 32% of farmers and 23% of nonfarmers approved of the idea of compensating farmers for damage caused by sandhill cranes (McIvor and Conover 1994a). Kellert (1979) found that only 7% of cattle ranchers, 11% of sheep ranchers, and 26% of the general public approved of using tax monies to compensate ranchers for livestock killed by coyotes. Only 31% of Iowa farmers (Gabrey et al. 1993) and 9% of Wisconsin farmers (Craven 1989) were in favor of compensation for turkey damage to their crops. It is not surprising that members of the general public might not approve of their tax dollars being used to compensate ranchers and farmers for wildlife damage, but the low level of support among the potential beneficiaries is more surprising. This trend may result because most compensation programs require a great deal of paperwork and time and then only pay for a fraction of the actual cost of the damage. States and provinces often set aside a limited amount of money each year to compensate for damage, and there are usually more claims than funds. In some years, livestock owners in Utah who were eligible for compensation for losses from bears and mountain lions received less than 25% of the actual losses (Wagner et al. 1997). Utah's livestock producers also complain that they are only compensated for their losses when they can prove the animal was killed by a bear or lion but that, in most cases, the carcass is never found or is too badly decomposed to ascertain what predator killed it. To overcome this problem, Alberta compensates cattlemen 100% of the value of confirmed kills and 80% of the value of missing cattle (Bjorge and Gunson 1995).

Disagreements between agricultural producers and government agencies in charge of compensation programs are common because agency employees accuse farmers of overestimating losses, and farmers complain that agencies deliberately underestimate them. Disagreements over damage estimates can affect the relationship between the agricultural community and the wildlife agency. Neutral third parties, such as crop adjusters from insurance companies, could be used to estimate losses and minimize conflicts between producers and agencies.

ARE PEOPLE'S PERCEPTIONS ABOUT WILDLIFE DAMAGE ACCURATE?

One interesting question is whether perceptions of wildlife damage are accurate. That is, can people accurately estimate their losses to wildlife and, if not, do they generally overestimate or underestimate wildlife damage? Many wildlife biologists believe that farmers consistently overestimate wildlife damage, but few efforts have been made to document this (Conover and Decker 1991). There are several factors that reduce a producer's ability to estimate wildlife damage accurately.

One factor is variation in the conspicuousness of the damage. Bird damage to corn ears, for example, is very noticeable because the husks are shredded and discolored. Wakeley and Mitchell (1981) found that farmers' estimates of bird damage to their corn fields were four times higher than those found by field sampling because an ear's husk may be shredded, but few kernels consumed. Likewise, farmers overestimated damage to ripening grain fields by sandhill cranes (McIvor and Conover 1994b) and damage to ripening grain fields by turkeys (Gabrey et al. 1993) because these birds forage mostly along the edges of fields but rarely in the interior. Farmers normally survey their fields from the perimeter and consequently tend to overestimate damage by these species. For the same reason, farmers often underestimate damage caused by Canada geese in growing grain fields because geese forage mainly in the interior of fields (Conover 1988).

Farmers also are less likely to detect damage which occurs slowly over the course of weeks or months. This situation might occur if a small group of deer forages nightly in an alfalfa field. From one day to the next, there is little change in the appearance of the crop and, hence, damage may go unrecognized. In these cases, damage can only be determined if parts of the field are ungrazed, allowing the farmers to assess how tall the crop would have been if undamaged by wildlife (Conover 1988).

The conspicuousness of different wildlife species can also bias results. For instance, farmers blamed damage to their newly planted grain fields on turkeys because these birds are diurnal, gregarious, and conspicuous. In reality, much of the damage was caused by less visible species: squirrels and deer (Gabrey et al. 1993). In fact, most turkeys foraging in oat and alfalfa fields during the summer are eating insects and not damaging the crop (Wright et al. 1989). Fish hatchery managers may overestimate the amount of fish taken by birds that forage during the day and underestimate the amount of damage caused by nocturnal birds, such as the black-crowned night heron (Pitt and Conover 1996). In Pennsylvania, hatchery managers believed that great blue herons and belted kingfishers were responsible for most of their losses, because these birds were conspicuous and had a reputation as "fish thieves." However, most fish were actually taken by mallards and common grackles (Parkhurst et al. 1992). Coyotes hunt mostly during the night and ranchers markedly underestimated their losses of lambs and ewes to these predators (Stoddart et al. 2001). Damage caused by other species that forage at night, such as deer or elk, may be underestimated by farmers as well.

Some commodities are difficult to survey for signs of wildlife damage. In these cases, losses are estimated by counting the number of depredating animals foraging on a crop. This method of estimating losses may result in biased estimates. For instance, a fish farmer cannot count his fish while they are growing and therefore must estimate bird predation by noting the number of birds at his ponds (Parkhurst et al. 1992). Many farmers assume that all of the animals present are meeting their nutritional needs from the fish. If this is not the case, damage will be less than perceived. In contrast, if there is a turnover in the animals during the day, farmers may underestimate the total number of birds that are feeding at their facility.

People tend to worry more about new threats than old, stable ones. New diseases are more alarming to us because of their unknown potential severity. The same is true for wildlife damage. Agricultural producers are more concerned with new or increasing threats and tend to overestimate their severity. For instance, McIvor and Conover (1994a) found that farmers were more concerned about damage by sandhill cranes than by deer, although the latter caused more damage. However, crane damage was a new problem for local farmers, whereas deer damage was stable; farmers had become accustomed to deer damage and worried less about it. Flyger and Thoerig (1962) and Sayre and Decker (1990) also noted that farmers were less tolerant of deer damage in areas where deer had only recently arrived. Despite difficulties in estimating losses, I believe that most agricultural producers are reasonably accurate in estimating their loses, a conclusion that Wywialowski (1996) also reached.

SUMMARY

In the last few decades, the goal of wildlife management has shifted from trying to maximize wildlife populations to a more difficult goal of trying to maximize wildlife values for society. A major difficulty in trying to achieve this goal is that the benefits and liabilities of wildlife fall unevenly upon different segments of society. This causes disagreements over what the ideal wildlife population really is and how wildlife should be managed. For this reason, each stakeholder group has its own "wildlife acceptance capacity." When combined, these become the "cultural carrying capacity" or the wildlife population which everyone can tolerate.

Important stakeholders in wildlife damage management often include agricultural producers, hunters, fur trappers, animal welfare activists, animal rights activists, rural residents, and metropolitan residents. Experiencing wildlife problems can change one's perception of wildlife and one's tolerance of wildlife damage, especially if the problem is serious and chronic. People suffering from wildlife damage usually are reasonably accurate in estimating the extent of wildlife damage. One way to alleviate wildlife damage problems is to increase the injured persons' tolerance for damage by increasing their appreciation of wildlife, expanding their knowledge, allowing them to reap economic benefits from wildlife, or compensating them for their losses.

LITERATURE CITED

Berg, N. A., USDA goals for strengthening private land management in the 1980s. *Trans. North Am. Wildl. Nat. Res. Conf.*, 46, 137–145, 1986.

Berryman, J. H., Needed now: an action program to maintain and manage wildlife habitat on private lands. P. 6–10 in R. T. Dumke, G. V. Burger, and J. R. March, Eds., *Wildlife Management on Private Lands*. Wisconsin Chapter, The Wildlife Society, Madison, 1981.

Bingham, G., Seeking consensus on resource management. *Trans. North Am. Wildl. Nat. Res. Conf.*, 62, 127–134, 1997.

Bjorge, R. R. and J. R. Gunson, Evaluation of wolf control to reduce cattle predation in Alberta. *J. Range Manage.*, 38, 483–487, 1985.

Brown, T. L., D. J. Decker, and C. P. Dawson, Willingness of New York farmers to incur white-tailed deer damage. *Wildl. Soc. Bull.*, 6, 235–239, 1978.

Burger, G. V. and J. G. Teer, Economic and socioeconomic issues influencing wildlife management on private lands. P. 252–278 in R. T. Dumke, G. V. Burger, and J. R. March, Eds., *Wildlife Management on Private Lands*. Wisconsin Chapter, The Wildlife Society, Madison, 1981.

Campa III, H., S. R. Winterstein, R. B. Peyton, G. R. Dudderar, and L. A. Leefers, An evaluation of a multidisciplinary problem: ecological and sociological factors influencing white-tailed deer damage to agricultural crops in Michigan. *Trans. North Am. Wildl. Nat. Res. Conf.*, 62, 431–440, 1997.

Conover, M. R., Effect of grazing by Canada geese on the winter growth of rye. *J. Wildl. Manage.*, 52, 76–80, 1988.

Conover, M. R., Wildlife management by metropolitan residents in the United States: practices, perceptions, costs, and values. *Wildl. Soc. Bull.*, 25, 306–311, 1997.

Conover, M. R., Perceptions of American agricultural producers about wildlife on their farms and ranches. *Wildl. Soc. Bull.*, 26, 597–604, 1998.

Conover, M. R. and D. J. Decker, Wildlife damage to crops: perceptions of agricultural and wildlife professionals in 1957 and 1987. *Wildl. Soc. Bull.*, 19, 46–52. 1991.

Craven, S. R., Farmer attitudes toward wild turkeys in southwestern Wisconsin. In *Proc. East. Wildl. Damage Control Conf.*, 4, 113–119, 1989.

Curtis, P. D., R. J. Stout, B. A. Knuth, L. A. Myers, and T. M. Rockwell, Selecting deer management options in a suburban environment: a case study from Rochester, New York. *Trans. North Am. Wildl. Nat. Res. Conf.*, 58, 102–116, 1993.

De Boer, W. F. and D. S. Baquete, Natural resource use, crop damage and attitudes of rural people in the vicinity of the Maputo Elephant Reserve, Mozambique. *Environ. Conserv.*, 25, 208–218, 1998.

Decker, D. J. and T. L. Brown, Fruit growers' vs. other farmers' attitudes toward deer in New York. *Wildl. Soc. Bull.*, 10, 150–155, 1982.

Decker, D. J. and T. L. Brown, How the animal rightists view the "wildlife management–hunting system." *Wildl. Soc. Bull.*, 15, 599–602, 1987.

Decker, D. J. and T. A. Gavin, Public attitudes toward a suburban deer herd. *Wildl. Soc. Bull.*, 15, 173–180, 1987.

Decker, D. J., G. F. Mattfeld, and T. L. Brown, Influence of deer damage on farmers' perceptions of deer population trends: important implications for managers. In *Proc. East. Wildl. Damage Control Conf.*, 1, 191–194, 1983.

Decker, D. J. and K. G. Purdy, Toward a concept of wildlife acceptance capacity in wildlife management. *Wildl. Soc. Bull.*, 16, 58–62, 1988.

Diefenbach, D. R., W. L. Palmer, and W. K. Shope, Attitudes of Pennsylvania sportsmen towards managing white-tailed deer to protect the ecological integrity of forests. *Wildl. Soc. Bull.*, 25, 244–251, 1997.

Doig, H. E., The importance of private lands to recreation. P. 7–10 in *Recreation on Private Lands: Issues and Opportunities*. President's Commission on Americans Outdoors, Washington D.C., 1986

Ehrenfeld, D., Conservation and the rights of animals. *Conserv. Biol.*, 5, 1–3, 1991.

Feder, G. and R. Slade, The role of public policy in the diffusion of improved agricultural technology. *Am. J. Agric. Econ.*, 67, 423–428, 1985.

Flyger, V. and T. Thoerig, Crop damage caused by the Maryland deer. In *Proc. Southeastern Assoc. Game Fish Commissioners*, 16, 45–52, 1962.

Gabrey, S. W., P. A. Vohs, and D. H. Jackson, Perceived and real crop damage by wild turkeys in Northeastern Iowa. *Wildl. Soc. Bull.*, 21, 39–45, 1993.

Green, D., G. R. Askins, and P. D. West, Public opinion: obstacle or aid to sound deer management? *Wildl. Soc. Bull.*, 25, 367–370, 1997.

Guynn, D. E., Miracle in Montana — Managing conflicts over private lands and public wildlife issues. *Trans. North Am. Wildl. Nat. Res. Conf.*, 62, 146–154, 1997.

Heinen, J. T., Park–people relations in Kosi Tappu Wildlife Reserve, Nepal: a socio-economic analysis. *Environ. Conserv.*, 20, 25–34, 1993.

Hewitt, D. G. and T. A. Messmer, Responsiveness of agencies and organizations to wildlife damage: policy process implications. *Wildl. Soc. Bull.*, 25, 418–428, 1997.

Hooper, J. K., Animal welfarists and rightists: insights into an expanding constituency for wildlife interpreters. *Legacy,* (Nov./Dec. 1992), 20–25, 1992.

Hygnstrom, S. E., R. M. Timm, and G. E. Larson, *Prevention and Control of Wildlife Damage*. University of Nebraska Cooperative Extension, Lincoln, 1994.

Kellert, S. R., Public attitudes toward critical wildlife and natural habitat issues. U.S. Fish and Wildlife Service Report, Washington, D.C., 1979.

Kellert, S. R., Americans' attitudes and knowledge of animals. *Trans. North Am. Wildl. Nat. Res. Conf.*, 45, 111–124, 1980a.

Kellert, S. R., American attitudes toward and knowledge of animals: an update. *Int. J. Study Anim. Problems*, 1, 87–119, 1980b.

Kellert, S. R., Contemporary values of wildlife in American society. P. 31–59 in W. W. Shaw and E. H. Zube, Eds., *Wildlife Values.* U.S. Forest Service., Rocky Mountain Forest and Range Experiment Station, Instructional Service Report 1, 1980c.

Kelly, R. G., Forests, arms, and wildlife in Vermont: a study of landowner values. P. 64–72 in R. T. Dumke, G. V. Burger, and J. R. March, Eds., *Wildlife Management on Private Lands*. Wisconsin Chapter, The Wildlife Society, Madison, 1981.

Kirby, S. B., K. M. Babcock, S. L. Sheriff, and D. J. Witter, Private land and wildlife in Missouri: a study of farm operator values. P. 88–101 in R. T. Dumke, G. V. Burger, and J. R. March, Eds., *Wildlife Management on Private Lands*. Wisconsin Chapter, The Wildlife Society, Madison, 1981.

Leopold, A., *Game Management*. Charles Scribner's Sons, New York, 1933.

Loker, C. A., D. J. Decker, and S. J. Schwager, Social acceptability of wildlife management actions in suburban areas: three cases from New York. *Wildl. Soc. Bull.*, 27, 152–159, 1999.

Matthews, O. P., Who owns wildlife? *Wildl. Soc. Bull.*, 14, 459–465, 1986.

McIvor, D. E. and M. R. Conover, Perceptions of farmers and non-farmers toward management of problem wildlife. *Wildl. Soc. Bull.*, 22, 212–219, 1994a.

McIvor, D. E. and M. R. Conover, Impact of greater sandhill cranes foraging on corn and barley crops. *Agric. Ecosystems Environ.*, 49, 233–237, 1994b.

Messmer, T. A., L. Cornicelli, K. J. Deehu, and D. G. Hewitt, Stakeholder acceptance of urban deer management techniques. *Wildl. Soc. Bull.*, 25, 360–366, 1997.

Messmer, T. A. and C. E. Dixon, Extension's role in achieving hunter, landowner, and wildlife agency objectives through Utah's big game posted hunting unit program. *Trans. North Am. Wildl. Nat. Res. Conf.*, 62, 47–56, 1997.

Messmer, T. A., C. A. Lively, D. MacDonald, and S. A. Schroeder, Motivating landowners to implement conservation practices using calendars. *Wildl. Soc. Bull.*, 24, 757–763, 1996.

Messmer, T. A. and S. A. Schroeder, Perceptions of Utah alfalfa growers about wildlife damage to their hay crops: implications for managing wildlife on private land. *Great Basin Nat.*, 56, 254–260, 1996.

Miller, J. E. and P. T. Bromley, Wildlife management on Conservation Reserve Program land: the farmer's view. *J. Soil Water Conserv.*, 44, 438–440, 1989.

Minnis, D. L. and R. B. Peyton, Cultural carrying capacity: modeling a notion. P. 19–34 in J. B. McAninch, Ed., *Urban Deer: A Manageable Resource?* North Central Section, The Wildlife Society, St. Louis, MO, 1995.

Musgrave, R. S. and M. A. Stein, *State Wildlife Laws Handbook.* Government Institutes, Rockville, MD, 1993.

Newmark, W. D., N. L. Leonard, H. I. Sariko, and D. G. M. Gamassa, Conservation attitudes of local people living adjacent to five protected areas in Tanzania. *Biol. Conserv.*, 63, 177–183, 1993.

Olsen, L., Compensation: giving a break to ranchers and bears. *West. Wildlands,* 17; 25–29, 1991.

Parkhurst, J. A., R. P. Brooks, and D. E. Arnold, Assessment of predation at trout hatcheries in central Pennsylvania. *Wildl. Soc. Bull.*, 20, 411–419, 1992.

Parry, D. and B. Campbell, Attitudes of rural communities to animal wildlife and its utilization in Chobe Enclave and Mababe Depression, Botswana. *Environ. Conserv.*, 23, 207–217, 1992.

Pease, J. L., Attitudes and behaviors of Iowa farmers toward wildlife. Ph.D. dissertation, Iowa State University, Ames, 1992.

Pitt, W. C. and M. R. Conover, Predation at Intermountain West fish hatcheries. *J. Wildl. Manage.*, 60, 616–624. 1996.

Richards, R. T. and R. S. Krannich, The ideology of the animal rights movement and activists' attitudes towards wildlife. *Trans. North Am. Wildl. Nat. Res. Conf.*, 56, 363–371, 1991.

Rimbey, N. R., R. L. Gardner, and P. E. Patterson, Wildlife depredation policy development. *Rangelands,* 13, 272–275, 1991.

Robel, R. J., A. D. Dayton, F. R. Henderson, R. L. Meduna, and C. W. Spaeth, Relationships between husbandry methods and sheep losses to canine predators. *J. Wildl. Manage.*, 45, 894–911, 1981.

Sayre, R. W. and D. J. Decker, Deer damage to the ornamental horticulture industry in suburban New York: extent, nature, and economic impact. Human Dimension Research Unit Series Report, 90–1, Department of Natural Resources, Cornell University, Ithaca, NY, 1990.

Schmidt, R. H., Animal welfare and wildlife management. *Trans. North Am. Wildl. Nat. Res. Conf.*, 54, 468–475, 1989.

Schmidt, R. H., Why do we debate animal rights? *Wildl. Soc. Bull.*, 18, 459–461, 1991a.

Schmidt, R. H., Influence of the animal rights movement on range management activities — productive directions. *Rangelands,* 13, 276, 1991b.

Svoboda, F. J., Minnesota landowner attitudes toward wildlife habitat management. *Trans. North Am. Wildl. Nat. Res. Conf.*, 49, 154–158, 1984.

Stoddart, L. C., R. E. Griffiths, and F. F. Knowlton, Coyote responses to changing jackrabbit abundance affect sheep predation. *J. Range Manage.*, 54, 15–20, 2001.

Stout, R. J., S. A. Knuth, and P. D. Curtis, Preferences of suburban landowners for deer management techniques: a step towards better communication. *Wildl. Soc. Bull.*, 25, 348–359, 1997.

Stout, R. J., R. C. Stedman, D. J. Decker, and B. A. Knuth, Perceptions of risk from deer-related vehicle accidents: implications for public preferences for deer herd size. *Wildl. Soc. Bull.*, 21, 237–249. 1993.

Tanner, G. and R. W. Dimmick, An assessment of farmers' attitudes toward deer and deer damage in west Tennessee. In *Proc. East. Wildl. Damage Control Conf.*, 1, 195–199, 1983.

Tchamba, M. N., History and present status of the human/elephant conflict in the Waza–Logone region, Cameroon, West Africa. *Biol. Conserv.*, 75, 35–41, 1996.

U. S. Bureau of the Census, *Statistical Abstract of the United States: 111th ed.* U.S. Government Printing Office, Washington, D.C., 1992.

U. S. Department of the Interior. 1991 *National Survey of Fishing, Hunting, and Wildlife-associated Recreation.* U.S. Government Printing Office, Washington, D.C., 1993.

Wagner, K. K., R. H. Schmidt, and M. R. Conover, Compensation programs for wildlife damage in North America. *Wildl. Soc. Bull.*, 25, 312–319, 1997.

Wakeley, J. S. and R. C. Mitchell, Blackbird damage to ripening field corn in Pennsylvania. *Wildl. Soc. Bull.*, 9, 52–55, 1981.

Wallace, M. S., H. L. Stribling, and H. A. Clouts, Factors influencing land access selection by hunters in Alabama. *Trans. North Am. Wildl. Nat. Res. Conf.*, 54, 183–189, 1989.

Warner, R. E., An adoption model for roadside habitat management by Illinois farmers. *Wildl. Soc. Bull.*, 11, 238–249, 1983.

Wigley, T. B. and M. A. Melchiors, State wildlife management programs for private land. *Wildl. Soc. Bull.*, 15, 580–584, 1987.

Wright, R. G., R. N. Paisley, and J. F. Kubisiak, Farmland habitat use by wild turkeys in Wisconsin. In *Proc. East. Wildl. Damage Control Conf.*, 4, 120–126, 1989.

Wywialowski, A. P., Wildlife damage to field corn in 1993. *Wildl. Soc. Bull.*, 24, 264–271, 1996.

Zinn, H. C. and W. F. Andelt, Attitudes of Fort Collins, Colorado, residents towards prairie dogs. *Wildl. Soc. Bull.*, 27, 1098–1106, 1999.

CHAPTER 16

Developing an Integrated Approach

"I adjure you, ye mice here present, that ye neither injure nor suffer another mouse to do so. I give you yonder field [the neighbor's] to use but if ever I catch you here again, by the mother of God I will rend you in seven pieces."

An ancient Greek "rat letter" written over 2000 years ago and left inside a building for rats to read

"There are no easy solutions to problems caused by wildlife. If there were, they wouldn't be problems."

Anonymous

"I have not failed, I have found 10,000 things that won't work."

Thomas Edison before he invented the light bulb

In the management of wildlife damage, there are no panaceas — no magic bullets — but there are techniques that can reduce wildlife damage and alleviate human–wildlife conflicts. By considering all potential solutions (Figure 16.1), wildlife damage managers can select the most effective, cost-efficient, humane, and socially acceptable. Usually, a combination of techniques will be more effective than any single one, especially if they are used in an integrated fashion. To illustrate this integrated approach, this chapter examines three vexing wildlife problems and explores potential methods to alleviate them.

REDUCING BLACKBIRD DAMAGE TO SUNFLOWERS

In the Great Plains of North America, sunflowers have become a popular cash crop. Unfortunately, sunflower seeds also are popular with red-winged blackbirds, yellow-headed blackbirds, and common grackles (Figure 16.2). Bird damage occurs

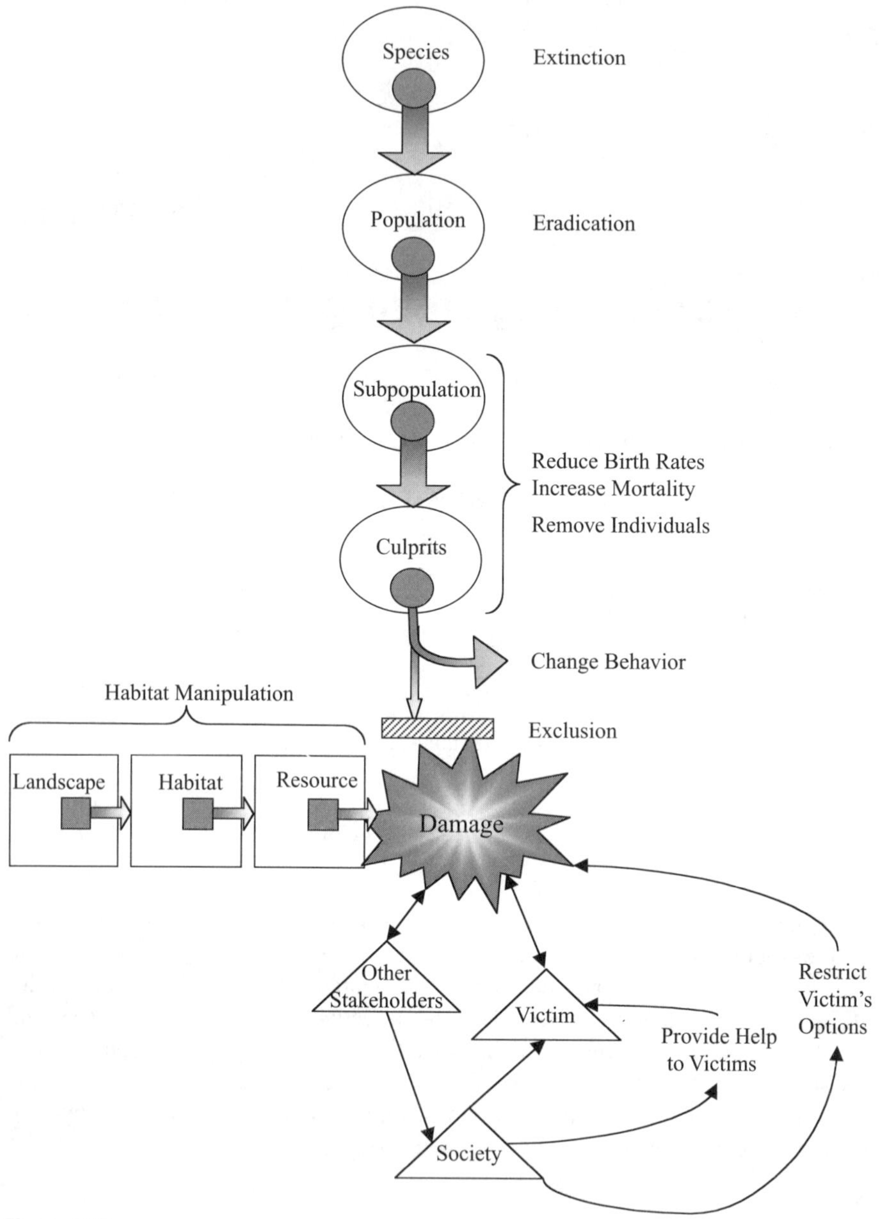

Figure 16.1

in the late summer or fall when sunflower seeds ripen and the blackbirds forage in flocks numbering in the thousands. A large flock can destroy an entire sunflower field in a few days. Each year, these birds damage $5 to 10 million worth of sunflower crops (Hothem et al. 1988). This loss only represents about 2% of the total sunflower crop, but damage is not evenly distributed across all fields. Instead, most sunflower fields have little damage, whereas a few suffer great losses (Figure 16.3).

Figure 16.2 Flocks of yellow-headed blackbirds (A) and red-winged blackbirds (B) forage in sunflower fields in the fall. (Courtesy of George Linz and U.S.D.A. Wildlife Services. With permission.)

Lethal Control

One method of alleviating blackbird damage to sunflowers is lethal control. Blackbirds are migratory birds that are protected by the federal government and fall under the jurisdiction of the Federal Migratory Bird Treaty Act, signed by the U.S., Canada, and Mexico. This treaty contains a provision allowing for the lethal control of blackbirds when they are "committing or about to commit depredations upon ornamental or shade trees, agricultural crops, livestock, or wildlife, or when con-

Figure 16.3 Large flocks of blackbirds can destroy a sunflower field in a few days. (Courtesy of George Linz and U.S.D.A. Wildlife Services. With permission.)

centrated in such numbers and manner as to constitute a health hazard or other nuisance" (Linz and Hanzel 1997).

Some farmers fire shotguns at blackbirds foraging in their sunflower fields. This method is more effective in scaring birds away than in reducing the blackbird population or eliminating culprits. Drawbacks with shooting blackbirds are that it is controversial, labor intensive, and dangerous, and that shotgun shells are expensive (Linz and Hanzel 1994).

Lethal toxicants, such as Avitrol®, can be used to reduce bird damage in sunflower fields (Besser et al. 1984; Knittle et al. 1988). Avitrol® consists of cracked-corn bait treated with 4-aminopyridine. When a blackbird consumes one of the treated baits, it gives distress calls, flies erratically, goes through convulsions, and dies within an hour. Because blackbirds forage in large flocks, other birds observe these distressed individuals, become alarmed, and leave the field. To avoid killing nontarget birds, Avitrol® should be applied only to the interiors of sunflower fields because most other avian species forage along the edges of these fields.

DRC-1339 (3-chloro-4-methylbenzamine HCl) is another toxicant sometimes used to kill blackbirds. It is usually sprayed on sunflower or corn seed baits and then distributed in sunflower fields. DRC-1339 can only be used by trained government employees; farmers cannot use it themselves. In some studies, distributing DRC-1339-treated bait in ripening sunflower fields was ineffective (Linz and Bergman 1996).

Fear-Provoking Stimuli

Fear-provoking stimuli can be used to scare blackbirds from sunflower fields. These stimuli should be employed before blackbird damage begins because it is much easier to prevent birds from starting to forage in the field than it is to stop them once they have established a foraging pattern. Propane cannons are the most popular method of scaring birds from sunflower fields, although scarecrows and the playback of bird distress calls are occasionally used (Linz et al. 1996a). Propane cannons should be elevated on stands above the crop and should be set to fire once every 5 to 10 minutes. If the cannon is moved every few days, it can protect 4 to 8 ha (Linz and Hanzel 1994). Shooting at a few birds with a shotgun can increase the effectiveness of propane cannons (Dolbeer 1994). Unfortunately, blackbirds habituate to propane cannons within a few days (see Chapter 10). This is a problem because sunflowers are vulnerable to bird damage for six to ten weeks (Linz and Hanzel 1997).

Harassment is also used to scare blackbirds out of sunflower fields. People patrolling fields can use pyrotechnic devices, such as cracker shells, and bird bangers. In some cases, the federal government uses airplanes to chase blackbirds from sunflower fields. Although expensive, this technique is effective in getting birds to move elsewhere (Linz et al. 1996a).

Chemical Repellents

One problem with trying to protect sunflower seeds with repellents is that blackbirds shell the husk off the sunflower seed and only eat the kernel inside. Hence, if a field were sprayed with a repellent, the chemical would land on the husks rather than on the seed kernels and would not be ingested (Figure 16.4). Currently, there are no repellents registered for use in sunflower fields.

Figure 16.4 It is difficult to protect sunflower seeds by spraying fields with a repellent because all of the repellent ends up on the seed husks which birds do not eat. (Photo courtesy of U.S.D.A. Wildlife Services.)

Diversion

One problem for the U.S. Fish and Wildlife Service is that blackbirds often roost in their national wildlife refuges in the hundreds of thousands. These birds leave the refuge during the day to forage in neighboring sunflower fields (Otis and Kilburn 1988). This upsets local farmers who view the refuges as the source of their problems. Hence, the federal government has planted diversionary fields in an effort to lure blackbirds from sunflower fields (Cummings et al. 1987). Unfortunately, sunflower seeds are so palatable to birds that the only effective lure crop for a sunflower field is another sunflower field. Therefore, diversion may not be as cost effective as allowing the birds to forage where they will and then compensating farmers for their losses. However, blackbirds prefer oilseed cultivars over confectionery cultivars and the latter sell for higher prices (Linz and Hanzel 1997). Hence, it makes economic sense to lure birds from confectionery sunflower fields to lure fields planted with oilseed sunflowers. In one novel approach to the use of diversion to protect sunflower fields from blackbirds, Linz and Hanzel (1997) recommended that plowing or tilling harvested sunflower fields should be delayed because such fields provide attractive feeding sites for blackbirds, luring them from unharvested fields.

Habitat Modification

The main factor influencing whether a sunflower field encounters bird damage is its proximity to marshes that blackbirds use for roosting (Otis and Kilburn 1988). One way to reduce bird damage to sunflowers is to plant fields that are near large roosts with crops that birds do not bother, such as soybeans (Linz and Hanzel 1997). Alternatively, if blackbirds can be convinced to roost elsewhere, the amount of damage in neighboring sunflower fields will decrease. Linz et al. (1992, 1996b) showed that blackbirds can be discouraged from using a marsh for roosting by spraying it with Rodeo®, a glyphosate herbicide that kills the cattails in large streaks (Figure 16.5). This makes the marsh more attractive to ducks seeking open water and less attractive to blackbirds.

Good weed control in sunflower fields can also help to reduce bird damage. Weed seeds mature before sunflowers and attract blackbirds to a field (Linz and Hanzel 1994). Once there, the birds are quick to realize when the sunflower seeds are ripe and then switch their feeding to sunflower seeds.

Isolated sunflower fields are more likely to suffer high loss rates from blackbirds because the birds from nearby roosts may concentrate their foraging in these fields. In contrast, large sunflower fields and large clusters of sunflower fields will suffer a lower proportion of damage because the damage caused by blackbirds from local roosts can be distributed over a large number of acres. So, it may be wise for farmers to coordinate their efforts and plant all sunflower crops in the same area.

Sunflower seeds are vulnerable to damage from the time they start to mature until the crop is harvested. Anything that shortens this period of vulnerability will reduce losses. Hence, farmers should consider harvesting fields vulnerable to blackbirds early. A harvesting delay of only a few days can make a big difference in the

Figure 16.5 Blackbirds can be discouraged from roosting in specific cattail marshes by an aerial application of a herbicide that kills cattails in broad streaks. (Courtesy of George Linz and U.S.D.A. Wildlife Services. With permission.)

amount of bird damage (Linz et al. 1996a). Desiccants cause plants to dry more quickly and their application makes it possible to harvest fields earlier.

During fall migration, blackbirds pass quickly through some areas, making it possible to time the planting of sunflower fields so that the sunflowers are vulnerable either before or after the migratory flocks are expected in the area. Also, all fields in an area should be planted so that they enter the period when they are most vulnerable to bird damage at the same time. That way, bird damage will be distributed across as many fields as possible (Linz et al. 1996a).

There has been some research focused on identifying and breeding bird-resistant sunflower hybrids (Hanzel and Gulya 1993). These resistant hybrids exhibit morphological traits that make it harder for birds to reach the seeds. For instance, the seed heads of resistant hybrids are more concave and more likely to face downward toward the ground (Linz et al. 1996a). These may be useful in areas where heavy levels of bird damage are expected.

Human Dimensions

Sunflower growers find damage by blackbirds frustrating, in part, because it occurs late in the growing season. Growers may be anticipating a large harvest, only to watch it be destroyed by blackbirds. One way to address this bird problem is to compensate growers who suffer losses above a certain threshold. Perhaps a better approach may be to offer cash incentives to sunflower growers located near large roosts, especially those on federal migratory bird refuges, to grow a diversionary crop or make habitat modifications that may reduce the vulnerability of sunflowers to bird damage.

Developing an Integrated Approach

Much of the bird damage to sunflower fields can be prevented by not growing sunflowers near large roosts, clustering sunflower fields together in both space and time (i.e., so they all ripen at the same time), or leaving harvested fields unplowed to provide an alternate place to forage. However, there are practical limitations to some of these approaches, making farmers less likely to implement them. For instance, farmers usually practice crop rotation and may need to plant sunflowers in vulnerable areas on occasion. Also, sunflowers can generate more money than other crops, causing farmers to be reluctant about growing less valuable crops. Farmers prefer to have their fields ripen over an extended period of time so they are less vulnerable to a drought or a period of bad weather. Other techniques, such as the use of fear-provoking stimuli or lethal control, may offer some help if a bird problem does occur, but they are not as effective as farmers would hope (Linz et al. 1996a). Hence, an integrated management approach that uses a combination of these methods offers the best course of action for alleviating problems. Research also continues to develop better and less expensive methods of protecting sunflower fields from birds. The recent discovery that blackbirds might be discouraged from roosting in a cattail marsh if vegetation is reduced may become an important way to alleviate this problem in an environmentally friendly and humane manner (Linz et al. 1996b).

REDUCING BIRD PREDATION AT FISH FARMS

Bird predation at fish farms is a widespread problem in the U.S., occurring at most fish farms or aquaculture facilities (Scanlon et al. 1978; Parkhurst 1989). Large economic losses are sustained by trout farms in the northern U.S., catfish farms in the South, crayfish farms in Louisiana, and minnow farms in Arkansas (Parkhurst et al. 1987, 1992; Stickley and Andrews 1989; Hoy et al. 1989; Pitt and Conover 1996). Similar problems occur in Europe (Draulans and van Vessem 1985) and Australia (Barlow and Bock 1984). In addition to consuming fish, birds can also introduce diseases and parasites to the facilities (Swanson 1984; Peters and Neukirch 1986).

Birds that catch fish while swimming (e.g., double-crested cormorants) are the main problem at most catfish farms in the South (Figure 16.6). Trout are commonly produced in narrow raceways where there is moving water. These narrow confines and moving water prevent cormorants from easily foraging in raceways, but can produce ideal foraging opportunities for wading birds (e.g., great blue herons, black-crowned night herons; Figure 16.7). Which techniques should be employed to protect fish farms depend upon whether the facility contains raceways or ponds, the species of fish produced, the size of the facility, and the number and species of birds responsible for the damage. Time and cost factors also play important roles in determining the control method(s) to be utilized.

Lethal Control

All fish-eating birds in North America are considered to be migratory and, hence, are protected by the Migratory Bird Treaty Act. Therefore, permits must be obtained from the U.S. Fish and Wildlife Service before these birds may be trapped or killed.

Figure 16.6 Pelicans (A) and cormorants (B) are a problem at aquaculture facilities where fish are raised in large ponds.

Figure 16.7 Great blue herons (A), black-crowned night herons (B), and other wading birds pose problems at aquaculture facilities that raise trout in concrete raceways.

Many states have additional restrictions on the killing of these birds and require that state permits be obtained as well.

Populations of double-crested cormorants and colonial wading birds have increased dramatically in North America since the 1970s as the populations have responded to a ban on the sale of DDT (dichloro-diphenyl-trichloroethane), protection from human persecution, and an increasing food supply resulting from the growth of the aquaculture industry (Fleury and Sherry 1995). Hence, many bird problems facing fish farmers are of recent origin. For instance, during the last two decades, the catfish industry expanded rapidly in the Mississippi Delta. During the same period, the wintering population of double-crested cormorants in the area increased to over 50,000 birds. This was due to an increased winter survival of cormorants that feed at fish farms and a migratory shift in winter range to where the catfish farms are located (Glahn and Stickley 1995; Jackson and Jackson 1995). Each winter, 50,000 cormorants are capable of eating 940 metric tons of catfish or 20 million catfish fingerlings (Glahn and Brugger 1995; Glahn et al. 1999).

Programs to reduce local populations of double-crested cormorants have been implemented in areas where the birds have caused environmental damage. For example, the cormorant population in the St. Lawrence River estuary in Quebec was reduced over a four-year period from 17,000 to 10,000 pairs through a combination of egg oiling and shooting adults (Bedard et al. 1999). During 1992 and 1993, about 8000 double-crested cormorants were killed in the U.S. under depredation permits (Reinhold and Sloan 1999). In 1998, federal laws were changed so that commercial fish farms and government fish hatcheries located in 13 states could kill double-crested cormorants without a federal permit (Reinhold and Sloan 1999).

Lethal methods are most successful at fish farms when small numbers of birds are responsible for the damage. Usually, when lethal means are employed, the birds are shot. Before initiating a lethal control program, careful observations are required to determine which species of birds are responsible for the damage. Sometimes, fish farmers blame conspicuous fish-eating birds that forage during the day (e.g., kingfishers) for the damage when the real culprits are night feeders, such as black-crowned night herons (Parkhurst et al. 1992). Lethal techniques are most beneficial when used in an integrated program to enhance the effectiveness of nonlethal methods. For instance, some fish farmers use lethal control only to remove those individual birds that stop responding to nonlethal techniques.

Fear-Provoking Stimuli

Frightening techniques are most applicable for problems lasting only a few days or weeks because birds quickly habituate to them. Because fish farms have bird problems that last months or years, frightening techniques may be of limited effectiveness when used alone. The success of frightening devices depends upon the bird species causing damage, how long birds have been foraging at the site, and the proximity of alternative food sources.

Success occasionally results from using just one fear-provoking stimuli; however, success is more likely if combinations of fear-provoking stimuli are employed and if their location is frequently altered. Bird numbers at a fish hatchery can be reduced if birds are harassed as soon as they arrive at the hatchery and not allowed time to forage. Birds can be scared away by people patrolling the fish farm on foot, in vehicles, or in boats (Draulans and van Vessem 1985). Radio-controlled scale models of airplanes or boats and dogs can also be used to harass bird.

Many noise-making devices have been used to scare birds away from fish farms (Gorenzel et al. 1994). Noise originating from the shore may not scare birds that are foraging in the middle of the pond. Hence, cracker shells, whistle bombs, and screamers are more effective for larger ponds because they can be fired to explode over the middle of the pond. Propane cannons are particularly effective if used in combination with permits that allow a few birds to be shot (Littauer 1990a).

Scarecrows, especially when dressed like hunters or facility personnel, have shown some success in scaring birds that forage by day (Figure 16.8). Pop-up

Figure 16.8 Scarecrows can be used as fear-provoking stimuli. (From Hygnstrom et al. 1994.)

scarecrows (i.e., Scarey Man®) are available and are more effective than stationary ones (Stickley and King 1995; Stickley et al. 1995). Lights can be used to confuse, frighten, and temporarily blind night-feeding predators (Littauer 1990a). Bright flood lights should be placed so that they shine horizontally across a pond or raceway. As with noise-making devices, the effectiveness of visual scare devices is often short term as birds may quickly become accustomed to them (Draulans and van Vessem 1985).

One approach to reducing problems caused by cormorants is to disperse them from roosts located near fish farms. At night, double-crested cormorants roost by the hundreds in the tops of trees. During the day, cormorants leave these roosts to forage in local fish ponds. The number of birds using a roost can be reduced by harassing the birds at sunset when they return to the roost site. By forcing birds to abandon local roosts, the number of birds foraging at fish farms can be reduced by more than half (Mott et al. 1998; Reinhold and Sloan 1999).

Diversion

Fish are grown at much higher densities (5000 to 150,000 per hectare for catfish) at fish farms than occur in nature because farmers provide feed for the fish and, in some cases, oxygenate the water. Unfortunately, these high fish densities make it easier for birds to catch fish at fish farms than elsewhere. Farm-raised fish are also highly nutritious, making it difficult to divert fish-eating birds from fish farms. It is possible that birds can be diverted from eating valuable fish to eating crayfish or shad, which are easier to catch and swallow (Barlow and Bock 1984). By draining water bodies to concentrate crayfish and shad in shallow water, short-term diversions can be established.

Exclusion

Two types of physical barriers can be used for controlling bird predation at fish farms: complete exclosures (which totally exclude birds from gaining access to fish), and partial exclosures (which inhibit a bird's ability to land and feed). Complete exclosures (Figure 16.9) are very effective against bird damage and are created by erecting large nets over the ponds or raceways and supporting them with sturdy poles and wires (Salmon and Conte 1981; Littauer 1990b). Nets should be constructed high enough to allow people to work beneath them. Total exclusion is impractical for most large ponds due to the difficulties of spanning large distances with netting. Exclosures are expensive but can still be cost effective if they provide protection for many years (Littauer 1990b; Martin and Hagar 1990).

Partial exclusion involves placing overhead wires above the fish farms (Figure 16.9). They are less effective than total exclusion systems but are less

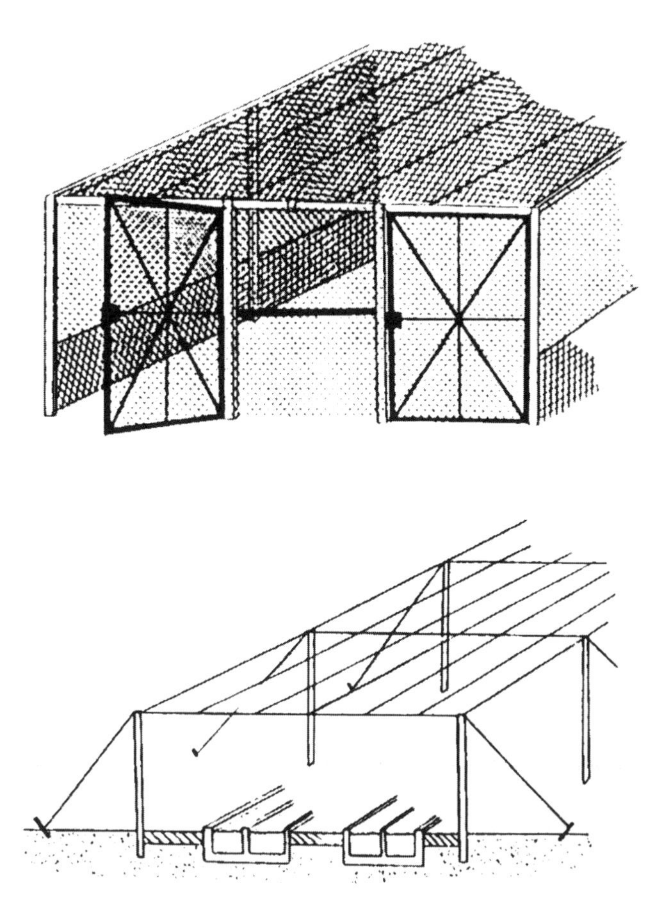

Figure 16.9 Bird predation at fish farms can be reduced by netting the facilities or erecting overhead wires. (From Hygnstrom et al. 1994.)

expensive to construct (Ostergaard 1981; Gorenzel et al. 1994). Overhead wires are most effective against birds that hunt from the air, such as terns, gulls, kingfishers, and ospreys, than those which catch prey while wading (e.g., great blue herons and black-crowned night herons) or swimming (e.g., cormorants). To guard against wading birds, the sides of the hatchery should be fenced to keep birds from walking to the water's edge (Gorenzel et al. 1994; Curtis et al. 1996).

Habitat Modification

Many bird depredation problems could be avoided by locating and designing fish farms to minimize bird problems. Predation problems can be expected when fish farms are constructed in areas where fish-eating birds are known to congregate. Care should be taken to avoid such areas or to take this factor into consideration when designing the facility. For example, the severity of bird depredation can be reduced if several fish farms are clustered together. Small facilities suffer a higher proportional loss than large ones, which are protected by the economics of scale. Other techniques include increasing the water depth of raceways and ponds to 1 m, which inhibits some wading birds, and increasing the height of sidewalls above the water level (Figure 16.10), which may prevent wading birds from fishing from the sidewalls (Salmon and Conte 1981). Problems can also be alleviated by creating ponds with steep sides, providing less room for wading birds to fish, and by removing perch sites near ponds or raceways. When they cannot be removed, perch sites can be covered with metal spines, porcupine wire, or other sharp objects, preventing

Figure 16.10 Bird predation at trout farms using concrete raceways can be reduced by keeping the water level low enough that wading birds standing on the concrete sides cannot easily reach down far enough to catch a fish (also note the use of overhead wires above the raceway).

birds from landing. Vegetation along pond banks should be kept short so that it will not provide cover for birds (Salmon and Conte 1981).

Adjustments to everyday operations may further help to reduce bird predation problems. More valuable or vulnerable fish, such as fry or fingerlings, should be located near buildings or areas of human activity (Moerbeek et al. 1987). Losses can be reduced by modifying stocking rates, size of fish stocked, and stocking time (Barlow and Bock 1984). Fish that feed at the water's surface are more susceptible to predation than those that feed below the surface. Thus, the use of floating feed may increase predation problems. Fish that are fed by hand may be conditioned to approach an overhead movement, which can be a fatal decision if the movement is caused by a fish-eating bird. Increasing the water's turbidity may help protect fish from avian predators because fish are harder to see and capture when the water is opaque (Van Vuren 1998).

Human Dimensions

Many fish farmers enjoy seeing birds at their facilities and are willing to tolerate low levels of damage. When losses exceed their tolerance, however, fish farmers can become frustrated due to government regulations that prevent farmers from taking action they believe will solve the problem (i.e., killing the birds). The problem is that, while a fish farmer may consider these birds to be pests and want to kill them, other people consider them to be a source of wonder and beauty. Hence, the use of lethal control of birds at fish farms is controversial (Williams 1992).

One way to reduce the reliance of fish farmers on lethal control is by providing them with accurate information about nonlethal techniques capable of alleviating their losses. Fish farmers often lack the money to construct bird-proof exclosures around their facilities. A solution to this financial problem is to provide zero-interest loans or cash incentives to fish farmers who are willing to enclose their facilities with bird-proof netting in lieu of using lethal techniques.

Developing an Integrated Approach

By using an integrated approach, fish farmers can obtain some relief from losses caused by avian predators. Ideally, fish farms should be located in areas where there is a concentration of fish farms rather than a concentration of fish-eating birds. Fish farms require considerable money to build and are designed to last for decades. Consequently, it often makes financial sense to enclose them in bird-proof exclosures rather than suffer losses from birds every year. If exclosures are not going to be used, fish ponds or raceways should be constructed to reduce their vulnerability to bird predation by having steep sides and high walls to discourage wading birds; nearby roost sites also should be removed. Some problems can be reduced by changing the density or size of fish to be stocked or the stocking season. People, radio-controlled airplanes or boats, trained dogs, and pyrotechnics can be used to harass birds. If all else fails, shooting birds may provide some relief at fish farms that have problems with a small number of birds. Shooting birds may also increase the effectiveness of fear-provoking stimuli.

PROTECTING GROUND-NESTING BIRDS FROM MAMMALIAN PREDATORS

High predation rates on ground-nesting birds and their eggs are a serious problem in areas where human modifications of the environment have resulted in a loss of high-quality nesting habitat (Sugden and Beyersbergen 1984; Cowardin et al. 1985; Greenwood et al. 1995) or when populations of red fox, raccoon, or striped skunk have increased (Sargeant et al. 1984, 1993). This combination has resulted in a long-term decline in nesting success of ducks. Many studies on upland nesting ducks conducted in the Prairie Potholes of North America have found that predation by red fox (Figure 16.11), striped skunk, raccoon, and American crow is the leading cause of nest failure (Cowardin et al. 1985; Greenwood et al. 1987; Beauchamp et al. 1996). Wildlife managers often are given the task of increasing the nesting success of birds in areas where there are high densities of predators. Fortunately, these managers have several techniques at their disposal.

Lethal Control

Many of the mammalian predators of ground-nesting birds in the Prairie Potholes region (e.g., raccoon, fox, coyote, and mink) are trapped for their fur. Low pelt prices and a decrease in the number of fur trappers have reduced the ability of fur trapping to control predator populations. In response, several studies have examined whether removing predators with traps (Figure 16.12) or toxicants can increase the nesting success of ducks (Balser et al. 1968; Duebbert and Kantrud 1974; Duebbert and Lokemoen 1980). Nesting success generally increases when all or most of the predators in an area are removed. Efforts to remove only one predator species from an area while leaving the others alone have generally been ineffective. For example,

Figure 16.11 Red foxes pose a great danger to upland-nesting ducks.

Figure 16.12 Predators that find their way into gull colonies can be trapped to improve nesting success.

Greenwood (1986) removed striped skunks, and Clark et al. (1995) removed American crows, but neither increased the nesting success of ducks, presumably due to the presence of other predator species.

Lethal control works best when applied to large areas, because immigrating predators then need to travel longer distances to replace those predators that were removed. When predators were trapped from large parcels of land (over 4000 ha) duck nesting success doubled (Garrettson et al. 1996), but trapping had less impact when used on small parcels (61 to 301 ha; Sargeant et al. 1995). Toxicants can be used to control predators more cheaply and over larger areas than traps, but they pose a greater risk to nontarget species.

Fertility Control

One potential way to reduce predator populations is through the use of reproduction inhibitors. Consumption of diethylstilbestrol in baits during a 19-day period around the first mating caused pregnancy termination in captive foxes (Linhart and Enders 1964), but field tests were unsuccessful in reducing reproductive success of foxes (Allen 1982). One problem is that red foxes, striped skunks, and raccoons all have high reproduction rates and/or high dispersal rates, which help these species overcome any population reduction efforts. Despite many years of research, a practical method to reduce fertility in wildlife populations has not been developed. To accomplish this goal, a team approach to fertility research would involve molecular biologists, reproductive physiologists, and wildlife biologists (Warren 1995). Such a team approach has been implemented by Australia's Vertebrate Biocontrol Centre to develop an effective means of reducing fertility of the introduced red fox.

Exclosures

Nesting success of ground-nesting birds can be improved by building a predator-proof fence around individual nests (Post and Greenlaw 1989; Deblinger et al. 1992; Melvin et al. 1992), nesting colonies (Forster 1975; Patterson 1977; Minsky 1980), and habitat patches used by nesting waterfowl (Lokemoen et al. 1982; Greenwood et al. 1990; LaGrange et al. 1995). Fences, although costly to construct, require little maintenance and can endure for several years (Figure 16.13) When costs are amortized over the expected life of the fence, this method often is more cost effective than other techniques (Lokemoen 1984; Goodrich and Buskirk 1995). One drawback with fences is that they are not very effective against avian predators or small mammalian predators, such as mink or ground squirrels. Fences also can trap precocial and semiprecocial young inside, unless the fences are equipped with special exits (Pietz and Krapu 1994; Trottier et al. 1994; Howerter et al. 1996). Another way to protect nesting birds from mammalian predators is to provide elevated nesting platforms (Figure 16.14), earth-filled culverts (Ball and Ball 1991), or islands (Lokemoen and Messmer 1993).

Repellents

Several studies have investigated whether conditioned taste aversions can be used to teach predators not to eat eggs. Sheaffer and Drobney (1986) tried to reduce predation on waterfowl nests by distributing eggs injected with lithium chloride, a chemical that causes nausea. Although consumption of treated eggs declined during their study, there was no evidence that predators reduced their predation on untreated waterfowl nests. Avery and Decker (1994) placed chemically treated eggs within the territories of crows and ravens. These birds became ill after consuming the treated eggs and averted from them. Additionally, predation on tern nests located near these

Figure 16.13 Predators can be excluded from patches of nesting habitat by erecting predator-proof fences.

Figure 16.14 Elevated nesting structures can be used to protect mallards from most mammalian predators.

nesting ravens decreased (Avery et al. 1995). Efforts to teach mammalian predators to stop depredating nests have been less successful.

A major problem with conditioned food aversions is that predators often learn to discriminate between treated and untreated eggs (Conover 1990, 1997; Avery et al. 1995). When this happens, predators stop eating treated eggs but continue to depredate nests. Due to inconsistent results, conditioned taste aversions cannot be recommended for use in protecting nesting birds. Furthermore, the U.S. Food and Drug Administration has not approved any chemicals for this use in the U.S.

Diversion

It may be possible to reduce predation on nests by providing alternative prey for local predators while birds are nesting. Several studies have shown a correlation between the abundance of alternate prey and the nesting success of oldsquaw (Pehrsson 1986), brant (Summers 1986; Anthony et al. 1991), and blue-winged teal (Byers 1974; Weller 1979). Two studies investigated the effect of providing supplemental food for predators on nesting success of birds. Crabtree and Wolfe (1988) increased nesting success of ducks in Utah by distributing dead fish in duck-nesting areas. On the other hand, supplemental food provided in areas of the Prairie Potholes region failed to increase nesting success (Greenwood et al. 1998).

Habitat Modification

One proposed method of reducing predation on ducks is to encourage coyotes to establish territories in areas where ducks nest (Sovada et al. 1995). The rationale is that coyotes defend their territories against foxes, which are a much greater threat to nesting birds (Sargeant et al. 1987). Coyotes also may suppress raccoon densities (Johnson et al. 1989).

Nesting success can also be improved by increasing the quantity of nesting cover at the landscape scale. In the prairies, waterfowl nest success is positively correlated to the amount of grassland in the area and negatively correlated to the amount of farmland (Greenwood et al. 1987, 1995; Ball 1996). These findings indicate that preserving large tracts of grasslands can aid breeding success of ground-nesting birds. In the U.S., there has been a landscape-scale program since 1985 (the U.S. Department of Agriculture's Conservation Reserve Program, or CRP) which has converted millions of hectares of marginal farmland to perennial grassland by providing cash incentives to farmers that enroll land in the program. In the states of North Dakota, South Dakota, Montana, and Minnesota, over 3 million ha of farmland were enrolled in the CRP (Greenwood and Sovada 1996). Ducks often nest successfully on CRP land, and this program has greatly increased the quantity and quality of nesting habitat for birds in the U.S. portion of the Prairie Potholes region (Reynolds et al. 1994).

A major tenet of the North American Waterfowl Management Plan is that nesting success can be improved by increasing the amount of dense nesting cover. One way to accomplish this is by periodically burning the grasslands (Kirsch et al. 1973; Higgins 1986), although the benefits vary among avian species (Johnson et al. 1994; Kruse and Bowen 1996). Livestock grazing can also be used to enhance nesting cover by stimulating plant growth (Barker et al. 1990; Kantrud and Higgins 1992; Bowen and Kruse 1993). In many cases, however, the creation of dense nesting cover did not improve the nesting success of ducks (Garrettson et al. 1996). Dense cover conceals nests from visually oriented predators, such as crows and gulls, but does not protect nests against predators that rely on olfaction, such as snakes, skunks, and foxes (Clark and Nudds 1991). It also may increase predation from ground squirrels, which prefer dense cover (Choromanski-Norris et al. 1989).

Human Dimensions

Many effective techniques in protecting nesting birds from predators (e.g., erecting nest structures, constructing islands, managing habitat) are labor intensive and can only be completed on small areas. If these programs are going to be implemented, landowners or other local individuals must take the initiative. Extension programs, which provide information and a sense of mission, can help motivate landowners to help nesting birds (Messmer et al. 1996).

Most landowners do not benefit financially if their lands produce ducks. Instead, their rewards are mostly intangible (e.g., the sense of satisfaction at seeing birds). Many farmers simply do not have the time or money to manage their property for the benefit of birds. One innovative program designed to overcome these problems is the Delta Waterfowl Foundation's Adopt-A-Pothole Project. This program matches farmers with suitable nesting habitat, but limited finances, with other people who have money, but not land. Through teamwork, these partnerships are able to make habitat improvements that neither partner could do by himself.

Developing an Integrated Approach

Wildlife managers whose mission is to enhance the recruitment of waterfowl face a difficult dilemma. Techniques relying on exclusion or habitat modification often are so expensive that they cannot be justified on economic grounds (i.e., too few ducks produced for the cost). Lethal control of predators is a more cost-effective way to increase duck production, but it is controversial and may lack public support. Additionally, lethal control may not generate the desired results. Most nonlethal techniques to protect birds from mammalian predators can only be applied under limited conditions and are only effective against certain predators. For small areas, nesting success can be improved by building nesting platforms, islands, or predator-proof fences. For larger areas, predation can be reduced by improving nesting habitat and encouraging coyotes to settle in the area. A well-designed management plan, which integrates different techniques and is adjusted based on the bird species needing protection and the local predators, can help boost avian recruitment. The results of the management plan should be monitored to determine how well the plan is actually increasing avian recruitment.

SUMMARY

Alleviating human–wildlife conflicts requires a careful balance between the needs of humans and wildlife. Wildlife damage managers often face vexing problems with no easy solutions. But there are techniques that can be employed to reduce the severity of a human–wildlife conflict. However, some proposed techniques are too expensive, unreliable, or not cost effective. Hence, wildlife damage managers often have to sift through the multitude of potential techniques to find one or, more likely, a combination of several techniques that can reduce wildlife damage to acceptable levels. Solutions may involve changing wildlife behavior, excluding animals, eliminating culprits, or reducing local subpopulations (Figure 16.1). Other solutions may involve making habitat changes or modifying the resource that needs protection. Human perceptions can be changed through education or by increasing human tolerance for wildlife, which involves increasing the value of the wildlife to the person suffering damage. The science of wildlife damage management has not progressed to the point where there is a panacea or magic bullet for every problem, but it has progressed to the point where there is no wildlife damage problem — no human–wildlife conflict — where we cannot do something to reduce the conflict and alleviate wildlife damage. It is incumbent upon wildlife damage managers to use their knowledge to identify an integrated solution which can reduce damage caused by wildlife to tolerable levels in a manner which is humane, environmentally benign, and socially acceptable. The goal is to create a world where humans and wildlife can coexist and have as little negative impact on each other as possible.

LITERATURE CITED

Allen, S. H., Bait consumption and diethylstilbestrol influence on North Dakota red fox reproductive performance. *Wildl. Soc. Bull.*, 10, 370–374, 1982.

Anthony, R. M., P. L. Flint, and J. S. Sedinger, Arctic fox removal improves nest success of black brant. *Wildl. Soc. Bull.*, 19, 176–184, 1991.

Avery, M. L. and D. G. Decker, Responses of captive fish crows to eggs treated with chemical repellents. *J. Wildl. Manage.*, 58, 261–266, 1994.

Avery, M. L., M. A. Pavelka, D. L. Bergman, D. G. Decker, C. E. Knittle, and G. M. Linz, Aversive conditioning to reduce raven predation on California least tern eggs. *Colonial Waterbird,* 18, 131–138, 1995.

Ball, I. J., Managing habitat to enhance avian recruitment. *Trans. North Am. Wildl. Nat. Res. Conf.*, 61, 109–117, 1996.

Ball, I. J., and S. K. Ball, Earth-filled culverts as nest sites for waterfowl. *Prairie Nat.,* 23, 85–88, 1991.

Balser, D. S., H. H. Dill, and H. K. Nelson, Effect of predator reduction on waterfowl nesting success. *J. Wildl. Manage.*, 32, 669–682, 1968.

Barker, W. T., K. K. Sedivec, T. A. Messmer, K. F. Higgins, and D. R. Hertel, Effects of specialized grazing systems on waterfowl production in south central North Dakota. *Trans. North Am. Wildl. Nat. Res. Conf.*, 55, 462–474, 1990.

Barlow, C. G. and K. Bock, Predation of fish in farm dams by cormorants, *Phalacrocorax* spp. *Aust. Wildl. Res.*, 11, 559–566, 1984.

Beauchamp, W. D., R. R. Koford, T. D. Nudds, R. G. Clark, and D. H. Johnson, Long-term declines in nest success of prairie ducks. *J. Wildl. Manage.*, 60, 247–257, 1996.

Bedard, J., A. Nadeau, and M. Lepage, Double-crested cormorant culling in the St. Lawrence River estuary: results of a five-year program. P. 147–154 in M. E. Tobin, Ed., *Symposium on Double-Crested Cormorants: Population Status and Management Issues in the Midwest.* U.S. Department of Agriculture, Animal and Plant Health Inspection Service, Technical Bulletin 1879, 1999.

Besser, J. F., D. J. Brady, T. L. Burst, and T. P. Funderberg, 4-Aminopyridine baits on baiting lanes protect sunflower fields from blackbirds. *Agric. Ecosystems Environ.*, 11, 281–290, 1984.

Bowen, B. S. and A. D. Kruse, Effects of grazing on nesting by upland sandpipers in south central North Dakota. *J. Wildl. Manage.*, 57, 291–301, 1993.

Byers, S. M., Predator–prey relationships on an Iowa waterfowl nesting area. *Trans. North Am. Wildl. Nat. Res. Conf.*, 39, 223–229, 1974.

Choromanski-Norris J., E. K. Fritzell, and A. B. Sargeant, Movements and habitat use of Franklin's ground squirrels in duck-nesting habitat. *J. Wildl. Manage.*, 53, 324–331, 1989.

Clark, R. G. and T. D. Nudds, Habitat patch size and duck nesting success: the crucial experiments have not been performed. *Wildl. Soc. Bull.*, 19, 534–543, 1991.

Clark, R. G., D. E. Meger, and J. B. Ignatiuk, Removing American crows and duck nesting success. *Can. J. Zoo.*, 73, 518–522, 1995.

Conover, M. R., Reducing mammalian predation on eggs by using a conditioned taste aversion to deceive predators. *J. Wildl. Manage.*, 54, 360–365, 1990.

Conover, M. R., Behavioral principles governing conditioned food aversions based on deception. P. 29–41 in J. R. Mason, Ed., *Repellents in Wildlife Management*. Colorado State University, Fort Collins, 1997.

Cowardin, L. M., D. S. Gilmer, and C. W. Shaiffer, Mallard recruitment in the agricultural environment of North Dakota. *Wildl. Monogr.,* 92, 1985.

Crabtree, R. L. and M. L. Wolfe, Effects of alternate prey on skunk predation of waterfowl nests. *Wildl. Soc. Bull.*, 16, 163–169, 1988.

Cummings, J. L., J. L. Guarino, C. E. Knittle, and W. C. Royall, Jr., Decoy plantings for reducing blackbird damage to nearby commercial sunflower fields. *Crop Protect.*, 6, 56–60, 1987.

Curtis, K. S., W. C. Pitt, and M. R. Conover, Overview of techniques for reducing bird predation at aquaculture facilities. Jack H. Berryman Institute Publication 12, Utah State University, Logan, 1996.

Deblinger, R. D., J. J. Vaske, and D. W. Rimmer, An evaluation of different predator exclosures used to protect Atlantic coast piping plover nests. *Wildl. Soc. Bull.*, 20, 274–279, 1992.

Dolbeer, R. A., Blackbirds. P. E25–E32 in S. E. Hygnstrom, R. M. Timm, and G. E. Larson, Eds., *Prevention and Control of Wildlife Damage*. University of Nebraska Cooperative Extension, Lincoln, 1994.

Draulans, D. and J. van Vessem, The effect of disturbance on nocturnal abundance and behaviour of gray herons (*Ardea cinerea*) at a fish-farm in winter. *J. App. Ecol.*, 22, 19–27, 1985.

Duebbert, H. F. and H. A. Kantrud, Upland duck nesting related to land use and predator reduction. *J. Wildl. Manage.*, 38, 257–265, 1974.

Duebbert, H. F. and J. T. Lokemoen, High duck nesting success in a predator-reduced environment. *J. Wildl. Manage.*, 44, 428–437, 1980.

Fleury, B. E. and T. W. Sherry, Long-term population trends of colonial wading birds in the southern United States: the impact of crayfish aquaculture on Louisiana populations. *Auk*, 112, 613–632, 1995.

Forster, J. A., Electric fencing to protect sandwich terns against foxes. *Biol. Conserv.*, 7, 85, 1975.

Garrettson, P. R., F. C. Rohwer, J. M. Zimmer, B. J. Mense, and N. Dion, Effects of mammalian predator removal on waterfowl and non-game birds in North Dakota. *Trans. North Am. Wildl. Nat. Res. Conf.*, 61, 94–101, 1996.

Glahn, J. F. and K. E. Brugger, The impact of double-crested cormorants on the Mississippi Delta catfish industry: a bioenergetics model. *Colonial Waterbirds,* 18, 168–175, 1995.

Glahn, J. F. and A. R. Stickley, Jr., Wintering double-crested cormorants in the Delta region of Mississippi: population levels and their impact on the catfish industry. *Colonial Waterbirds*, 18, 137–142, 1995.

Glahn, J. F., M. E. Tobin, and J. B. Harrel, Possible effects of catfish exploitation on overwinter body condition of double-crested cormorants. P. 107–113 in M. E. Tobin, Ed., *Symposium on Double-Crested Cormorants: Population Status and Management Issues in the Midwest*. U.S. Department of Agriculture, Animal and Plant Health Inspection Service Technical Bulletin 1879, 1999.

Goodrich, J. M. and S. W. Buskirk, Control of abundant native vertebrates for conservation of endangered species. *Conserv. Biol.*, 9, 1357–1364, 1995.

Gorenzel, W. P., F. S. Conte, and T. P. Salmon, Bird damage at aquaculture facilities. P. E5–E18 in S. E. Hygnstrom, R. M. Timm, and G. E. Larson, Eds., *Prevention and Control of Wildlife Damage*. University of Nebraska Cooperative Extension, Lincoln, 1994.

Greenwood, R. J., Influence of striped skunk removal on upland duck nest success in North Dakota. *Wildl. Soc. Bull.*, 14, 6–11, 1986.

Greenwood, R. J., P. M. Arnold, and M. G. McGuire, Protecting duck nests from mammalian predators with fences, traps, and a toxicant. *Wildl. Soc. Bull.*, 18, 75–82, 1990.

Greenwood, R. J., D. G. Pietruszewski, and R. D. Crawford, Effects of food supplementation on depredation of duck nests in upland habitat. *Wildl. Soc. Bull.*, 26, 219–226, 1998.

Greenwood, R. J., A. B. Sargeant, D. H. Johnson, L. M. Cowardin, and T. L. Shaffer, Mallard nest success and recruitment in prairie Canada. *Trans. North Am. Wildl. Nat. Res. Conf.*, 52, 299–309, 1987.

Greenwood, R. J., A. B. Sargeant, D. H. Johnson, L. M. Cowardin, and T. L. Shaffer, Factors associated with duck nest success in the Prairie Pothole region of Canada. *Wildl. Monogr.,* 59, 1995.

Greenwood, R. J. and M. A. Sovada, Prairie duck populations and management. *Trans. North Am. Wildl. Nat. Res. Conf.*, 61, 31–42, 1996.

Hanzel, J. J. and T. J. Gulya, Registration of two bird-resistant oilseed sunflower germplasm lines. *Crop Sci.,* 33, 1419–1420, 1993.

Higgins, K. F., A comparison of burn season effects on nesting birds in North Dakota mixed-grass prairie. *Prairie Nat.,* 18, 219–228, 1986.

Hothem, R. L., R. W. DeHaven, and S. D. Fairaizl, Bird damage to sunflowers in North Dakota, South Dakota, and Minnesota, 1979–1981. U.S. Fish and Wildlife Service, Fish and Wildlife Technical Report 15, Washington, D.C., 1988.

Howerter, D. W., R. B. Emery, B. L. Joynt, and K. L. Guyn, Mortality of mallard ducklings exiting from electrified predator exclosures. *Wildl. Soc. Bull.*, 24, 667–672, 1996.

Hoy, M., J. Jones, and A. Bivins, Economic impact and control of wading birds at Arkansas minnow ponds. In *Proc. East. Wildl. Damage Control Conf.*, 4, 109–112, 1989.

Hygnstrom, S. E., R. M. Timm, and G. E. Larson, Eds., *Prevention and Control of Wildlife Damage*. University of Nebraska, Cooperation Extension, Lincoln, 1994.

Jackson, J. A. and B. J. S. Jackson, The double-crested cormorant in the southeastern United States: habitat and population changes of a feathered pariah. *Colonial Waterbirds,* 18, 118–130, 1995.

Johnson, D. H., R. L. Kreil, G. B. Berkey, R. D. Crawford, D. O. Lamberth, and S. F. Galipeau, Influences of waterfowl management on nongame birds: the North Dakota experience. *Trans. North Am. Wildl. Nat. Res. Conf.*, 59, 293–302, 1994.

Johnson, D. H., A. B. Sargeant, and R. J. Greenwood, Importance of individual species of predators on nesting success of ducks in the Canadian Prairie Pothole region. *Can. J. Zoo.*, 67, 291–297, 1989.

Kantrud, H. A. and K. F. Higgins, Nest and nest site characteristics of some ground-nesting, non-passerine birds in northern grasslands. *Prairie Natur.,* 24, 67–84, 1992.

Kirsch, L. M., A. T. Klett, and H. W. Miller, Land use and prairie grouse population relationships in North Dakota. *J. Wildl. Manage.*, 37, 449–453, 1973.

Knittle, C. E., J. L. Cummings, G. M. Linz, and J. F. Besser, An evaluation of modified 4-aminopyridine baits for protecting sunflowers from blackbird damage. In *Proc. Vert. Pest Conf.*, 13, 248–253, 1988.

Kruse, A. D. and B. S. Bowen, Effects of grazing and burning on densities and habitats of breeding ducks in North Dakota. *J. Wildl. Manage.*, 60, 233–246, 1996.

LaGrange, T. G., J. L. Hansen, R. D. Andrews, A. W. Hancock, and J. M. Kienzler, Electric fence predator exclosure to enhance duck nesting: a long-term case study in Iowa. *Wildl. Soc. Bull.*, 23, 256–260, 1995.

Linhart, S. B. and R. K. Enders, Some effects of diethylstilbestrol on reproduction in captive red foxes. *J. Wildl. Manage.*, 28, 358–363, 1964.

Linz, G. M. and D. L. Bergman, DRC-1339 avicide fails to protect ripening sunflowers. *Crop Protect.*, 15, 307–310, 1996.

Linz, G. M. and J. J. Hanzel, Birds. P. 69–72 in D. R. Berglund, Ed., Sunflower Production. North Dakota State University Extension Service, Fargo, 1994.

Linz, G. M. and J. J. Hanzel, Birds and sunflowers. P. 381–394 in *Sunflower Technology and Production. Agron. Monogr.*, 35, 1997.

Linz, G. M., D. L. Bergman, and W. J. Bleier, Progress on managing cattail marshes with Rodeo® herbicide to disperse roosting blackbirds. In *Proc. Vert. Pest Conf.*, 15, 56–61, 1992.

Linz, G. M., R. A. Dolbeer, J. J. Hanzel, and L. E. Huffman, Controlling blackbird damage to sunflower and grain crops in the northern Great Plains. U.S. Department of Agriculture, Animal Plant Health Inspection Service, Agriculture Information Bulletin 679, Washington, D.C., 1996a.

Linz, G. M., D. L. Bergman, H. J. Homan, and W. J. Bleier, Effects of herbicide-induced habitat alterations on blackbird damage to sunflowers. *Crop Protect.*, 15, 625–629, 1996b.

Littauer, G., Avian predators: frightening techniques for reducing bird damage at aquaculture facilities. U.S. Department of Agriculture, Cooperative Extension Service, Southern Regional Aquaculture Center Publication No. 401, 1990a.

Littauer, G., Control of bird predation at aquaculture facilities: strategies and cost estimates. U.S. Department of Agriculture, Cooperative Extension Service, Southern Regional Aquaculture Center Publication No. 402., 1990b.

Lokemoen, J. T., Examining economic efficiency of management practices that enhance waterfowl production. *Trans. North Am. Wildl. Nat. Res. Conf.*, 49, 584–607, 1984.

Lokemoen, J. T. and T. A. Messmer, Locating, constructing, and managing islands for nesting waterfowl. Jack H. Berryman Institute Publication 5, Logan, UT, 1993.

Lokemoen, J. T. and T. A. Messmer, Locating and managing peninsulas for nesting ducks. Jack H. Berryman Institute Publication 6, Utah State University, Logan, 1994.

Lokemoen, J. T., H. A. Doty, D. E. Sharp, and J. E. Neaville, Electric fences to reduce mammalian predation on waterfowl nests. *Wildl. Soc. Bull.*, 10, 318–323, 1982.

Martin, L. R., and S. Hagar, Bird control on containment pond sites. In *Proc. Vert. Pest Conf.*, 14, 307–310, 1990.

Melvin, S. M., L. H. MacIvor, and C. R. Griffin, Predator exclosures: a technique to reduce predation at piping plover nests. *Wildl. Soc. Bull.*, 20, 143–148, 1992.

Messmer, T. A., C. A. Lively, D. D. MacDonald, and S. S. Schroeder, Motivating landowners to implement wildlife conservation practices using calendars. *Wildl. Soc. Bull.*, 24, 757–763, 1996.

Minsky, D., Preventing fox predation at a least tern colony with an electric fence. *J. Field Ornithol.*, 51, 17–18, 1980.

Moerbeek, D. J., W. H. van Dobben, E. R. Osiech, G. C. Boere, and C. M. Bungenberg de Jong, Cormorant damage prevention at a fish farm in the Netherlands. *Biol. Conserv.*, 39:23–38, 1987.

Mott, D. F., J. F. Glahn, P. L. Smith, D. S. Reinhold, K. J. Bruce, and C. A. Sloan, An evaluation of winter roost harassment for dispersing double-crested cormorants away from catfish production areas in Mississippi. *Wildl. Soc. Bull.*, 26, 584–591, 1998.

Ostergaard, D. E., Use of monofilament fishing line as a gull control. *Progr. Fish Culturist* 43:134, 1981.

Otis, D. L. and C. M. Kilburn, Influence of environmental factors on blackbird damage to sunflower. U.S. Fish and Wildlife Service, Fish and Wildlife Technical Report 16, Washington, D.C, 1988.

Parkhurst, J. A., Assessment and management of wildlife depredation at fish-rearing facilities in central Pennsylvania. Ph.D. dissertation, Pennsylvania State University, University Park, 1989.

Parkhurst, J. A., R. P. Brooks, and D. E. Arnold, A survey of wildlife depredation and control techniques at fish-rearing facilities. *Wildl. Soc. Bull.*, 15, 386–394, 1987.

Parkhurst, J. A., R. P. Brooks, and D. E. Arnold, Assessment of predation at trout hatcheries in central Pennsylvania. *Wildl. Soc. Bull.*, 20, 411–419, 1992.

Patterson, I. J., The control of fox movement by electric fencing. *Biol. Conserv.*, 11, 267–278, 1977.

Peters, F. and M. Neukirch, Transmission of some fish pathogenic viruses by the heron *Ardea cinerea*. *J. Fish Dis.*, 9, 539–544, 1986.

Phersson, O., Duckling production of the oldsquaw in relation to spring weather and small-rodent fluctuations. *Can. J. Zoo.*, 64, 1835–1841, 1986.

Pietz, P. J. and G. L. Krapu, Effects of predator exclosure design on duck brood movements. *Wildl. Soc. Bull.*, 22, 26–33, 1994.

Pitt, W. C. and M. R. Conover, Predation at Intermountain West fish hatcheries. *J. Wildl. Manage.*, 60, 616–624, 1996.

Post, W. and J. S. Greenlaw, Metal barriers protect near-ground nests from predators. *J. Field Ornithol.*, 60, 102–103, 1989.

Reinhold, D. S. and C. A. Sloan, Strategies to reduce double-crested cormorant depredation at aquaculture facilities in Mississippi. P. 99–105 in M. E. Tobin, Ed., *Symposium on Double-Crested Cormorants: Population Status and Management Issues in the Midwest*. U.S. Department of Agriculture, Animal and Plant Health Inspection Service Technical Bulletin 1879, 1999.

Reynolds, R. E., T. L. Shagger, J. R. Sauer, and B. G. Peterjohn, Conservation Reserve Program: benefit for grassland birds in the Northern Plains. *Trans. North Am. Wildl. Nat. Res. Conf.*, 59, 328–336, 1994.

Salmon, T. P. and F. S. Conte, Control of bird damage at aquaculture facilities. U.S. Fish and Wildlife Service Wildlife Leaflet 475, Washington, D.C., 1981.

Sargeant, A. B., S. H. Allen, and R. T. Eberhardt, Red fox predation on breeding ducks in midcontinent North America. *Wildl. Monogr.*, 89, 1–41, 1984.

Sargeant, A. B., S. H. Allen, and J. O. Hastings, Spatial relationships between sympatric coyotes and red foxes in North Dakota. *J. Wildl. Manage.*, 51, 285–293, 1987.

Sargeant, A. B., R. J. Greenwood, M. A. Sovada, and T. L. Shaffer, Distribution and abundance of predators that affect duck production — Prairie Pothole region. U.S. Fish and Wildlife Service, Resource Publication 194, Washington, D.C., 1993.

Sargeant, A. B., M. A. Sovada, and T. L. Shaffer, Seasonal predator removal relative to hatch rate of duck nests in waterfowl production areas. *Wildl. Soc. Bull.*, 23, 507–513, 1995.

Scanlon, P. F., L. A. Helfrich, and R. E. Stultz, Extent and severity of avian predation at federal fish hatcheries in the United States. In *Proc. Ann. Conf. Southeastern Assoc. Fish Wildl. Agencies*, 32, 470–473, 1978.

Sheaffer, S. E. and R. D. Drobney, Effectiveness of lithium chloride induced taste aversions in reducing waterfowl nest predation. *Trans. Missouri Academy Sci.*, 20, 59–63, 1986.

Sovada, M. A., A. B. Sargent, and J. W. Grier, Differential effects of coyotes and red foxes on duck nest success. *J. Wildl. Manage.*, 59, 1–9, 1995.

Stickley, A. R., Jr. and K. J. Andrews, Survey of Mississippi catfish farmers on means, effort, and costs to repel fish-eating birds from ponds. In *Proc. East. Wildl. Damage Control Conf.*, 4, 105–108, 1989.

Stickley, A. R. Jr. and J. O. King, Long-term trial of an inflatable effigy scare device for repelling cormorants from catfish ponds. In *Proc. East. Wildl. Damage Control Conf.*, 6, 89–92, 1995.

Stickley, A. R. Jr., D. F. Mott, and J. O. King, Short-term effects of an inflatable effigy on cormorants at catfish farms. *Wildl. Soc. Bull.*, 23, 73–77, 1995.

Sugden, L. G. and G. W. Beyersbergen, Farming intensity on waterfowl breeding grounds in Saskatchewan parklands. *Wildl. Soc. Bull.*, 12, 22–26, 1984.
Summers, R. W., Breeding populations of dark-bellied brant geese *Branta bernicla* in relation to lemming cycles. *Bird Study*, 33, 105–108, 1986.
Swanson, G. A., Dissemination of amphipods by waterfowl. *J. Wildl. Manage.*, 48, 988–991, 1984.
Trottier, G. C., D. C. Duncan, and S. C. Lee, Electric predator fences delay mallard brood movements to water. *Wildl. Soc. Bull.*, 22, 22–26, 1994.
Van Vuren, D., Manipulating habitat quality to manage vertebrate pests. In *Proc. Vert. Pest Conf.*, 18, 383–390, 1998.
Warren, R. J., Should wildlife biologists be involved in wildlife contraception research and management? *Wildl. Soc. Bull.*, 23, 441–444, 1995.
Weller, M. W., Density and habitat relationships of blue-winged teal nesting in northwestern Iowa. *J. Wildl. Manage.*, 43, 367–374, 1979.
Williams, T., Killer fish farms. *Audubon*, 94(2), 14–22, 1992.

Appendix

Latin Names for Species Mentioned in the Text

alder, green (*Alnus viridis*)
alligator, American (*Alligator mississippiensis*)
antelope, pronghorn (*Antilocapra americana*)
apple (*Malus domestica*)
armadillo (*Dasypus novemcinctus*)
ash, white (*Fraxinus americana*)
aspen (*Populus tremuloides*)
aspen, quaking (*Populus tremuloides*)
baboon, olive (*Papio anubis*)
badger, American (*Taxidea taxus*)
badger, European (*Meles meles*)
bamboo (*Phyllostachys* spp.)
barley (*Hordeum vulgare*)
bat, flying fox (*Pteropus* spp.)
bat, giant fruit (*Pteropus giganteus*)
bear, Asiatic black (*Ursus thibetanus*)
bear, black (*Ursus americanus*)
bear, brown (*Ursus arctos*)
bear, grizzly (*Ursus arctos horribilis*)
bear, polar (*Ursus maritimus*)
bear, sloth (*Melursus ursinus*)
beaver (*Castor canadensis*)
beaver, mountain (*Aplodontia rufa*)
beech (*Fagus sylvatica*)
beech, American (*Fagus grandifolia*)
bison (*Bison bison*)
bitterbrush (*Purshia tridentata*)
blackbird, red-winged (*Agelaius phoeniceus*)
blackbird, yellow-headed (*Xanthocephalus xanthocephalus*)
blackbush (*Coleogyne ramosissima*)
blueberry (*Vaccinium* spp.)
bluefish (*Pomatomus saltatrix*)
bluegrass, Kentucky (*Poa pratensis*)
brant, black (*Branta bernicla nigricans*)
boar, wild (*Sus scrofa*)
bobcat (*Lynx rufus*)
buffalo (*Bison bison*)
buffalo, water (*Bubalus bubalis*)
bullfinch (*Pyrrhula pyrrhula*)
burro (*Equus asinus*)
butterfly, monarch (*Danaus plexippus*)
camel (*Camelus dromedarius*)
cardinal, northern (*Cardinalis cardinalis*)
carp (*Cyprinus carpio*)
cat, feral (*Felis catus*)
catbird (*Dumetella carolinensis*)
catfish (*Ictalurus* spp.)
cattail (*Typha* spp.)
cattle, feral (*Bos taurus*)
cedar, northern white (*Thuja occidentalis*)
cedar, western red (*Thuja plicata*)
chamois (*Rupicapra rupricapra*)
cherry (*Prunus* spp.)
cherry, black (*Prunus serotina*)
cherry, choke (*Prunus virginiana*)
chimpanzee (*Pan troglodytes*)
chipmunk (*Tamias* spp.)
cockatoo, sulphur-crested (*Cacatua galerita*)
cocoa (*Theobroma cacao*)

condor, California (*Gymnogyps californianus*)
cormorant, great (*Phalacrocorax carbo*)
corn (*Zea mays*)
corn borer, European (*Ostrinia nubilalis*)
cottontail, desert (*Sylvilagus audubonii*)
cottontail, eastern (*Syvilagus floridanus*)
cougar (*Felis concolor*)
cowbird, brown-headed (*Molothrus ater*)
coyote (*Canis latrans*)
crane, sandhill (*Grus canadensis*)
crane, whooping (*Grus americana*)
crow, American (*Corvus brachyrhynchos*)
crow, carrion (*Corvus corone*)
crow, fish (*Corvus ossifragus*)
deer, black-tailed (*Odocoileus hemionus columbianus*)
deer, fallow (*Dama dama*)
deer, mule (*Odocoileus hemionus*)
deer, muntjac (*Muntiacus* spp.)
deer, red (*Cervus elaphus*)
deer, roe (*Capreolus capreolus*)
deer, sika (*Cervus nippon*)
deer, white-tailed (*Odocoileus virginianus*)
dingo (*Canis familiaris dingo*)
dog, feral (*Canis familiaris*)
donkey (*Equus asinus*)
dove, eared (*Zenaida auriculata*)
dove, Inca (*Columbina inca*)
dove, mourning (*Zenaida macroura*)
dove, red-eyed turtle (*Streptopelia decaocto*)
duck, wood (*Aix sponsa*)
dunlin (*Calidris alpina*)
eagle, golden (*Aquila chrysaetos*)
egret, cattle (*Bubulcus ibis*)
eider, common (*Somateria mollissima*)
elephant, African (*Loxodonta africana*)
elephant, Asian (*Elephas maximus*)
elk (*Cervus elaphus*)
falcon, peregrine (*Falco peregrinus*)
ferret, black-footed (*Mustela nigripes*)
fescue, tall (*Festuca arundinaceae*)
finch, African weaver (*Quelea quelea*)
finch, house (*Carpodacus mexicanus*)
fir, Douglas (*Pseudotsuga menziesii*)
fir, Fraser (*Abies fraseri*)
fody, red (*Foudia madagascariensis*)
fox, arctic (*Alopex lagopus*)
fox, gray (*Urocyon cinereoargenteus*)
fox, red (*Vulpes vulpes*)
fox, San Joaquin kit (*Vulpes macrotis mutica*)
foxglove (*Digitalis purpurea*)
gadwall (*Anas strepera*)
galah (*Eolophus roseicapillus*)
gaur (*Bos gaurus*)
goat, feral (*Capra hircus*)
goat, mountain (*Oreamnos americanus*)
goose, Aleutian Canada (*Branta canadensis leucopareia*)
goose, brant (*Branta bernicla*)
goose, Canada (*Branta canadensis*)
goose, Hawaiian (*Nesochen sandvicensis*)
goose, pink-footed (*Anser brachyrhynchus*)
goose, Ross' (*Chen rossii*)
goose, snow (*Chen caerulescens*)
gopher, pocket (Geomyidae)
goshawk, northern (*Accipiter gentilis*)
grackle, boat-tailed (*Quiscalus major*)
grackle, common (*Quiscalus quiscula*)
grouse, ruffed (*Bonasa umbellus*)
grouse, sharp-tailed (*Tympanuchus phasianellus*)
gull, laughing (*Larus atricilla*)
gyrfalcon (*Falco rusticolus*)
hare, snowshoe (*Lepus americanus*)
hawk, red-shouldered (*Buteo lineatus*)
hemlock, eastern (*Tsuga canadensis*)
heron, black-crowned night (*Nycticorax nycticorax*)
heron, gray (*Ardea cinerea*)
heron, great blue (*Ardea herodias*)
hippopotamus (*Hippopotamus amphibius*)
holly, American (*Ilex opaca*)
holly, China girl (*Ilex X meserveae*)
holly, Japanese (*Ilex crenata*)
horse, feral (*Equus caballus*)
horse, Przewalski's (*Equus caballus przewalskii*)
hummingbird, black-chinned (*Archilochus alexandri*)
hyaena (*Hyaena hyaena*)
ibex, alpine (*Capra ibex*)
jackdaw (*Corvus monedula*)

jackrabbit, black-tailed (*Lepus californicus*)
jaguar (*Panthera onca*)
jay, blue (*Cyanocitta cristata*)
junglefowl, red (*Gallus gallus*)
kangaroo (*Macropus* spp.)
kestrel, American (*Falco sparverius*)
killdeer (*Charadrius vociferus*)
kingfisher, belted (*Ceryle alcyon*)
koala (*Phascolarctos cinereus*)
lapwing (*Vanellus vanellus*)
leopard (*Panthera pardus*)
lion, African (*Panthera leo*)
lion, Asiatic (*Panthera leo persica*)
lion, mountain (*Felis concolor*)
lizard, western fence (*Sceloporus occidentalis*)
llama (*Lama glama*)
mallard (*Anas platyrhynchos*)
maple, sugar (*Acer saccharum*)
mange, sarcoptic (*Sarcoptes scabiei*)
marmot, yellow-bellied (*Marmota flaviventris*)
mink (*Mustela vison*)
mockingbird (*Mimus polyglottos*)
moles (Talpidae)
mongoose, Indian (*Herpestes javanicus*)
monkey, red-tailed (*Cercopithecus ascanius*)
moose (*Alces alces*)
mouse, house (*Mus musculus*)
mouse, striped field (*Apodemus agrarius*)
mouse, white-footed (*Peromyscus leucopus*)
mouse, yellow-necked (*Apodemus flavicollis*)
muskrat (*Ondatra zibethicus*)
night-heron, black-crowned (*Nycticorax nycticorax*)
nilgai (*Boselaphus tragocamelus*)
nutria (*Myocastor coypus*)
oldsquaw (*Clangula hyemalis*)
oleander (*Nerium* spp.)
opossum (*Didelphis virginiana*)
oriole, northern (*Icterus galbula*)
osprey (*Pandion haliaetus*)
owl, barn (*Tyto alba*)
owl, great-horned (*Bubo virginianus*)
owl, spotted (*Strix occidentalis*)
oystercatcher (*Haematopus ostralegus*)
palm, oil (*Elaeis guineensis*)
parakeet, Mauritius (*Psittacula echo*)
parakeet, monk (*Myiopsitta monachus*)
parakeet, rose-ringed (*Psittacula krameri*)
pear (*Pyrus communis*)
peccary (*Pecari tajacu*)
petrel, dark-rumped (*Pterodroma phaeopygia*)
pheasant, ring-necked (*Phasianus colchicus*)
pika (*Ochotona princeps*)
pig, feral (*Sus scrofa*)
pigeon (*Columba livia*)
pigeon, passenger (*Ectopistes migratorius*)
pigeon, Picazuro (*Columba picazuro*)
pigeon, spot-winged (*Columba maculosa*)
pigeon, wood (*Columba palumbus*)
pine, Austrian (*Pinus nigra*)
pine, lodgepole (*Pinus contorta*)
pine, Ponderosa (*Pinus ponderosa*)
pine, Scotch (*Pinus sylvestris*)
plover, piping (*Charadrius melodus*)
pocket gopher (Geomyidae)
pocket gopher, Mazama (*Thomomys mazama*)
poplar (*Populus* spp.)
porcupine (*Erethizon dorsatum*)
possum, brush-tailed (*Trichosurus vulpecula*)
prairie-chicken, lesser (*Tympanuchus pallidicinctus*)
prairie dog, black-tailed (*Cynomys ludovicianus*)
prairie dog, white-tailed (*Cynomys leucurus*)
ptarmigan, willow (*Lagopus lagopus*)
quelea, red-billed (*Quelea quelea*)
rabbit, cottontail (*Sylvilagus floridanus*)
rabbit, European (*Oryctolagus cuniculus*)
raccoon (*Procyon lotor*)
rat, black (*Rattus rattus*)
rat, canefield (*Rattus sordidus*)
rat, cotton (*Sigmodon hispidus*)
rat, kangaroo (*Dipodomys* spp.)
rat, Malayan wood (*Rattus tiomanicus*)

rat, Norway (*Rattus norvegicus*)
rat, Polynesian (*Rattus exulans*)
rat, rice (*Oryzomys palustris*)
rat, roof (*Rattus rattus*)
rattlesnake (*Crotalus* spp.)
reindeer (*Rangifer tarandus*)
robin, American (*Turdus migratorius*)
rock-wallaby, brush-tailed (*Petrogale penicillata*)
rock-wallaby, yellow-footed (*Petrogale xanthopus*)
rook (*Corvus frugilegus*)
rootworm, northern corn (*Diabrotica longicornis*)
rowan (*Sorbus aucuparia*)
salal (*Gaultheria shallon*)
sandpiper, pectoral (*Calidris melanotos*)
seal, common (*Phoca vitulina*)
seal, harp (*Phoca groenlandica*)
sheep, domestic (*Ovis aries*)
shrew, short-tailed (*Blarina brevicauda*)
silvereye (*Zosterops lateralis*)
skunk, striped (*Mephitis mephitis*)
snake, brown tree (*Boiga irregularis*)
snake, garter (*Thamnophis sirtalis*)
snake, rat (*Elaphe obsoleta*)
sorghum, grain (*Sorghum vulgare*)
soybean (*Glycine max*)
sparrow, house (*Passer domesticus*)
sparrow, seaside (*Ammodramus maritimus*)
sparrow, Spanish (*Passer hispaniolensis*)
spruce, white (*Picea glauca*)
squirrel, Belding's ground (*Spermophilus beldingi*)
squirrel, California ground (*Spermophilus beecheyi*)
squirrel, Douglas (*Tamiasciurus douglasii*)
squirrel, golden-mantled ground (*Spermophilus lateralis*)
squirrel, gray (*Sciurus carolinensis*)
squirrel, ground (*Spermophilus* spp.)
squirrel, red (*Tamiasciurus hudsonicus*)
starling (*Sturnus vulgaris*)
starling, European (*Sturnus vulgaris*)
starling, glossy (*Lamprotornis chalybaeus*)
sunflower (*Helianthus annuus*)
stoat (*Mustela erminea*)
swallow, barn (*Hirundo rustica*)
swan, mute (*Cygnus olor*)
sycamore (*Acer pseudoplatanus*)
Tasmanian devil (*Sarcophilus harrisii*)
teal, blue-winged (*Anas discors*)
tern, common (*Sterna hirundo*)
tern, least (*Sterna antillarum*)
tern, sandwich (*Sterna sandvicensis*)
thylacine (*Thylacinus cynocephalus*)
tiger (*Panthera tigris*)
tiger, Siberian (*Panthera tigris altaica*)
toad, cane (*Bufo marinus*)
tortoise, desert (*Gopherus agassizii*)
turkey (*Meleagris gallopavo*)
vole, bank (*Clethrionomys glareolus*)
vole, meadow (*Microtus pennsylvanicus*)
vole, pine (*Microtus pinetorum*)
vole, southern red-backed (*Clethrionomys gapperi*)
vulture, black (*Coragyps atratus*)
wallaby, tammar (*Macropus eugenii*)
warbler, black-and-white (*Mniotilta varia*)
warbler, black-throated green (*Dendroica virens*)
warbler, Kirtland's (*Dendroica kirtlandii*)
waxwing, cedar (*Bombycilla cedrorum*)
weasel, short-tailed (*Mustela erminea*)
weaver, black-headed (*Ploceus melanocephalus melanocephalus*)
weaver, village (*Ploceus cucullatus*)
weaver, yellow-backed (*Ploceus melanocephalus capitalis*)
wedelia (*Wedelia trilobata*)
wheat (*Triticum aestivum*)
white-eye, Japanese (*Zosterops japonicus*)
wildebeast (*Connochaetes taurinus*)
wolf (*Canis lupus*)
wolf, gray (*Canis lupus*)
wolfsbane (*Aconitum napellus*)
woodchuck (*Marmota monax*)
woodpecker, red-cockaded (*Picoides borealis*)
wood-pigeon (*Columba palumbus*)
yew, Japanese (*Taxus cuspidata*)
zebra (*Equus* spp.)

Index

A

B

C

D

E

F

G

H

I

J

K

L

M

N

O

P

Q

R

S

T

U

V

W

Z